73 Topics in Current Chemistry
Fortschritte der Chemischen Forschung

Organic Chemistry

Springer-Verlag
Berlin Heidelberg GmbH 1978

This series presents critical reviews of the present position and future trends in modern chemical research. It is addressed to all research and industrial chemists who wish to keep abreast of advances in their subject.

As a rule, contributions are specially commissioned. The editors and publishers will, however, always be pleased to receive suggestions and supplementary information. Papers are accepted for "Topics in Current Chemistry" in English.

ISBN 978-3-662-15828-9 ISBN 978-3-540-37037-6 (eBook)
DOI 10.1007/978-3-540-37037-6

Library of Congress Cataloging in Publication Data. Main entry under title: Organic chemistry. (Topics in current chemistry ; 73) Bibliography: p. Includes index. 1. Chemistry, Organic–Addresses, essays, lectures. I. Series. QD1.F58 vol. 73 [QD255] 540'.8s [547] 77-14137

Originally published by Springer-Verlag Berlin Heidelberg New York in 1978
Softcover reprint of the hardcover 1st edition 1978

Triltsch, Graphischer Betrieb, 8700 Würzburg
2152/3140-543210

Contents

Springer-Verlag	Postfach 105 280 · D-6900 Heidelberg 1 Telephone (0 62 21) 4 87-1 · Telex 04-61 723 Heidelberger Platz 3 · D-1000 Berlin 33 Telephone (0 30) 82 20 01 · Telex 01-83319
Springer-Verlag New York Inc.	175, Fifth Avenue · New York, NY 100 10 Telephone 4 77-82 00

Stereochemistry of Penta- and Hexacoordinate Phosphorus Derivatives

Dr. William S. Sheldrick

Gesellschaft für Biotechnologische Forschung mbH, Mascheroder Weg 1,
3300 Braunschweig-Stöckheim, Germany

Table of Contents

1. Introduction

Phosphorus is a stereochemically versatile element. It may be found coordinated to between one and six ligands in various compounds, but the vast majority of known phosphorus derivatives have coordination numbers of [3] and [4]. In contrast to nitrogen, the chemistry of phosphorus is characterised by higher coordination numbers [4–6], and is, therefore, in this respect more analogous to that of the heavier Group V elements, arsenic and antimony. The tricoordinate derivatives of phosphorus constitute a large class of compounds with nitrogen counterparts. However, the bond configuration around tricoordinate nitrogen may vary between trigonal pyramidal (*e.g.* NH_3, NMe_3) and planar [*e.g.* $N(SiH_3)_3$][1)]. On the other hand, the analogous phosphorus derivatives are exclusively trigonal pyramidal, with valence angles normally in the range 90–100°. In spectroscopic studies of members of the series, $M(SiH_3)_3$, M = P, As, Sb, the inference was drawn that the molecular skeleton is planar, as in the nitrogen analogue[2, 3)]. However, subsequent electron diffraction studies have shown these molecules to have the expected pyramidal structure[4, 5)]. The stereochemical insensitivity of tricoordinate phosphorus to the chemical nature of its substituents may be ascribed to the fact that, in contrast to nitrogen, it shows no tendency to form multiple bonds in which its p_π orbitals are involved.

A coordination number of [4] is most characteristic of the phosphorus atom. It is displayed in naturally occurring phosphates and nucleotides, and in the numerous coordination complexes of phosphine ligands, characterised over the last 20 years. According to classical valence theory the pentavalent phosphorus in the former class of tetracoordinate derivatives must be formally double bonded to one ligand (usually oxygen) in non-ionised structures. The distribution of the substituents at phosphorus is basically tetrahedral but, owing to the presence of bonds of differing nature and multiplicity within an individual structure, the valence angles can vary within wide limits (100–120°). The importance of π-bonding involving the phosphorus $3d$-orbitals in PO_4^{3-} ions has been discussed by Cruickshank[6)].

The existence of two possible geometrical configurations of very similar energies, namely the trigonal bipyramid and the square (or rectangular) pyramid, bestows the stereochemistry of pentacoordinate phosphorus with a particular fascination. The stereochemical and electronic structural aspects of pentacoordination in both main group elements and in transition metals have been the subject of a number of reviews[7–9)]. Both of the possible geometries are characterised by non-equivalent bonding; in the former there are three equatorial and two axial, and in the latter one apicel and four basal substituents. Theoretical considerations based on molecular orbital and electrostatic calculations have consistently predicted that the trigonal bipyramidal configuration should be energetically slightly more favourable (ca. 5–10 kcal/mol) for acyclic pentacoordinate phosphorus derivatives, and these findings have been borne out by electron and X-ray diffraction studies. However, in view of the relatively small energy difference between the two modifications, it is possible that a particular arrangement of ligands of differing electronegativities or the inclusion of phosphorus in a strained small 4–5 membered ring can lead to a relative stabilisation of the square (or rectangular) pyramid. Rectangular pyramidal structures have indeed been confirmed by X-ray diffraction for a number of spirobicyclic der-

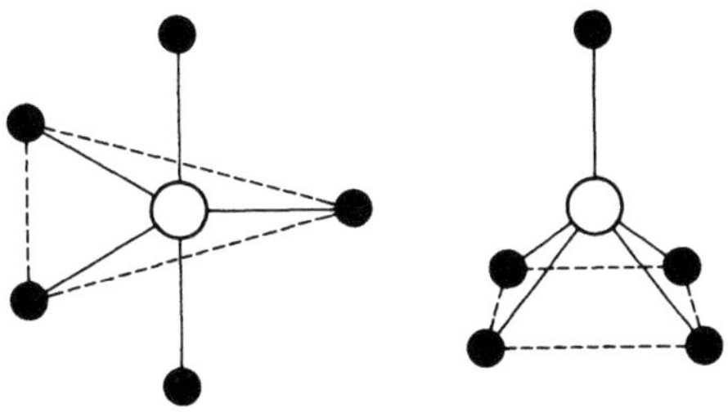

Fig. 1.1. Trigonal bipyramidal and square pyramidal geometries

ivatives of pentacoordinate phosphorus. The significance of the $3d$-orbitals in the bonding of pentacoordinate phosphorus is still contested. The general features of molecular geometry can be explained in a qualitative manner without consideration of these orbitals, and their absolute role will only be better understood as more structural parameters become available.

The dynamic stereochemistry of pentacoordinate phosphorus has received very considerable recent interest in view of the role of transitory pentacoordinate species in mechanisms advanced to account for the hydrolysis of phosphate esters[10)] or nucleophilic displacements in phosphonium systems[11)]. Rapid intramolecular ligand exchange takes place readily in many phosphoranes. The ^{19}F NMR-spectrum of PF_5 contains, for instance, only one fluorine resonance[12)], indicating that the non-equivalent axial and equatorial sites are scrambled by molecular inversions during the relatively long time scale of the NMR-measurements. Mechanisms involving the rearrangement of the substituent ligands in the trigonal bipyramid by internal rotation (the so called turnstile rotatiom mechanism[13)]), or by a pseudorotation pathway through an intermediate of C_{4v} symmetry as suggested by Berry[14)], have been advanced to account for such phenomena.

Phosphorus may make use of its $3d$-orbitals, to some extent, to increase its coordination number from five to six, as for instance in the adducts of PF_5 with nitrogen bases (*e.g.* $PF_5 \cdot$ pyridine) or in anions of the type PF_6^-. The neutral complexes display a distorted octahedral geometry with a very long coordinate bond. Regular octahedral geometry with equivalent bond distances is observed in the PF_6^- anion, for which the hybridisation may be described as sp^3d^2.

The most interesting recent developments in the structural chemistry of phosphorus have involved penta- and hexacoordinate derivatives. Little structural information was available until recently owing, in the main, to the instability and volatility of many of these compounds. In the last decade improved methods of analysis by electron and X-ray diffraction and the availability of increasing numbers of phosphoranes, which owe their stability to the inclusion of phosphorus in one or two small rings, have led to a substantial growth in the fund of determined structures. At the end of 1976 more than 50 diffraction studies of pentacoordinate phosphorus derivatives had been published, 14 of these being for acyclic molecules. This information has in its turn led to a better theoretical understanding of the nature of pentacoordination at phosphorus. The present state of knowledge of the static stereochemistry of penta- and hexacoordinate phosphorus derivatives as determined by diffraction methods will be discussed in this review. Although a detailed consideration of the dynamic stereochemistry of pentacoordinate phosphorus is outside the scope of this article, it is nevertheless self-evident that many of the factors governing the stability of trigonal bipyramidal species, which may be inferred from studies of static geometries, are of immediate relevance to the development of ligand exchange mechanisms and to proposed reaction mechanisms involving transitory pentacoordinate intermediates. Furthermore, distortions in the regular trigonal bipyramidal geometry at phosphorus have been interpreted as providing evidence for the onset of just such a ligand exchange. The stereochemistry of tri- and tetracoordinate phosphorus derivatives have been extensively reviewed elsewhere[15–17].

2. Stereochemistry of Pentacoordinate Phosphorus Derivatives

2.1. Static and Dynamic Configurations

Before we proceed to a detailed discussion of the significance of individual structural studies, it is apposite to summarise those features which characterise the stereochemistry of the simplest class of compounds containing pentacoordinate phosphorus, namely the acyclic derivatives. In the later sections we shall see how the inclusion of phosphorus in one or more strained small ring system (4–5 atoms) may lead to a modification of these structural principles. Structural data on compounds containing acyclic pentacoordinate phosphorus are however relatively limited, owing to the high volatility and reactivity of these species. Therefore, of particular relevance to the development of bonding theory has been the determination of accurate structural parameters by electron diffraction for the series of methyl substituted fluorophosphoranes $PF_{5-n}Me_n$, $n = 0–3$[18–20]. Previously the structure of PF_5 had been the subject of some controversy as a result of early electron diffraction studies carried out in 1937–38, which had suggested that the PF_5 molecules are trigonal bipyràmids in which the axial and equatorial P-F bonds are equal in length[21,22]. A similar analysis on the related molecule PCl_5, carried out two years later demonstrated, however, that the P-Cl axial bonds are significantly longer than the equatorial bonds (0.15 ± 0.06 Å) in this likewise trigonal bipyramidal molecule[23]. The situation was finally clarified in 1965 when Hansen and Bartell[18] were able to demonstrate, using modern electron diffraction techniques, that the P-F bond distances in PF_5 actually are significantly different. They obtained values of 1.577(5) for the axial and 1.534(4) Å for the equatorial bonds (see Fig. 2.1.1.). Any theory of bonding should be capable of explaining, at least in a qualitative manner, the stereochemical features observed in the fluorophosphorane series (Fig. 2.1.1.), which are representative for all acyclic derivatives:

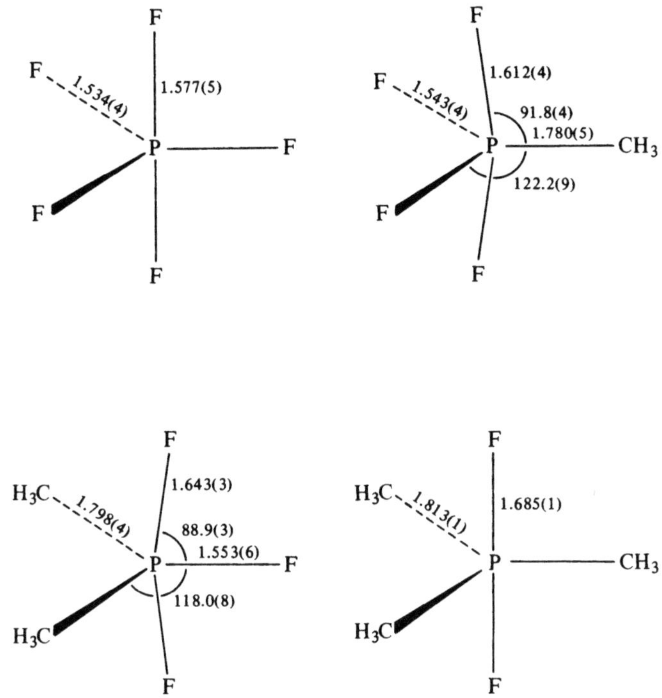

Fig. 2.1.1. Results of electron diffraction studies on the series $PF_{5-n}Me_n$, $n = 0-3$

1. The molecules are trigonal bipyramids or distorted trigonal bipyramids rather than the alternative square pyramid.
2. The more electronegative substituents (fluorine) preferentially occupy the axial sites in the trigonal bipyramid.
3. Axial bonds are longer than the corresponding equatorial bonds.
4. All the corresponding bond lengths and the ratio of the length of the axial bonds to that of the equatorial bonds, r_{ax}/r_{eq}, increase, the former linearly, as the number of methyl substituents increases.
5. Methyl substitution causes the P-F bonds to be bent away from the methyl groups [*e.g.* C_{eq}-P-F_{ax} = 91.8(4) in $MePF_4$].

The preferential occupation of axial sites by more electronegative substituents had also been anticipated by Muetterties *et al.*[24, 25] on the basis of IR and NMR studies. The observation of axial bonds which are significantly longer than the corresponding equatorial bonds suggests that the former are weaker than the latter. This has been confirmed for phosphorus and other analogous non-transition metal derivatives by investigation of vibrational stretching frequencies. These have been shown to be 100–200 cm^{-1} higher for equatorial bonds[26]. The stretching force constants are correspondingly lower for axial as compared to equatorial bonds[27, 28].

The NMR-equivalence of all fluorine atoms in PF_5 at temperatures above $-100°$ implies very rapid interchange of axial and equatorial ligands. This phenomenon, which has since been observed for numerous acyclic and cyclic phosphoranes prompted Berry in 1960 in propose the pseudorotation exchange mechanism, which has since been named after him[14]. He suggested that a pairwise exchange of the two axial ligands with two equatorial ligands may take place through synchronous vibrational motions, as illustrated in Fig. 2.1.2. mechanism 6. The third equatorial ligand (3), which may be regarded as functioning as a pivot, is not involved in the exchange retains its original position. As may be seen from mechanism 6, molecular C_{2v}- symmetry isretained throughout the exchange pathway, which also passes through a C_{4v} square pyramidal intermediate state.

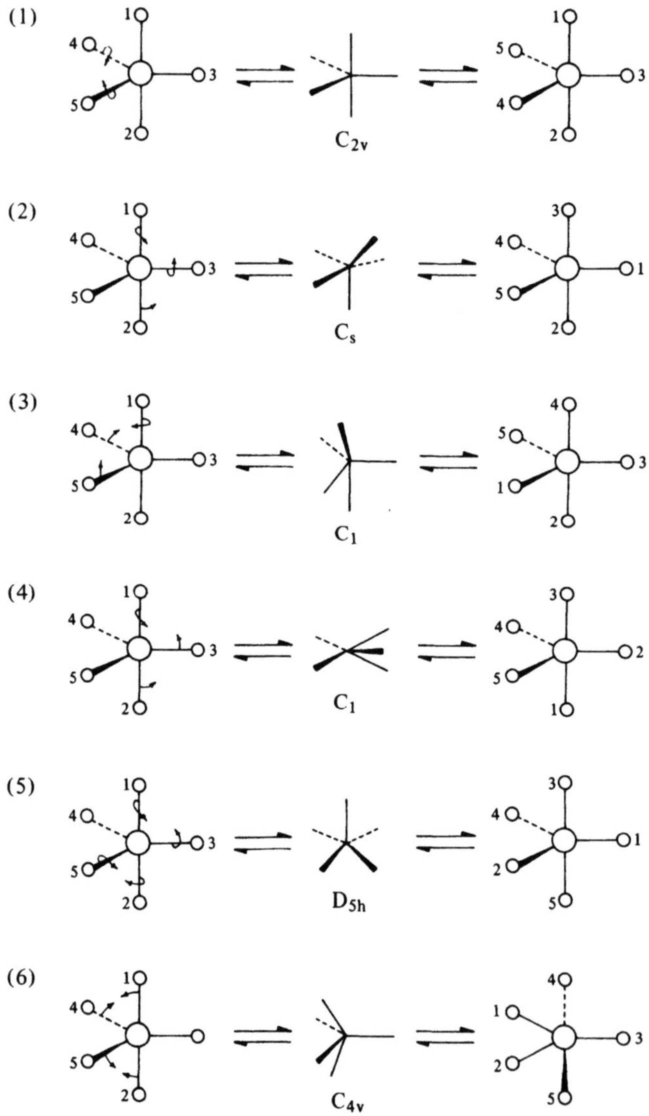

Fig. 2.1.2. Possible mechanisms for the isomerisation of trigonal bipyramidal molecules as proposed by Mutterties[29, 30]

Muetterties attempted to symbolise the conceivable mechanisms and symmetries of intermediate states for the regular pentatopal isomerisation of trigonal bipyramidal molecules [Fig. 2.1.2. mechanisms (1)–(6)][29, 30]. A further physically reasonable model, which was not discussed by Muetterties, is the turnstile rotation mechanism developed by Ugi, Ramirez *et al.*[31–33], which is depicted in Fig. 2.1.3. This mechanism corresponds to an internal rotation of one axial and one equatorial ligand with a pair angle α_2 about an approximate local C_2 axis, combined with a simultaneous opposite rotation of the three remaining ligands (trio angle α_3) about an approximate local C_3 axis. It has been envisaged as involving the pathway illustrated in Fig. 2.1.3. As a prerequisite to internal rotation, the diequatorial angle (4)-P-(5) contracts from

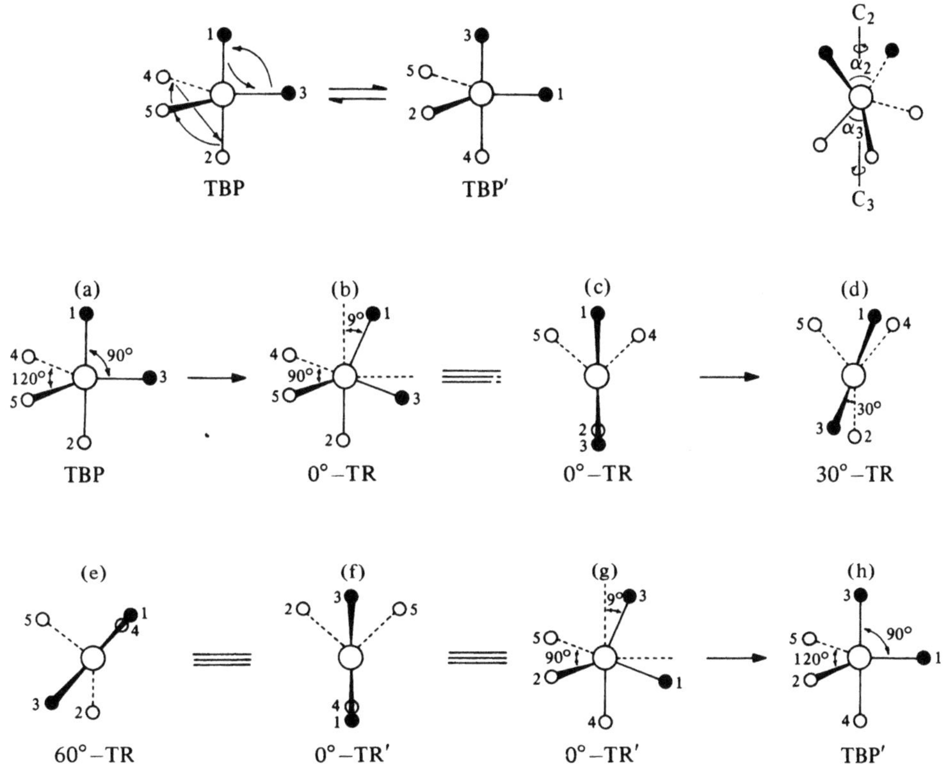

Fig. 2.1.3. The turnstile rotation mechanism for pentatopal isomerisation

120° to 90° and the ligand pair (1)(3) tilts 9° in the plane P(1)(3)(2) towards the axial ligand (2) to yield (b), which is termed the 0°-TR configuration (TR = turnstile rotation). The ligand pair (1)(3) rotates against the trio (2)(4)(5) leading to the 30°-TR configuration (d), which is the halfway intermediate between the two isomeric trigonal bipyramidal geometries. Further rotation leads to the 60°-TR configuration (e), which through tilting of the pair (3)(1) and expansion of the angle (5)-P-(2) generates the new trigonal bipyramidal geometry TBP′. It should be noted that ligand (5) retains an equatorial position in the new trigonal bipyramid. Assuming a staggered 30°-TR configuration, optimisation of the angles α_2 and α_3 for PF_5 has been carried out by CNDO/2[31, 32], extended Hückel[34] and *ab initio*[35] MO-calculations. The lowest barrier for a turnstile rotation was found by all these methods for values of $\alpha_2 \approx 85°$ and $\alpha_3 \approx 95°$.

The mechanistic alternatives 1–5 considered by Muetterties have since been demonstrated to be in disagreement with experimental evidence and have therefore been discarded. This may be illustrated by reference to an elegant study by Whitesides and Mitchell[36] of the temperature dependence of the ^{19}F-NMR spectrum of the substituted fluorophosphorane $(CH_3)_2NPF_4$. At temperatures below −100°

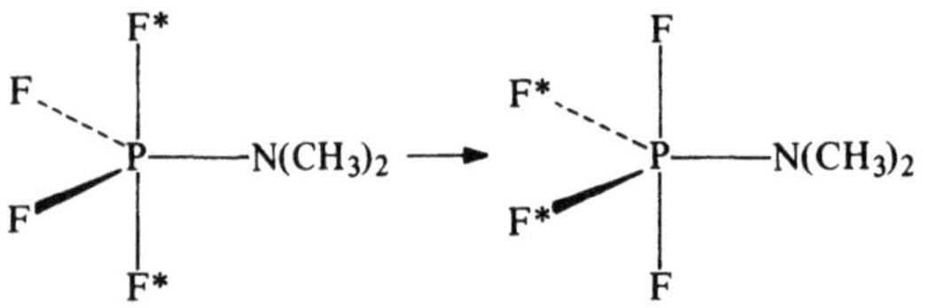

there is no exchange of equatorial and axial fluorines. Above −50° all four fluorines are NMR-equivalent, but there is an intermediate temperature range in which the equatorial fluorine pair replaces the axial pair *via* a single concerted step. F* and F do not, however, exchange with one another. Only the Berry pseudorotation and turnstile rotation mechanisms are capable of explaining this phenomenon. MO-Calculations (to be discussed in Section 2.2.) indicate that the Berry pseudorotation mechanism is energetically more favourable than the turnstile rotation mechanism for the regular pentatopal isomerisation of acyclic phosphoranes. Extended Hückel[34] and *ab initio*[35] MO-calculations on PF_5 have produced energy barriers of 1.4 and 4.8 kcal/mol for the pseudorotation and 10.0 and 18.1 kcal/mol for the turnstile rotation mechanism. The high values for the latter energy barrier would seem to rule out turnstile rotation as a possible mechanism for the interconversion of PF_5. On the basis of CNDO/2 calculations, however, which yield respective barriers of 3.5 and 9.1 kcal/mol for the two mechanisms in discussion, Ugi, Ramirez *et al.*[31, 32] concluded that neither of the mechanisms is quantum mechanically impossible. They also pointed out that, as there are four possible turnstile rotation pathways which lead to the same isomerisation as one Berry pseudorotation, the involvement of the former mechanism in interconversion processes becomes relatively more likely from a statistical point of view. Furthermore, in some cyclic oxyphosphoranes. for which pentatopal isomerisation has been confirmed by NMR spectroscopy there may be very high energy barriers to Berry pseudorotation. Variable temperature NMR studies[37] on a polycyclic oxyphosphorane derived from a phosphatrioxaadamantane serve to illustrate such a state of affairs. From the ^{19}F- and ^1H-NMR data, it would appear that there is a permutational isomerisation in this molecule. Solvent dependent ^{31}P-NMR data indicate that this isomerisation is regular *i.e.* it occurs without bond breaking and re-formation. Examination of molecular models reveals that any pentatopal isomerisation by Berry pseudorotation must traverse a prohibitively high energy barrier, but that the experimental data may be nicely explained by the turnstile rotation mechanism. This may be readily appreciated

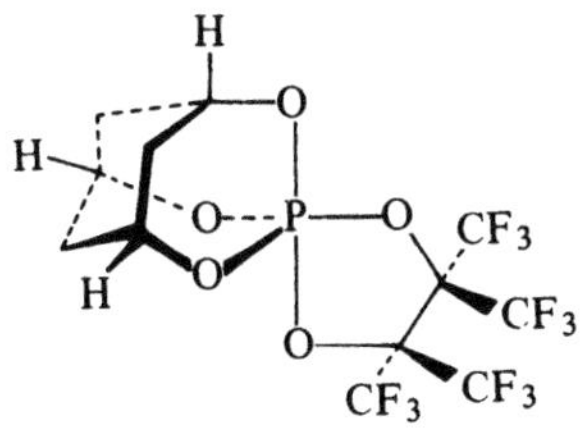

by observing that the five-membered ring oxygens may provide the pair and the adamantoid oxygens the trio for such a process. Therefore, in conclusion, it seems probable that isomerisation of acyclic phosphoranes involves a pseudorotation process, but that for some cyclic derivatives simple or multiple turnstile rotations must be invoked to explain experimental observations.

As mentioned previously, it is outside the scope of this review to pursue this topic in detail. Its relevance to an appreciation of the static stereochemistry of pentacoordinate phosphorus derivatives lies in the fact that, as will be seen in later sections, deviations from idealised trigonal bipyramidal (and square pyramidal) geometries in cyclic derivatives have often been interpreted as being in accordance with a partial pseudorotation or turnstile rotation of the molecule in question. For instance, Ramirez has interpreted X-ray analyses of caged adamantoid polycyclic oxyphosphoranes as supporting the latter mechanism for the permutational isomerisation of such species[37]. He has incorporated the turnstile rotation mechanism in its turn into his "oxyphosphorane concept", which proposes that metastable oxyphosphoranes are intermediates in the nucleophilic displacement reactions of tetracoordinate phosphorus compounds[37]. This concept has been invoked in the design of new types of phosphorylating agents suitable for the synthesis of oligodesoxyribonucleotides.

2.2. Theories of Bonding

Many theories of bonding of pentacoordinate phosphorus have been advanced over the last 20 years, which purport to explain, at least in a qualitative manner, those features of molecular geometry, which have been characterised by recent structural investigations using spectroscopic and diffraction techniques. One of the most fascinating aspects of this area of bonding theory is how models which start from totally opposite conceptual standpoints give rise to closely parallel predictions for the external aspects of molecular structure. For instance, as will be discussed subsequently, a simple Hückel model[38], which altogether neglects explicit electron repulsions, explains the experimentally observed trends in molecular geometry of the trigonal-bipyramidal series $PF_{5-n}Me_n$, $n = 0$–3, which are also unambiguously predicted by the valence shell electron pair repulsion model (VSEPR), which has been developed by Gillespie[39].

The shape of AB_5 complexes with fixed A-B distances of equal length has been investigated in terms of electrostatic repulsion forces[40]. It was found that the trigonal bipyramid is energetically slightly more favourable than an optimised square pyramid with a B_{ap}-A-B_{ba} (ap = apicel, ba = basal) angle of 104.1 and B_{ba}-A-B_{ba}

angles of 86.6 and 151.9°. Variation of the angles ϕ_x and ϕ_y, whilst preserving local C_{2v}-symmetry along the axis A-B^3, showed that the two equivalent trigonal bipyramids TBP and TBP′ (ϕ_x = 90, ϕ_y = 120 and ϕ_x = 120, ϕ_y = 90°) lie at opposite ends of a trough in the potential energy diagram of AB_5, which has an intermediate square pyramidal form SP ($\phi_x = \phi_y$ = 104.1°) at a flat minimum. This is, of course, exactly the proposed pathway for molecular inversion between trigonal bipyramids by the Berry pseudorotation mechanism[14)].

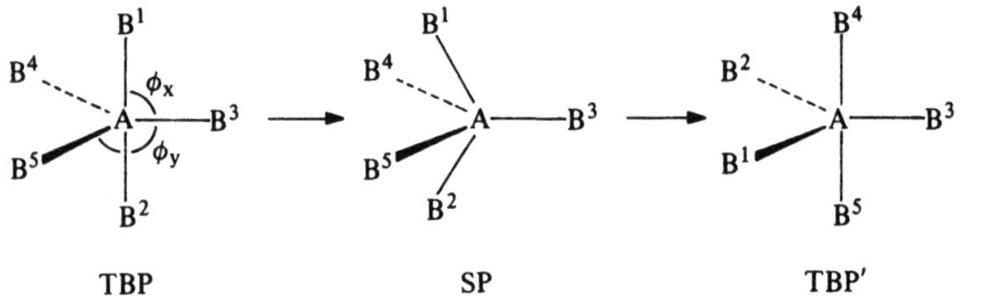

Analogous results were obtained by Foppl[41)], who investigated the optimum arrangement of 2–9 particles on the surface of a sphere (equivalent to introducing fixed bond lengths in an AB_n system, n = 2–9) for an inverse square force law. Zeman[40)] also found that the introduction of non-Coulombic repulsion terms into his calculations served to further increase the relative energetic stabilisation of the trigonal bipyramid.

According to the VSEPR model of Gillespie[39, 42–47)], which has augmented and developed the original simple electrostatic model of Sidgwick and Powell[48)], pairs of electrons in a valence shell, irrespective of whether they are bonding or non-bonding, are always arranged so as to minimise the repulsions between themselves. Gillespie demonstrated that the mutual interactions of the valence shell electrons arising from the Pauli exclusion principle are, in general, more important than those due to electrostatic repulsions. The magnitude of this Pauli exchange force increases rapidly with increasing overlap of the orbitals of two electrons with the same spin and may be represented as being proportional to $1/r^n$, where n is large. The optimum arrangement of particles on the surface of a sphere under such a force law has been investigated by Shutte and Van der Waerden[49)]. They obtained the same results as had been given by Foppl for an inverse square law[41)] in all cases (n = 2–9) except that of seven particles. Thus for AB_5 systems a trigonal bipyramidal rather than a square pyramidal shape was found to be most favourable under any inverse force law.

The VSEPR model provides an unambiguous explanation of all the trends observed in the trigonal bipyramidal methyl-substituted fluorophosphoranes (see Fig. 2.1.1.). Because an axial bonding electron pair has three nearest neighbouring pairs at 90°, while the equatorial pairs have only two such neighbours, minimisation of the repulsions can only be attained if the axial pairs are at a greater distance from the nucleus than the equatorial pairs. Therefore, axial bonds in a trigonal bipyramid must be longer than the corresponding equatorial bonds. Now electron pairs bonding electronegative ligands must be drawn further away from the central nucleus than those bonding less electronegative ligands, and as a consequence will take up less space on the surface of the central atom. Thus electronegative ligands with small electron pairs will tend to occupy the axial positions in order to minimise interactions with other electron pairs, while the larger electron pairs will occupy the equatorial positions where there is more space available for them. Substitution of an equatorial fluorine by a methyl group leads to a decrease in the effective electronegativity of the phosphorus and allows all the bonding pairs to move away from the phosphorus slightly, thereby increasing all the bond lengths. At the same time, there will be increased axial-equatorial repulsions, because of the size of the electron pair bond to carbon, and this will lead to the $P-F_{ax}$ bonds being both further lengthened and bent away from the methyl groups. Thus the VSEPR model is capable of accounting for all the observed structural features of the methyl-substituted fluorophosphoranes without having to make any arbitrary assumptions about the atomic orbitals which take part in the bonding (*e.g.* the role of the phosphorus $3d$-orbitals). The success of such an inherently simple model hints at an underlying simplicity in nature. It cannot, however, tell us anything about the geometrical preferences of π-electron donors and acceptors, independent of their electronegativities, which we shall see are predicted for pentacoordinate phosphorus derivatives by a recent MO-description.

The earliest descriptions of the hybrid orbitals for PF_5 always assumed significant involvement of the phosphorus $3d$-orbitals, which are of low energy and about the same size as the $3s$ and $3p$. Cotton[50)] constructed an orthonormal set of trigonal bipyramidal hybrid orbitals from an sp^3d set of orthonormal atomic orbitals in order to investigate the relative strengths of the axial and equatorial bonds in PF_5 and PCl_5, using the overlap integral criterion of Mulliken[51)]. Calculations, using overlap integrals computed from the Slater orbitals, over the entire range of distribution of *s*- and *d*-character between the axial and equatorial bonds, showed the latter to be stronger over wide ranges and particularly at values corresponding to greatest total overlap. Another model based on overlap considerations, due to Craig *et al.*[52)], which also assumed that a least one $3d$-orbital was of low enough energy to participate in the bonding, showed that the maximum overlap of a substituent orbital with an equatorial orbital of the central atom occurs at a smaller internuclear distance than for an axial orbital.

In total contrast to these approaches was that of electron-rich three-centre bonding proposed by Rundle[53–56)], which suggests that a good first order representation of the structures of PX_5 species may be constructed without an appreciable contribution from the phosphorus $3d$-orbitals. The equatorial P—X bonds may be regarded as normal in this model, while the delocalised three-centre orbitals are constructed for the axial P—X bonding from the phosphorus $3p_z$ and the axial ligand *s*- and *p*-functions. The interaction diagram for the construction of such molecular orbitals for the unknown model compound PH_5 is shown in Fig. 2.2.1. The non-bonding a_1' orbital of this three-centre set is localised on the axial hydrogens. Upon mixing with the equatorial P—H σ and σ^* levels, this a_1' orbital transfers a small but significant part of its electron density onto the equatorial atoms. The P—H axial bonds will naturally be longer than the P—H equatorial bonds in this interpretation, as the linear H—P—H bonds contain one bonding and one non-bonding pair, whereas the latter are electron pair bonds. The character of the non-bonding orbital favours increasing the electron density on the axial atoms, thereby explaining why electro-

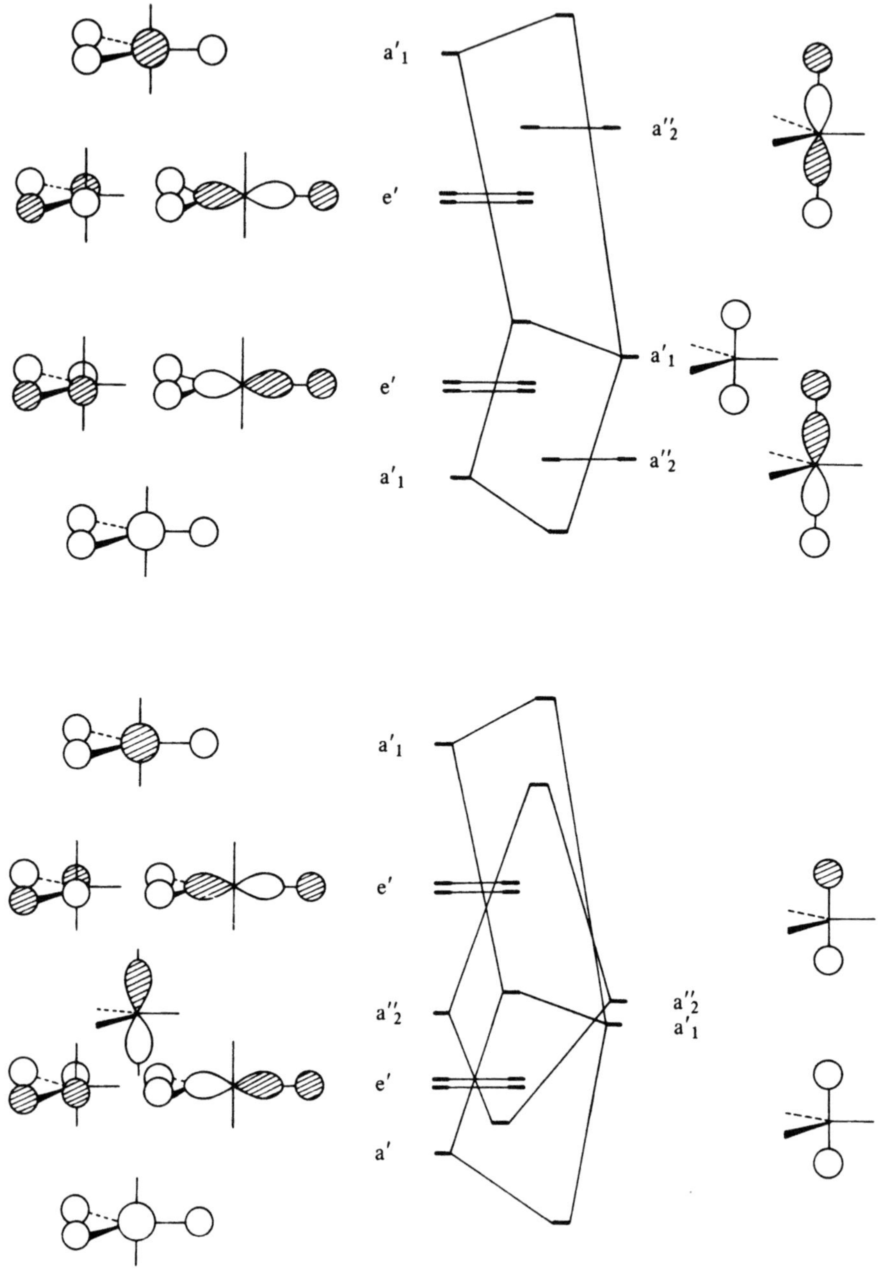

Fig. 2.2.1. Interaction diagrams for the construction of the molecular orbitals of trigonal bipyramidal PH_5.
(a) Semilocalised orbitals as calculated using Rundle's electron-rich three-centre bonding model.
(b) Fully delocalised orbitals as calculated using a semiempirical extended Hückel model (after Hoffmann *et al.*[34])

negative substituents will preferentially occupy these positions. Using a simple Hückel MO model, which considered only σ orbitals in the valence shell, Bartell was able to demonstrate that Rundle's model could explain the trends observed in the fluorophosphoranes[38)]. As he commented, it is somewhat remarkable that both the simple Hückel MO model, which totally neglects explicit electron repulsions, and the VSEPR model, which only considers such repulsions, can be invoked to interpret the same experimental results.

Various semiempirical (extended Hückel MO[34, 57–60)], modified extended Hückel MO[61)], CNDO/2 [32, 62)] VESCF-MO[63)]) calculations on PF_5[32, 34, 57, 60–63)], PH_5[32, 34, 58, 60)], oxyphosphoranes[32, 59)], and the series PF_nCl_{5-n}, $0 \leqslant n \leqslant 5$[57)], have been performed. More recently, these studies have been augmented by non-empirical LCAO-SCF-MO calculations on PH_5[64)], PF_5[35, 65)], and a series of substituted phosphoranes PF_nR_{5-n}[35)]. The resultant MO wave functions have been discussed with particular regard to the following aspects:

1. Qualitative understanding of the differences in bonding between trigonal bipyramidally coordinated phosphorus and axial and equatorial substituents.
2. The role and significance of phosphorus $3d$-orbitals in the fully delocalised molecular orbitals.
3. Possible interaction between π-electron donors and acceptors and the framework σ-orbitals or the phosphorus $3d$-orbitals.
4. Construction of a potential surface for permutational isomerisation by the Berry pseudorotation or turnstile rotation mechanisms.

The conclusions which have been arrived upon may be summarised as follows:

1. The bonding in the hypothetical model compound PH_5 (D_{3h} or C_{4v} geometry) has been discussed in terms of a set of five valence molecular orbitals[34, 58, 64)]. These are, in order of increasing energy (Fig. 2.2.2.), a nodeless orbital (a_1' or a_1), three singly noded orbitals of similar energy ($a_2'' + e'$ or $a_1 + e'$), and one doubly noded high-lying orbital (a_1' or b_1), which is separated by a large gap from the other bonding orbitals. The *ab initio* calculations[64)] led to an inversion of the levels of the a_2'' and e' orbitals, which is however of no consequence to a qualitative understanding of the bonding. The two geometries differ in that for D_{3h} there are no low-lying unoccupied orbitals, whereas for C_{4v} the $3a_1$ molecular orbital is not too high in energy. Of particular interest for an understanding of the bonding in trigonal bipyramidal derivatives is the non-bonding $2a_1'$ orbital, which is essentially distributed only over the hydrogens with very little contribution from the phosphorus $3s$. Inspection of the interaction diagram for the construction of the D_{3h} molecular orbitals (Fig. 2.2.1.) shows that this orbital is derived primarily from the a_1' axial hydrogen combination mixing itself into the planar PH_3 σ and $\sigma^* a'$ orbitals. Cancellation at P $3s$ occurs and comparison of the resultant delocalised molecular orbital shows it to be very similar to that derived from Rundle's semilocalised three-centre electron-rich model. The axial bonding is therefore weaker than the equatorial and the corresponding hydrogens more electronegative. A similar reasoning applies to the basal hydrogens in the square pyramid. Electronegative substituents will prefer to occupy those positions of highest electron density – the axial positions of a trigonal bipyramid, the basal positions of a square pyramid. The description of bonding in the model compound PH_5 appears to be equally valid in describing, in a general manner, the bond-

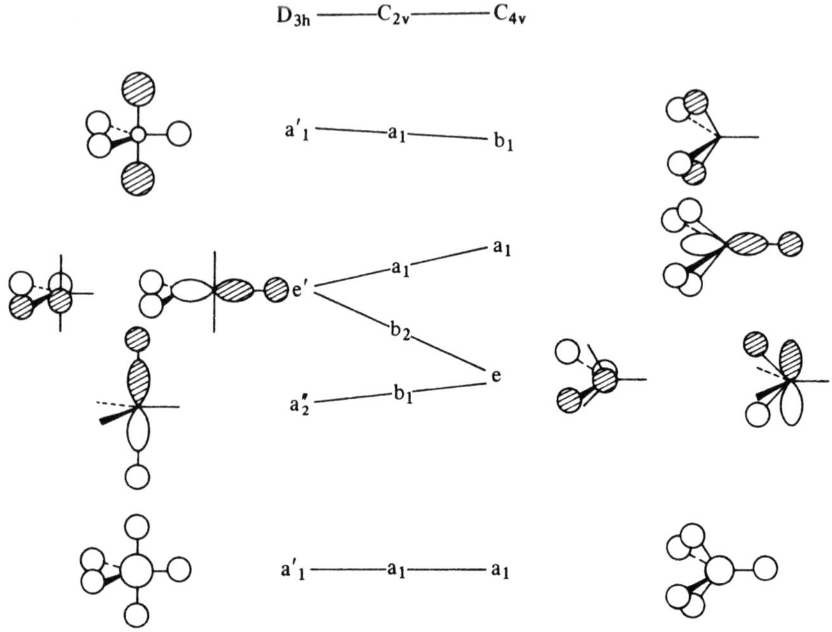

Fig. 2.2.2. Occupied molecular orbitals for D_{3h} and $C_{4v}PH_5$ with the correlation diagram for the Berry pseudorotation transformation

ing in PF_5, with the F $2p$ σ atomic orbitals now playing the role of the H 1*s*-orbitals. Although all MO-calculations have indicated, in agreement with the electrostatic and VSEPR considerations mentioned earlier, that the trigonal bipyramidal geometry should be energetically more favourable for acyclic phosphorus (V) derivatives than the alternative square pyramid, it is worth noting that for particular stoichiometries, distortions in the regular D_{3h} skeleton might reasonably be expected. For PX_3Y_2 molecules, where X is more electronegative than Y, one X ligand must enter an equatorial position. In a skeleton of C_s symmetry with local C_2 and C_3 axes there are three sites of high electron density and it is conceivable that the equilibrium geometry of such a molecule could be distorted from D_{3h} towards C_s, or that the latter could serve as a transition state for permutational isomerisation (*i.e.* that a turnstile rotation mechanism would be involved rather than a Berry pseudorotation). In cyclic PX_3Y_2 derivatives, where both Y atoms are included in a small ring system, one of them is forced to take up an axial position in order to reduce ring strain (see Section 2.4.). For such derivatives a distortion towards C_s symmetry, equivalent to a partial turnstile rotation would be predicted. In PX_4Y derivatives, where X is more electronegative than Y, a C_{2v} distortion from D_{3h} towards C_{4v} skeletal geometry could be favoured (*i.e.* a "frozen" partial Berry pseudorotation might be observed).

2. There is still considerable disagreement about the role of the phosphorus 3*d*-orbitals in bonding. Hoffmann *et al.* in their semiempirical calculations on PH_5 concluded, partly on the basis of pseudosymmetry observations, that the participation of P 3*d*-orbitals will effectively be limited to the stabilisation of the doubly-noded non-bonding orbital[34]. This stabilisation is also apparent in the *ab initio*

calculation[64)], but is simultaneously accompanied by a destabilisation of the same order of magnitude for all other orbitals. The semiempirical study, which contained a $3d$-orbital Slater exponent of 1.4 in good agreement with that obtained from an SCF-calculation on phosphorus[60)], found that the stabilisation is accompanied by a transfer of electron density of approximately 0.6 of an electron from the hydrogens to phosphorus. A more conservative figure of 0.27 of an electron was obtained from the *ab initio* study. The $3d$-orbitals might reasonably be expected to play a more significant role in PF_5, where the possibility of back bonding would be associated with the presence of the more electronegative fluorines. From the results of the semiempirical Hückel calculations, however, the conclusion was drawn that d-orbitals make only minor contributions to the bonding[34, 57)]. These findings contrast sharply with those from CNDO/2 studies by Ramirez *et al.*[31, 32, 37)], which predict substantial participation of the phosphorus $3d$-orbitals in the molecular bonding orbitals. According to their calculations a 25–35% increase in the stability of pentacoordinate phosphorus derivatives is provided by $(p \rightarrow d)\,\pi$ back donation of electron density from the substituents onto the central atom. The axial positions of a trigonal bipyramid will be less involved in d-orbital interaction than the equatorial positions and will consequently tend to be occupied by the most electronegative substituents, with the corollary that the axial bonds will be longer than comparable equatorial bonds. In terms of this bonding model, the assignment of ligand positions will, however, be significantly influenced by d-orbital effects, in addition to electronegativity considerations and ligand-ligand steric interactions. This has led to the introduction of the term "apicophilicity" to describe the proclivity of a particular ligand to assume an axial position in a trigonal bipyramidal phosphorane[32, 37)]. At present, however, there is no unambiguous example of a trigonal bipyramidal phosphorus (V) derivative in which the ligand positional assignment cannot also be explained in terms of the straightforward "polarity rule" based on electronegativity considerations (see Sections 2.3. for further details).

A further concept introduced by Ramirez *et al.* on the basis of the significant bonding role of the phosphorus $3d$-orbitals predicted by their CNDO/2 calculations is that of "stabilisation of ligand-set symmetry", an electronic effect which must be taken into account when comparing the relative stabilities of phosphoranes with different substituent patterns. An overall weakening of the bond system occurs when either the equatorial ligand subset, or the axial ligand subset, contains two different types of substituents. It appears that the mean contribution of a ligand to

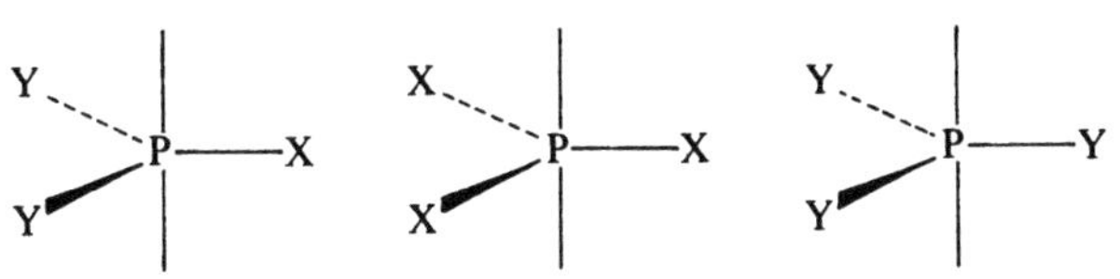

(A) (B) (C)

the total chemical binding energy of a phosphorane is less in a "heterosystem" [*e.g.* (A)] than the mean contribution of the same ligand in a "homosystem" [*e.g.* (B) or (C)]. An analogous, though less pronounced effect is obtained for the axial subset [*i.e.* heterosystem (D) *vs* homosystems (E) and (F)].

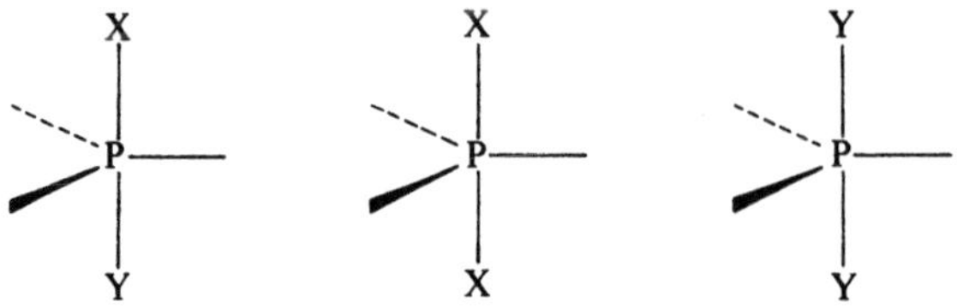

(D) (E) (F)

3. A methodical examination of the geometrical preferences of π-electron donors and acceptors in phosphoranes independent of their electronegativity has been made by Hoffmann *et al.* [34]. Their predictions are based on a theoretical consideration of symmetry restrictions and on computational studies of model donors and acceptors. If $3d$-orbitals are not of importance, the the only chemically significant π-interactions of a substituent will be with the occupied skeletal orbitals – for acceptors stabilising, for donors inherently destabilising. Those axial and equatorial interactions which are allowed by symmetry are shown below for the model compound PH_5. Interactions e_x and e_y are identical by symmetry, whereas b_2 is weaker than its axial counterpart e_y, where the p-type hydrogens are also involved. Calculations on model donors and acceptors showed the b_1 interaction to be of comparable strength to e_x and e_y

e_y 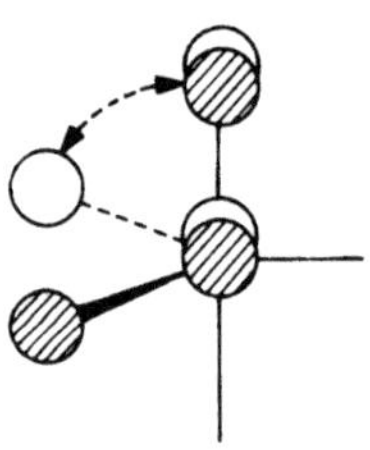

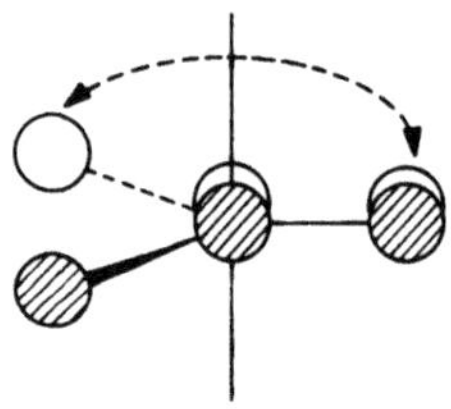 b_2

e_x 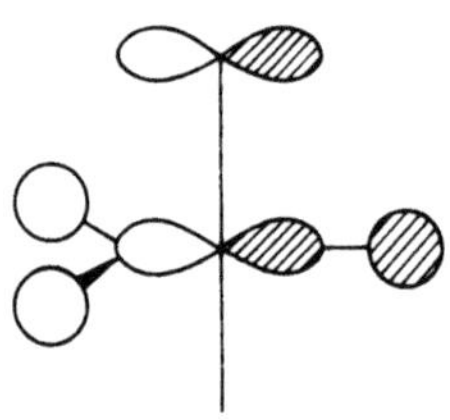

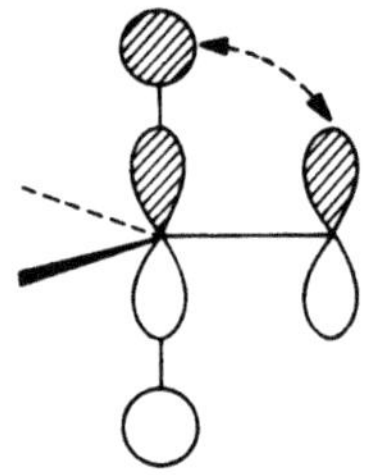 b_1

in every case, allowing the following conclusions to be drawn for trigonal bipyramidal geometry:

a) If $3d$-orbitals are not important, interactions between donor orbitals and framework σ-orbitals are stronger for axial than for equatorial substitution. Thus π-acceptors will prefer axial sites, π-donors equatorial sites.

b) An equatorial single π-system will prefer to have its acceptor orbital perpendicular to the equatorial plane, or its donor orbital in that plane.

The role of the phosphorus $3d$-orbitals, which could interact with donor π-orbitals in a stabilising manner, was investigated for the two extreme cases of high and low overlap. The following less confident conclusions were reached:

a) There is a $(p \rightarrow d)\,\pi$ bonding component with significant effects on computed bond orders over a wide range of d-orbital participation.

b) The difference between axial and equatorial interactions is dominated by the interaction with the framework σ-orbitals rather than with the $3d$-orbitals, except in the case of maximum d-orbital participation.

c) Equatorial $(p \rightarrow d)\,\pi$ bonding is more efficient than axial $(p \rightarrow d)\,\pi$ bonding. The latter finding is in agreement with the CNDO/2 calculations of Ramirez *et al.*, although, of course, they predicted a far more significant role for $(p \rightarrow d)\,\pi$ bonding in phosphoranes.

A similar analysis of donor and acceptor interactions for a square pyramidal geometry reached the following conclusions:

a) In a C_{4v} geometry π-donors will prefer the apicel positions, π-acceptors the basal sites.

b) If the basal substituent carries a single π-system, it will preferentially align that system in the basal plane for an acceptor, but parallel with the apicel bond for a donor.

4. All MO calculations, both semiempirical and *ab initio*, have demonstrated that the barrier to a D_{3h}-C_{4v}-D_{3h}' interconversion, the Berry pseudorotation process, is small for acyclic phosphoranes (see also Section 2.1.). Estimates of 1.4 (extended Hückel MO[34]), 4.8 (*ab initio*[35]) and 3.5 kcal/mol (CNDO/2[31]) have been obtained for PF_5. The barrier computed from the *ab initio* calculation increases to 8.5 kcal/mol if the $3d$-orbitals are not included. The turnstile rotation mechanism traverses a much higher energy barrier: 10.0 (extended Hückel MO), 18.1 (*ab initio*) and 9.1 kcal/mol (CNDO/2). On the basis of these calculations, the Berry pseudorotation mechanism must be the preferred explanation of pentatopal isomerisation in acyclic phosphoranes (see Section 2.1.).

2.3. Acyclic Derivatives

The results of 14 structural determinations of acyclic pentacoordinate phosphorus derivatives using diffraction and microwave techniques are listed in Table 2.3.1. Electron diffraction studies on members of the series $PF_{5-n}Me_n$, $n = 0$–3, the results of which are illustrated in Figs. 2.1.1. and 2.3.1., were extensively commented upon in Sections 2.1. and 2.2., on account of their relevance to the development of bonding theory for pentacoordinate phosphorus. The following general stereochemical features of acyclic derivatives may be recognised on the basis of the presently available information:

1. All the molecules are trigonal bipyramids. No considerable distortion towards a C_{4v} structure for PX_4Y or a C_s structure for PX_3Y_2 derivatives (X more electronegative than Y) is observed.

Table 2.3.1. Bond lengths (Å) in acyclic derivatives of pentacoordinate phosphorus

Compound	r(P-C) Axial	Equatorial	X	r(P-X) Axial	Equatorial	Method of investigation	Ref.
PF_5	–	–	F	1.577(5)	1.534(4)	ED[1])	18)
$MePF_4$	–	1.780(5)	F	1.612(4)	1.543(4)	ED	19)
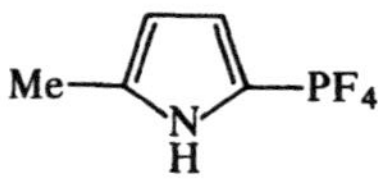	–	1.740(11)	F	1.622(11)	1.555(7)	X-ray[2])	67)
HPF_4	–	–	F	1.594(5)	1.55(3)	MW	68)
Me_2PF_3	–	1.798(4)	F	1.643(3)	1.553(6)	ED	19)
Me_3PF_2	–	1.813(1)	F	1.685(1)	–	ED	20)
$(C_6F_5)_3PF_2$	–	1.819(4)	F	1.638(2)	–	X-ray[2])	69)
$(Me_2N)_3PF_2$	–	–	F	1.632(6)	–	ED	70)
	–	–	N	–	1.674(5)	ED	70)
Cl_2PF_3	–	–	F	–	1.59(3)	ED[3])	22)
	–	–	Cl	2.05(3)	–	ED	22)
PCl_5	–	–	Cl	2.124(9)	2.020(7)	ED	71)
$(CF_3)_2PCl_3$	1.950(11)	–	Cl	–	2.036(2)	ED	72)
$(CF_3)_3PCl_2$	1.946(14)	1.938(31)	Cl	–	2.053(6)	ED	72)
$(C_6H_5O)_5P$	–	–	O	1.663(2)	1.572(5), 1.601(5)	X-ray[4])	73)
$(C_6H_5)_5P$	1.987(6)	1.850(20)	–	–	–	X-ray	74)

[1]) ED = Electron diffraction, X-ray = X-ray diffraction, MW = Microwave spectroscopy.
[2]) Bond distances corrected for librational motion.
[3]) Results of an early ED study (1938).
[4]) The equatorial distances show a statistically significant difference.

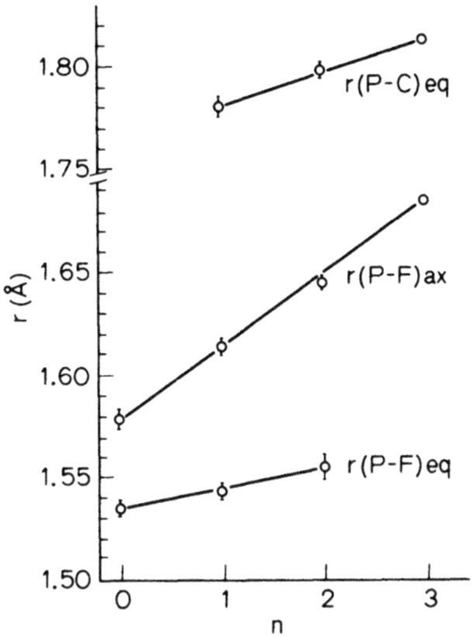

Fig. 2.3.1. Variation of bond distances in the fluorophosphoranes $PF_{5-n}Me_n$, $n = 0-3$, with increasing methyl substitution

2. The ligand distribution over the axial and equatorial sites of the trigonal bipyramid for all unambiguous structural determinations is in accordance with electronegativity predictions. An "apicophilicity" series which differs in order from that based on ligand electronegativities has not as yet been established on the basis of diffraction studies.

3. Axial bonds are longer than corresponding equatorial bonds. Both increase for corresponding bonds, the former more markedly, upon substitution of less electronegative ligands at phosphorus.

4. The orientation of the π-donor orbital of an equatorial substituent is dictated by a compromise between steric interactions and electronic energy and is not normally perpendicular to the axial plane of the trigonal bipyramid for bulky ligands.

The assignment of molecular point group symmetries for a series of phosphorus chlorofluorides PCl_nF_{5-n} by vibrational spectroscopy has been reviewed[75]. The results of an early electron diffraction study on Cl_2PF_3, which predicted that the chlorines are axially substituted[22], have been superseded by a more recent analysis of the vibrational spectrum, which shows the molecule to have the expected C_{2v} point group symmetry[76]. A recent electron diffraction study[72] on the two molecules $(CF_3)_2PCl_3$ and $(CF_3)_3PCl_2$ is of particular interest because NMR investigations by Cavell *et al.*[77–79] have been interpreted as suggesting that in such derivatives the chlorines preferentially occupy the axial sites in the trigonal bipyramid, despite the higher electronegativity of the CF_3 ligands ($\chi_{CF_3} = 3.46$, $\chi_{Cl} = 3.16$)[80]. This led to the conclusion that Cl is more "apicophilic" than CF_3. In contrast, however, the vibrational spectra[81,82] of CF_3PCl_4 and $(CF_3)_2PCl_3$ appeared to confirm the axial position for CF_3 as predicted by electronegativity considerations. The results of the electron diffraction study are in accordance with this latter conclusion, but also suggest that there is a very considerable C_{2v} distortion of the trigonal bipyramidal geometry in $(CF_3)_3PCl_2$ [C_{ax}-P-C_{eq} = 95.5(19), Cl_{eq}-P-Cl_{eq} = 133.0(17)°]. Contrary to the rule observed in all other acyclic derivatives, the P-C_{eq} bond is very long and not significantly shorter than the corresponding axial bond. However a C_{2v} distortion, which should lead to this equatorial carbon taking up some apicel character, will not be associated with a relative weakening of the P-C_{eq} bond. In view of the considerable simplifications and assumptions necessary in an electron diffraction study of a molecule of this complexity, these findings must be treated with some caution. For instance, no investigation was made of a possible distortion towards a C_s structure, which might possibly be expected for such a PX_3Y_2 system.

The preference of π-donor lone pairs on equatorially bonded heteroatoms to take up a position in the equatorial plane, as predicted by symmetry considerations[34), has been confirmed by NMR studies for sulphur in S-substituted thiotetrafluorophosphoranes[83) and for nitrogen in aminophosphoranes[84–87), where activation energies of 5–12 kcal/mol for rotation about the P-N bond have been reported. This activation energy may be regarded as a reasonable lower limit to the P-N π-interaction energy in these molecules and serves therefore as a general qualitative measure of this energy. The diffraction studies on $(C_5H_6N)PF_4$, $(C_6F_5)_3PF_2$ and $(Me_2N)_3PF_2$ provide experimental evidence for the preferred orientation of bulkier equatorially substituted π-donors, for which steric factors would be predicted to play a significant role. The pyrrole ring in $(C_5H_6N)PF_4$, which would be expected to behave as a single π-donor, is found to lie in the axial plane of the trigonal bipyramid, in accordance with theoretical prediction. The very short P-C_{eq} bond length of 1.740(11) Å is strongly suggestive of some degree of interaction between the pyrrole ring π-system and the phosphorus 3*d*-orbitals. In contrast, the NMe_2 groups in $(Me_2N)_3PF_2$ make a dihedral angle τ of 70(3)°, the (C_6F_5) groups in $(C_6F_5)_3PF_2$ of between 33.5 and 36.2° with the equatorial plane. It appears that the preferred orientation for potential π-donors is dictated by an optimum compromise between Coulombic steric repulsion forces and the electronic energy. The former forces would lead to an energy minimum at $\tau = 45°$, the latter at $\tau = 90°$. A weak polymerisation of the $(C_5H_6N)PF_4$ molecules through linear N-H···F_{ax} hydrogen bonds may account for the fact that $\tau = 90°$ for this compound in the crystalline state. The coordination at nitrogen in $(Me_2N)_3PF_2$ is virtually planar in accordance with theoretical predictions for equatorially substituted amino groups in fluorophosphoranes[35).

2.4. Monocyclic Derivatives

The introduction of a cyclic system imposes an additional constraint on the stereochemical permutations available in a trigonal bipyramidal configuration. Strained four- and five-membered rings are found to assume a conformation spanning axial-equatorial sites with the more electronegative ring component in a axial position if the ring ligands differ in character. It is to be expected that the minimisation of ring strain, which is achieved by the assumption of an axial-equatorial rather than an equatorial-equatorial arrangement, will be the dominating factor in the determination of which stereochemistry is energetically preferred, even at the expense of displacing a more electronegative acyclic substituent atom from an axial location. Bone *et al.* have proposed[88] that the energy difference between the isomers *1* and *2*, with an ae (axial/equatorial) and ee (diequatorial) placed five-membered ring respectively (X and Y are heteroatoms), is composed of three terms:

a) The increase in ring strain owing to the increase in the ring angle at phosphorus.

b) The energy required to rotate the lone pair on X from the equatorial to an axial plane.

c) The difference in "apicophilicity" between R and Y, when the lone pair on equatorial Y is constrained to an axial plane.

Their calculations suggested that the energy required to move five-membered rings containing heteroatoms bonded to phosphorus from an ae to an ee placement is considerably greater than is needed in the case of a phospholan ring (for Y=X=C, R=Ph, $\Delta G = 9$ kcal/mol; for Y=X=O, R=OPH, $\Delta G = 17$ kcal/mol). Furthermore, the energy will not only depend on the nature of the heteroatom which moves from an axial to an equatorial position but also on the nature of the atom which remains equatorial *i.e.* factor b). The energy difference between *1* and *2* was also estimated for a number of spirobicyclic derivatives from their dynamic NMR spectra, assuming

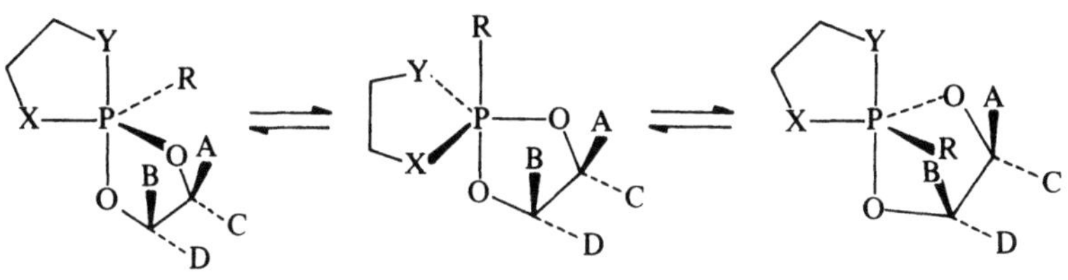

1 *2* *3*

that the high energy ee isomer is present as an intermediate in a pseudorotation between *1* and *3*. Good agreement was obtained between experimental and theoretical ΔG values.

It has, however, been postulated by Ramirez *et al.*[37, 89], in the development of their "oxyphosphorane" concept, that four-/five-membered rings are placed exclusively ae in the trigonal bipyramid even for transitory intermediates. According to their turnstile rotation mechanism an ee intermediate would not, therefore, be involved in the permutational isomerisation of such derivatives. However, a recent structural study of a tricyclic derivative [(I) in Fig. 2.7.1.] has demonstrated that the equatorial angle at phosphorus may be narrowed sufficiently to allow a four-membered ring to be placed ee. This enables one to accept the findings of Bone *et al.* more readily.

The following general stereochemical features of monocyclic derivatives may be recognised on the basis of the presently available information:

1. All the molecules may best be described as distorted bipyramids. No considerable distortion towards a C_s structure is observed in derivatives such as $[F_3PNMe]_2$ for which this might be expected.

2. Strained four- and five-membered rings are found to span axial-equatorial sites according to the following general structural representation.

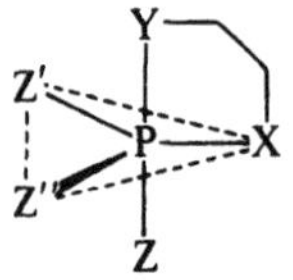

Electronegativities
Z ⩾ Z', Z''
Y ⩾ X
Z > < Y

3. Six-membered rings will prefer to assume a diequatorial placement in the trigonal bipyramidal skeleton where ring strain is in general insignificant and where ligand-ligand steric interactions are minimised. Although less favourable in terms of the latter interactions such rings can also occupy an axial-equatorial skeletal position if this is dictated by other factors *e.g.* electronegativities, π-interaction.
4. The heterocyclic five-membered rings of derivatives (III)-(VI) in Fig. 2.4.1. are planar. The shortness of the formally single C-O, C-N and N-N bonds in these rings indicates the presence of a significant degree of π-delocalisation over the members of the ring. The endocyclic P-O bonds in these derivatives are significantly longer than their exocyclic counterparts (Table 2.4.1.).
5. The diazadiphosphetidines (listed in Table 2.4.2.) are crystallographically centrosymmetric with, therefore, a planar $[PN]_2$ ring system.

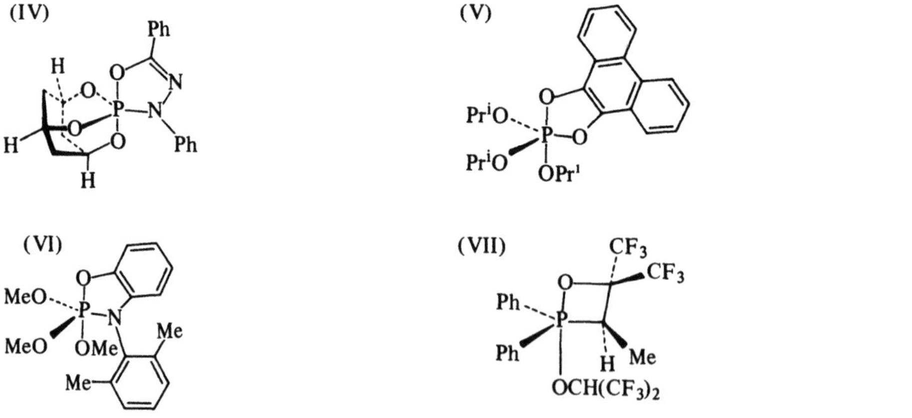

Fig. 2.4.1. Monocyclic and multicyclic pentacoordinate phosphorus derivatives which have been analysed by X-ray diffraction

Table 2.4.1. Bond lengths and angles in monocyclic derivatives of pentacoordinate phosphorus (Fig. 2.4.1.)

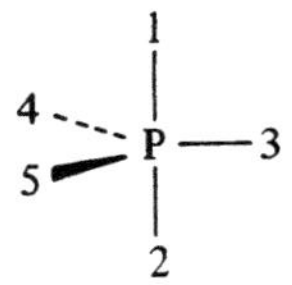

Compound	P-(1)[1]	Bond lengths (Å) P-(2)	P-(3)	P-(4)	P-(5)	(1)-P-(3)	Bond angles (°) (2)-P-(3)	(3)-P-(4)	(3)-P-(5)	(4)-P-(5)	Ref.
(I)	2.01(2)	1.85(2)	1.76(2)	1.90(2)	1.87(2)	71.3(10)	101.2(9)	125.8(9)	119.5(9)	106.8(9)	90)
(II)[2]	2.36	1.83	1.70	1.77	1.78	–	103	119	116	112	91)
(III)	1.749(6)	1.602(6)	1.949(6)	1.573(6)	1.549(6)	83.3(5)	85.2(5)	130.1(5)	121.7(5)	106.6(5)	92)
(IV)	1.754(4)	1.627(5)	1.671(5)	1.590(4)	1.601(5)	83.8(2)	86.4(2)	123.1(3)	129.5(3)	106.3(2)	93)
(V)[3]	1.751(7)	1.638(7)	1.633(7)	1.574(7)	1.588(8)	89.3(3)	88.6(4)	125.5(4)	117.2(4)	117.2(4)	94, 95)
(VI)	1.75(1)	1.61(1)	1.63(1)	1.57(1)	1.60(1)	86.7(7)	89.3(7)	124.3(7)	123.9(7)	111.8(6)	96)
(VII)	1.79(1)	1.71(1)	1.83(2)	1.75(2)	1.82(2)	75.5(6)	91.0(7)	119.3(8)	127.3(7)	112.7(8)	97)

1) This numbering system is used in all subsequent tables.
2) No standard deviations quoted; angle (1)-P-(3) not given.
3) The bond lengths quoted for (V) are those for the monoclinic form[95] and are not corrected for librational motions.

Table 2.4.2. Bond lengths and angles in diaza-diphosphetidines $[R^1R^4R^5P\text{-}NMe]_2$

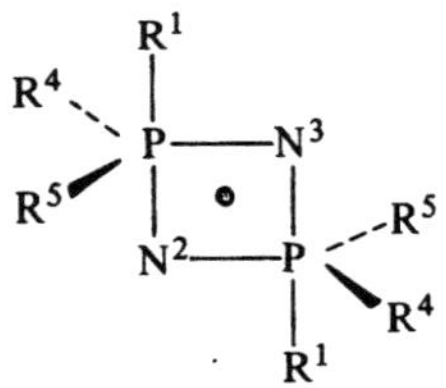

Compound	Bond lengths (Å) P-N^2	P-N^3	P-R^1	P-R^4	P-R^5	Bond angles (°) N^2-P-N^3	R^1-P-N^3	N^3-P-R^4	N^3-P-R^5	R^4-P-R^5	Ref.
R^1=R^4=R^5=Cl[1]	1.769(7)	1.635(7)	2.133(3)	2.029(5)	2.023(3)	80.5(4)	94.5(3)	125.0(3)	124.6(3)	110.0(2)	100, 101)
R^1=R^4=R^5=F[2]	1.735	1.595	1.605	1.570	1.570	77.9	99.1	127.5	127.5	103.9	102)
R^1=R^4=F, R^5=CCl_3	1.742(3)	1.621(3)	1.601(3)	1.559(3)	1.888(5)	79.9(2)	93.9(2)	134.3(2)	117.7(2)	107.9(2)	103)
R^1=R^4=F, R^5=C_6F_5	1.750(4)	1.631(4)	1.606(2)	1.555(3)	1.817(4)	79.8(2)	93.5(2)	126.7(2)	125.6(2)	107.6(2)	104)
R^1=R^4=F, R^5=Ph[3]	1.78(2)	1.64(2)	1.62(2)	1.57(2)	1.79(2)	80.6	92.5	128	122	109	105)
R^1=F, R^4=R^5=Ph[4]	1.780(3)	1.652(3)	1.683(3)	1.832(7)	1.839(4)	78.3(3)	90.9(2)	124.9(10)	122.2(11)	113.0(2)	106)

1) Results from Ref.[100].
2) Electron diffraction study; no standard deviations given.
3) No standard deviations in bond angles given.
4) Mean values for two crystallographically independent molecules.

The multicyclic adamantoid derivatives (III) and (IV) have been included in this section in view of the fact that the six-membered rings are relatively strainfree. Their ring oxygen atoms, one of which is axial and two equatorial in the trigonal bipyramid at phosphorus, may be regarded for classification purposes as displaying the characteristics of an acyclic substituent. Derivatives (I) and (II) have also been included in Fig. 2.4.1. although they both must contain considerable betaine character, in accordance with contributions from the resonance forms:

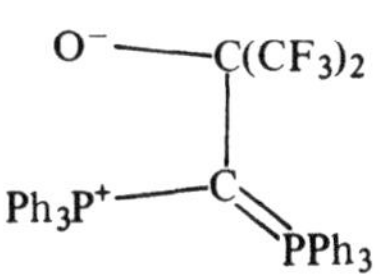

(I)

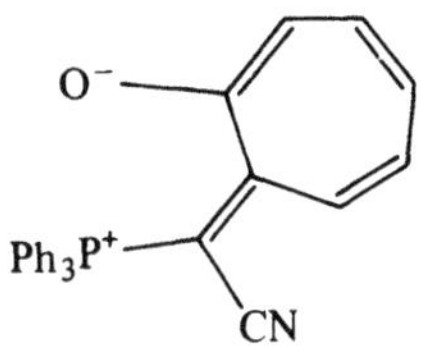

(II)

The axial P-O distance of 2.01(2) in (I), which was isolated as the proposed betaine intermediate of a Wittig reaction, in 0.22 Å longer than in the analogous four-membered ring of (VII). Even more noteworthy is the P-O bond length of 2.36 in (II), which lies between that of 1.75 Å in (III) – (VI) and the sum of the Van der Waals radii (3.30 Å)[98]. Very considerable distortion from a trigonal bipyramidal skeletal geometry is observed for both derivatives. In particular, the phosphorus atom in (II) may be regarded as exhibiting a geometry intermediate between this and that of a tetrahedral configuration (not involving the oxygen). The structural characterisation of derivative (I) further illustrates the role of pentacoordinate phosphorus intermediates in reaction mechanisms.

Ring strain energy calculations[99] on the planar benzodioxaphosph(V)ole five-membered ring system have indicated, in accordance with present experimental evidence, that the trigonal bipyramid is relatively more stable than the alternative square pyramid for such monocyclic derivatives. Stabilisation due to delocalised π-bonding over the constituent atoms of the five-membered ring (and the aromatic ring) was assumed to be of a greater magnitude than ring strain effects, so that planarity is maintained (as experimentally observed). Ring puckering would otherwise be expected, in order to minimise strain effects. Any strain introduced into such a planar ring system as a result of the unequal character of the axial and equatorial bonds in a trigonal bipyramid must be minimised by angle bonding and bond length alterations associated with the various ring atoms. The geometrical consequences for the benzodioxaphosph(V)ole five-membered ring system (where the aromatic C2-C3 double bond is assumed to remain constant) are illustrated in Fig. 2.4.2. The bond C3-04 will increase relative to O1-C2, the angles α_2 and α_4 will increase, whereas α_1 and α_3 decrease. Similar alterations in bond lengths and angles are observed in the ring systems of (III) – (VI) [(V) is illustrated in Fig. 2.4.2.]. It is also apparent from Table 2.4.1. that the endocyclic P-O bond distances in the planar five-membered rings of derivations (III) – (VI) are significantly longer than the corresponding exocyclic distances. Thus endocyclic P-O_{ax} and P-O_{eq} distances of 1.75 and 1.63 con-

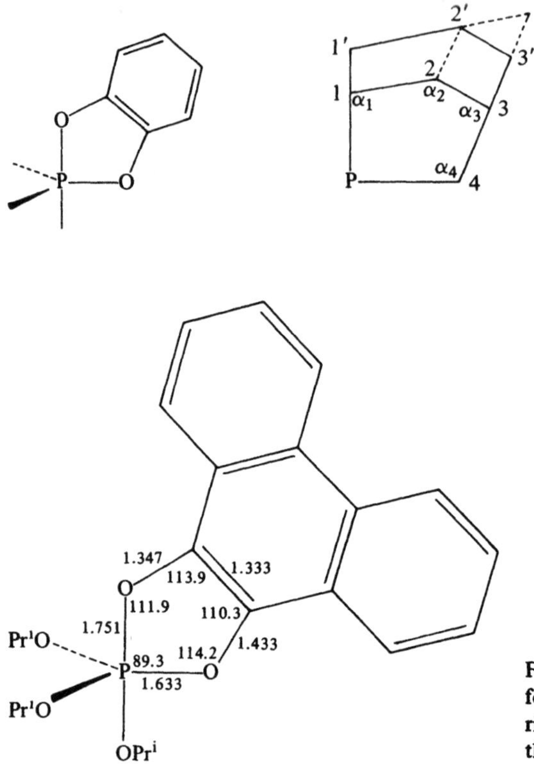

Fig. 2.4.2. The geometrical consequences for the planar benzodioxyphosph(V)ole ring system of the unequal character of the axial and equatorial P-O bonds

trast with exocyclic distances of 1.60–1.64 and 1.55–1.60 Å respectively. Two independent effects may be invoked to explain this phenomenon:

a) Lengthening of the ring axial and equatorial bonds may lead to a reduction in ring strain. Evidence for this effect is provided by the fact that the P-O_{ax} bond distance in the four membered ring of (VII) is even longer (1.79 Å).

b) Delocalisation over the atoms of the five-membered ring leads to a reduction in the degree of π-bonding in the ring axial and equatorial bonds. Further confirmation of this effect is provided by a comparison of the equatorial P-N bond distances of 1.67 and 1.68 in (IV) and (VI) with those in the diazadiphosphetidines (Table 2.4.2. 1.60–1.65 Å), where a higher P-N bond order would be expected.

It is interesting that no distortion towards a C_s structure is observed for the diazadiphosphetidine $[F_3PNMe]_2$, which has a PX_3Y_2 stoichiometry, and in which one of the less electronegative nitrogens is forced to take up an axial position in a trigonal bipyramid. The angular pattern at phosphorus approaches that predicted by MO calculations for an energetically most favourable C_s structure ($\alpha_2 \approx 85°$, $\alpha_3 \approx 95°$)[31, 34, 35]. Thus the equatorial F-P-F angle is narrowed to 103.9° as a result of reduced electron pair repulsion in the bonds to the highly electronegative fluorine, whilst the N-P-N angle is reduced to 77.9° in the four-membered ring system. However, trigonal bipyramidal character is confirmed by the very unequal character of the axial and equatorial P-N bonds (1.735 and 1.595 Å).

The structures of the caged oxyphosphoranes (III) and (IV), which also display very considerable distortions from the perfect perfect trigonal bipyramidal geometry, have been interpreted by Ugi and Ramirez[37] as providing evidence for the turnstile rotation mechanism as the preferred mode of permutational isomerisation in such species. The angular patterns observed do indeed correspond to those predicted for an incipient turnstile rotation *via* a C_s intermediate. The 1-P-3 angle is tilted towards O2 whilst the O4-P-O5 angle contracts as a result of the inclusion of atoms 4 and 5 in the adamantoid moiety. Furthermore, in (IV) the cage P-O axial and equatorial bond lengths are very similar to one another as required for a C_s intermediate. Whether one looks at the distortion in the ideal D_{3h} skeleton or in the ideal geometry of the adamantoid moiety, it may safely be concluded that the caged derivatives (III) and (IV) have structures of relatively high energy and that only minor geometrical modifications are necessary to reach the 30°-TR intermediate barrier configuration. These studies underline the fact that static molecular geometries can provide significant information for the understanding of dynamic processes. However, it would be unwise to carry over the conclusions of this work regarding polytopal isomerisations to other cyclic pentacoordinate phosphorus species.

2.5. Spirobicyclic Derivatives

Extension of the principle of axial-equatorial ring placement to spirobicyclic systems leads to the following general structural representation for a trigonal bipyramidal geometry:

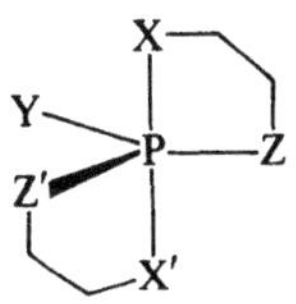

Electronegativities
X > Z, X' > Z'
Y > < X (X'), Z (Z')

A relative stabilisation of the square pyramidal geometry is particularly likely in spirobicyclic derivatives of the type PX_4Y (*i.e.* X = Z in the above general representation), where two of the more electronegative X substituents must occupy equatorial sites in a trigonal bipyramid. A square pyramidal configuration then allows the P-X linkages of each ring to assume more or less equal character and thereby minimise ring strain in comparison to a trigonal bipyramid, where the axial and equatorial linkages are of a substantially different nature. X-ray analyses on a series of spirobicyclic catechol derivatives (*1–5*) listed in Table 2.5.1., in which both electronegativity and ring strain considerations should favour a square pyramid, have demonstrated that this configuration is, in fact, observed. As a result of chelation, the endocyclic O-P-O angles are larger than the associated exocyclic angles, thereby reducing the symmetry of the pyramid base to rectangular rather than square (see Fig. 2.5.1.). However, the small sum of the squared angular deviations from an idealised square pyramidal, as calculated on the basis of the simple electrostatic model of Zemann[40] for an AB_5 system (X-P-X' = 86.6, 151.9, Y-P-X = 104.1°), indicates the basic correct-

Table 2.5.1. Bond lengths and angles in spirobicyclic derivatives of the type $(C_6H_4X_2)_2PY$. Comparison with idealised D_{3h} or C_{4v} geometries

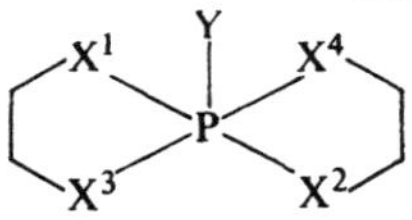

	D_{3h}	C_{4v}	*1* Y=Me, X=O	*2*[1] Y=Ad, X=O	*3* Y=Cl, X=O	*4* Y=OPh, X=O	*5* Y=Ph, X=O	*6* Y=F, X=O	*7* Y=Me, X=S	*8*[2] $(C_6H_5O)_5P$
$P\text{-}X^1$	Axial	Basal	1.667(3)	1.700(3)	1.662(3)	1.650(10)	1.691(4)	1.658(2)	2.204(8)	1.664(5)
$P\text{-}X^2$	Axial	Basal	1.674(3)	1.701(2)	1.658(3)	1.666(10	1.682(4)	1.659(2)	2.209(7)	1.661(5)
$P\text{-}X^3$	Equat	Basal	1.650(3)	1.667(2)	1.641(3)	1.666(10)	1.655(4)	1.625(2)	2.132(8)	1.601(5)
$P\text{-}X^4$	Equat	Basal	1.656(3)	1.663(2)	1.645(3)	1.666(10)	1.650(4)	1.629(2)	2.153(7)	1.572(5)
P-Y	Equat	Apical	1.776(3)	1.836(3)	2.031(1)	1.597(10)	1.775(6)	1.546(2)	1.82(2)	1.572(5)
$X^1\text{-}P\text{-}X^2$	180	151.9	156.6(1)	157.2(1)	162.9(1)	160.0(2)	160.0(2)	168.2(1)	158.0(4)	176.5(12)
$X^3\text{-}P\text{-}X^4$	120	151.9	147.9(1)	144.0(1)	149.8(1)	151.4(21)	145.4(2)	146.1(1)	143.5(4)	126.2(7)
$Y\text{-}P\text{-}X^3$	120	104.1	106.3(2)	108.1(1)	105.3(1)	105.4(10)	106.1(2)	106.6(1)	109.4(8)	118.9(6)
$Y\text{-}P\text{-}X^4$	120	104.1	105.8(2)	107.9(1)	104.9(1)	103.3(10)	108.5(2)	107.3(1)	107.2(8)	114.9(6)
$Y\text{-}P\text{-}X^1$	90	104.1	101.7(2)	100.9(1)	98.9(1)	98.0(9)	100.0(2)	95.8(1)	100.4(7)	91.3(4)
$Y\text{-}P\text{-}X^2$	90	104.1	101.8(2)	101.9(1)	98.3(1)	102.0(10)	100.0(2)	96.0(1)	101.6(8)	91.8(4)
$X^1\text{-}P\text{-}X^3$	90	86.6	90.4(1)	89.8(1)	90.4(1)	92.4(8)	89.9(2)	91.7(1)	91.4(3)	89.0(4)
$X^2\text{-}P\text{-}X^4$	90	86.6	89.9(1)	89.4(1)	91.0(1)	91.5(8)	90.0(2)	91.2(1)	90.6(3)	92.2(4)
$X^1\text{-}P\text{-}X^4$	90	86.6	83.3(1)	83.9(1)	85.1(1)	83.3(8)	84.0(2)	84.5(1)	83.0(3)	88.0(4)
$X^2\text{-}P\text{-}X^3$	90	86.6	83.6(1)	82.9(1)	84.6(1)	83.0(8)	84.2(2)	85.7(1)	81.2(3)	88.1(4)
$\Sigma\Delta^2(D_{3h})$	0	2756.7	2106.9	1749.0	1827.0	2218.1	1640.3	1284.0	1683.7	96.3
$\Sigma\Delta^2(C_{4v})$	2756.7	0	102.1	173.8	228.2	191.3	199.8	502.7	246.6	1957.8
Ref.	–	40)	107, 108)	109)	110)	111)	110)	107, 112)	113)	73)

1) Ad = adamantoid group.
2) This acyclic trigonal bipyramidal molecule is included for comparison purposes.

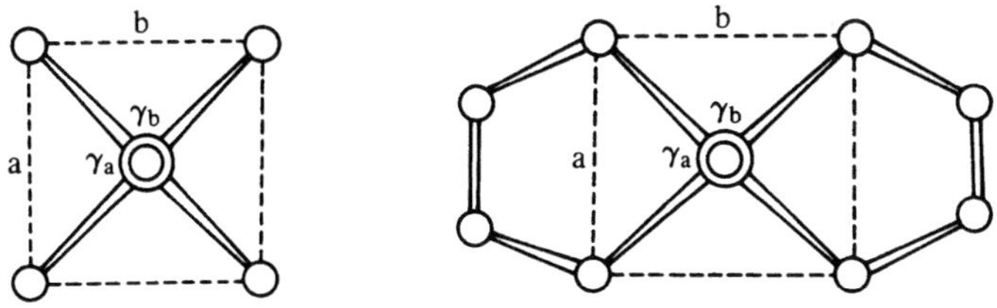

$a = b, \gamma_a = \gamma_b$ $a > b, \gamma_a > \gamma_b$

Fig. 2.5.1. (1) Square pyramidal and (2) chelate rectangular pyramidal configurations viewed along the Y-P bond

9.

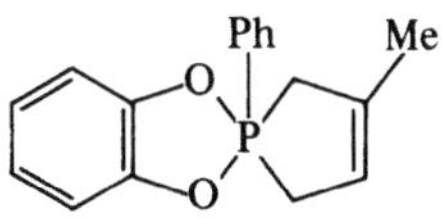

10.

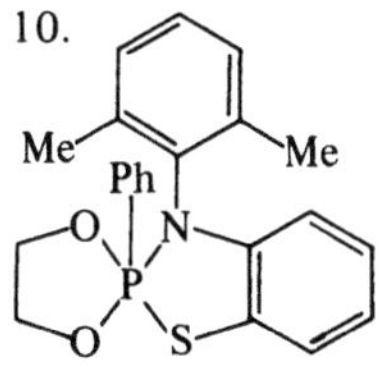

11.

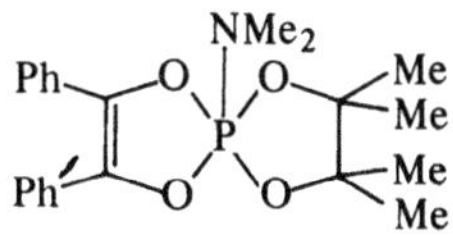

12.

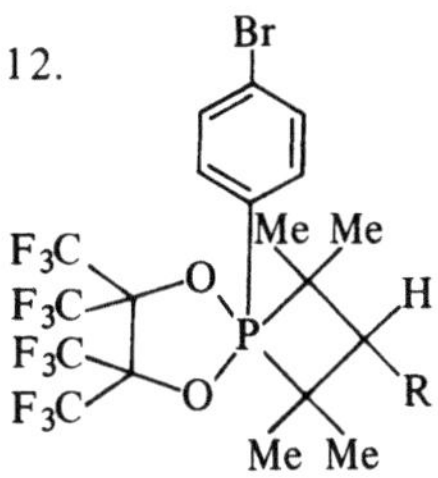

R = H, Me

13.

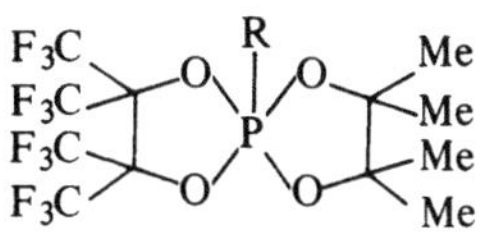

R = SPh, p–OC_6H_4Br

14.

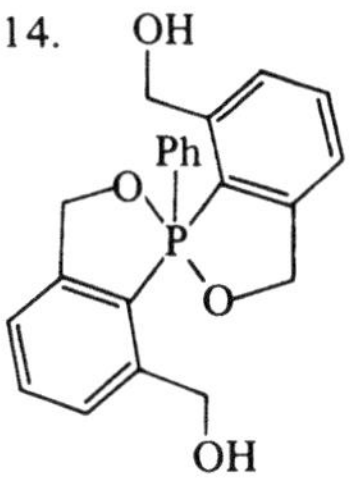

15.

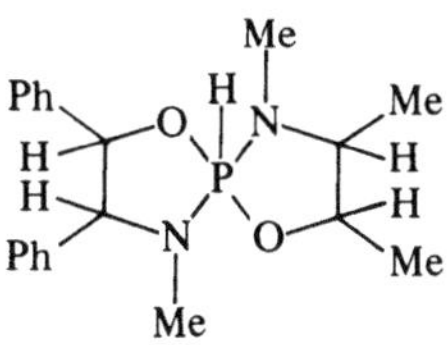

16.

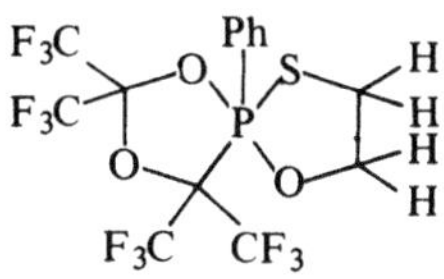

17.

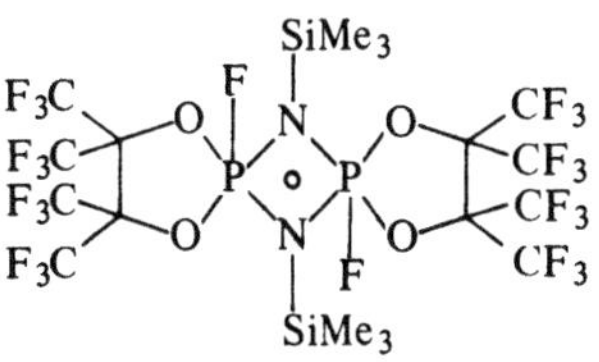

18.

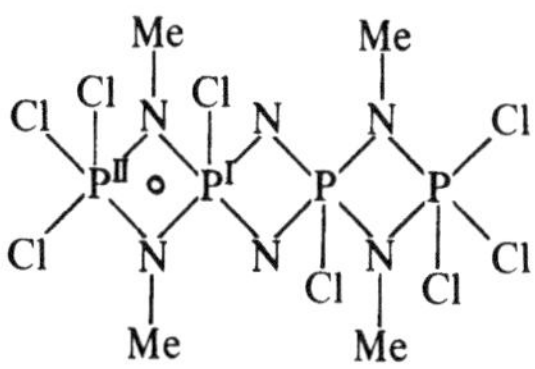

Fig. 2.5.2. Spirobicyclic pentacoordinate phosphorus derivatives which have been analysed by X-ray diffraction

ness of this configuration in describing the angular pattern in derivatives *1–5*. In general, the sum of the squared deviations $\Sigma\Delta^2$ should only be regarded as providing evidence about the degree of square pyramidal character in a pentacoordinate derivative when used in conjunction with that obtained for a trigonal bipyramid. Equivalence of the P-X_{ba} bonds and the X_{ba}-P-X'_{ba} and Y_{ap}-P-X_{ba} angles, (ba = basal, ap = apical), may be regarded as being characteristic for the square pyramidal geometry. It may be seen from Table 2.5.1. that the bond lengths and angles indicate that a small degree of trigonal bipyramidal character is preserved in derivatives *1–5*. This is evidenced, in particular, by the Y-P-X bond angles. A recent theoretical calculation of the ring strain in spirobicyclic benzodioxaphosph(V)ole derivatives has confirmed that for Y = O- the rectangular pyramid is about 5 kcal/mol more stable than the isomeric trigonal bipyramid[99]. The observation of rectangular pyramidal geometry in such catechol derivatives has led to a surge of interest over the last 2–3 years in the structural chemistry of spirobicyclic species, such that the large variety of molecules illustrated in Fig. 2.5.2. have now been characterised by C-ray diffraction. The following general stereochemical principles for spirobicyclic derivatives may now be recognised on the basis of presently available information:

1. Rectangular pyramidal geometry is observed, as a result of the following complementary effects, in spirobicyclic derivatives of the type $(C_6H_4O_2)_2PY$, where Y is less or equal in electronegativity to oxygen:

a) The four highly electronegative oxygens may occupy the electron-rich fourfold basal positions of the rectangular pyramid. In a trigonal bipyramid, two would be forced to take up an electronically unfavourable equatorial placement.

b) The rectangular pyramidal geometry possesses ligand subset symmetry (4 basal X, 1 apical Y) for such PX_4Y derivatives, which should lead to its stabilisation relative to the trigonal bipyramid, which must display a mixed equatorial ligand subset (2X, 1Y).

c) Stabilisation due to delocalised π-bonding over the constituent atoms of the five-membered ring (and the aromatic ring) is of a greater magnitude than any relaxation of ring strain effects which might be achieved by ring puckering. For the resultant planar five-membered ring system a greater reduction of ring strain may be achieved by the assumption of a rectangular pyramidal geometry, in which the P-O linkages are essentially equivalent.

It will be seen in the following discussion that a rectangular/square pyramidal geometry is only observed for spirobicyclic derivatives with highly strained ring systems, indicating that alleviation of ring strain is the determing factor in the relative stabilisation of this configuration. Any change in this combination of ligand electronegativity, ligand subset symmetry and (particularly) ring strain, would be expected to lead to a relative stabilisation of the trigonal bipyramid. Sufficient information is now available to allow analysis of the relative significance of these factors.

2. A significant relative increase in the electronegativity of Y in comparison to X (even when the X substituents are themselves highly electronegative), or a decrease in the electronegativity of the X substituents alone (in comparison to oxygen), in spirobicyclic derivatives of the type $(C_6H_4X_2)_2PY$, should result in a reduction in the relative stability of the rectangular pyramid. According to the VSEPR model such changes in the electronegativities of X and Y will lead to an increase in the

electron pair repulsions between the P-X bonds and therefore to a C_{2v} distortion towards a trigonal bipyramid, with the X substituents in both axial and equatorial positions, in order to minimise these forces. From an MO standpoint such a distortion would also be predicted on the following grounds:

a) If Y is much more electronegative than X then its forced occupation of the electronically less favourable apical site in a rectangular pyramid will lead to a relative destabilisation of this geometry.

b) If X is less electronegative than oxygen then the occupation by this substituent of the electron-rich fourfold basal positions in a rectangular pyramid will become relatively less electronically favourable than for oxygen. For derivative *6* (Y = F, X = O) a structure intermediate between the two isomeric geometries is indeed observed. In contrast the dithiophosphole *7* (Y = Me, X = S) still displays a basically rectangular pyramidal structure although the P-S bond distances now indicate some degree of trigonal bipyramidal character. However, ring strain differences between the two possible geometries would be expected to be lowered in this example, because of both longer ring distances and reduced force constants for the sulphur bonds. In addition, ring puckering is observed in *7*, thereby providing a further alleviation of strain. Unfortunately the necessary structural information is lacking on thio derivatives to allow an estimation of ring strain effects in such systems, as has been performed on the analogous oxyphosphoranes. But the observation of pronounced rectangular pyramidal character must indicate that despite the relieving factors mentioned above ring strain is still very significant in *7*.

3. A reduction in ligand subset symmetry to $PX_2X'_2Y$ in highly strained spirobicyclic derivations would likewise be expected to lead to a reduction in the relative stability of the rectangular/square pyramid even for derivatives containing highly electronegative heteroatoms. The presence of two sets of bond properties may well largely offset the differential ring strain effect favouring the rectangular/square pyramid. Increasing trigonal bipyramidal character would be predicted for the series with mixed heteroatoms depicted in Fig. 2.5.3. Experimental data is lacking on such mixed strained spirobicyclics with the exception of derivative *9* for which a near square pyramid is observed despite the reduced electronegativity of carbon relative to oxygen. In such mixed derivatives the pyramid base will lose its rectangular symmetry and so the geometry is best described in terms of its approach to a perfect square pyramid. It is interesting, however, that the two P-C distances are very different and display values typical for axial and equatorial bonds [1.966(14) and

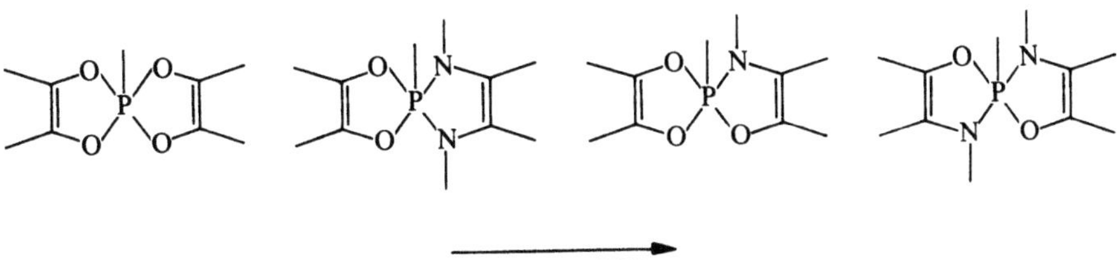

Fig. 2.5.3. Spirobicyclic derivatives with unsaturated ring systems containing heteroatoms bonded to phosphorus. Postulated order of increasing trigonal bipyramidal character

1.822(11) Å], whereas the P-O distances are identical, though longer than in the rectangular pyramidal derivatives *1–5*.

4. Trigonal bipyramidal geometry is observed for spirobicyclic derivatives with two saturated five-membered ring systems (or for unsaturated systems when extended π-delocalisation over the ring atoms is not possible *e.g.* derivative *14*), even when the ring substituents at phosphorus are highly electronegative. Ring puckering will produce a reduction in the ring strain in such derivatives, which is energetically more favourable than the stretching or compressing of bonds. Derivatives *13–16* all display a basically trigonal bipyramidal geometry, as does *11* (albeit distorted) which also includes one unsaturated heteroatom ring system. The geometry of 10 is intermediate between the two configurations.

5. The presence of four-membered rings in a spirobicyclic derivative will lead to an increased ring strain differential and should therefore enhance the relative stabilisation enjoyed by the square pyramid for a wider range of substituent electronegativities. A near square pyramid is indeed observed for the derivatives *12a* and *12b*, which contain saturated four- and five-membered rings. This feature serves to emphasise the decisive role of ring strain in determining the equilibrium geometry of spirobicyclic derivatives.

6. Distorted trigonal bipyramidal geometry is observed for the extended spirobicyclic diazadiphosphetidines *17* and *18*. This is to be expected, as the presence of two different P-N distances does not lead to additional ring strain in the planar diphosphetidine ring system because of its C_i symmetry. From Table 2.5.2. it may be seen that the extended spirobicyclic system leads to considerable angular distortion at phosphorus in these two derivatives.

Sarma *et al.*[111)] have argued that the structure of derivative *4* is better described as a 15°-TR configuration rather than a slightly distorted rectangular pyramid as suggested here. However, the equivalence of all four P-X bond lengths is difficult to explain in terms of this model. Furthermore, the shortness of the exocyclic P-O bond is characteristic for the rectangular pyramidal configuration. The angular pattern also approximates well to that of the analogous derivatives in Table 2.5.1. It is, therefore, not necessary to introduce the basically dynamic turnstile rotation concept as a means of describing the static geometry of such a spirobicyclic derivative.

2.6. Fused Bicyclic Derivatives

Extension of the principle of axial-equatorial ring placement to fused bicyclic systems leads to two possible structural representations for a trigonal bipyramidal geometry, namely with either equatorial or axial annelation, which are capable of

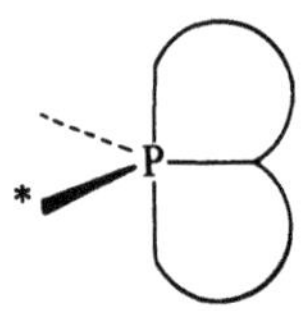

aea

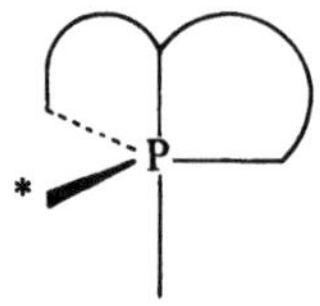

eae

Table 2.5.2. Bond lengths and angles in spirobicyclic derivatives 9–18 (Fig. 2.5.2.)

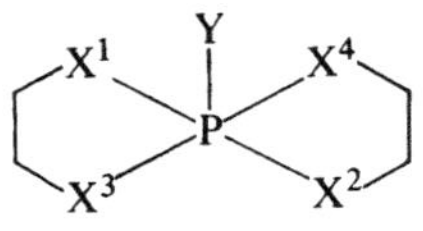

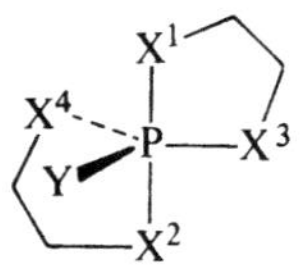

	9	*10*[2])	*11*	*12(a)*	*12(b)*	*14*	*15*[1])	*16*	*17*	*18*(I)[2])	*18*(II)[2])
Structure	SP	TBP/SP	TBP	SP	SP	TBP	TBP	TBP	TBP	TBP	TBP
P-X^1	1.724(2)	1.65	1.661(1)	1.74(2)	–	1.715(4)	1.703(7)	1.772(4)	1.704(7)	1.74	1.71
P-X^2	1.966(14)	2.187	1.714(1)	1.96(2)	–	1.737(4)	1.697(8)	1.632(4)	1.737(3)	1.77	2.16
P-X^3	1.724(2)	1.63	1.660(1)	1.68(2)	–	1.846(6)	1.698(10)	1.928(4)	1.648(6)	1.64	1.66
P-X^4	1.822(11)	1.72	1.673(1)	1.89(2)	–	1.823(5)	1.681(10)	2.080(2)	1.626(8)	1.67	2.05
P-Y	1.807(4)	1.81	1.652(2)	1.82(2)	–	1.866(5)	–	1.805(5)	1.543(6)	2.10	2.05
X^1-P-X^2	150.0(4)	163	169.7(1)	154.2[2])	155.3[2])	178.1(2)	176.6(4)	168.3(2)	173.5(3)	177.1	174.3
X^3-P-X^4	152.5(4)	139	129.9(1)	147.7	148.1	133.9(2)	123.3(2)	127.3(2)	133.1(4)	125.3	123.8
Y-P-X^3	102.2(1)	109	119.2(1)	101.6	100.6	113.0(2)	–	117.1(2)	110.3(3)	116.8	126.7
Y-P-X^4	104.9(3)	112	110.8(1)	110.7	111.3	113.1(2)	–	114.9(2)	116.5(4)	117.8	109.4
Y-P-X^1	102.2(1)	97	95.2(1)	100.0	99.5	90.9(2)	–	92.6(2)	91.9(3)	93.5	93.6
Y-P-X^2	105.8(4)	99	95.1(1)	105.6	105.0	90.9(2)	–	98.8(2)	94.5(3)	89.4	89.3
X^1-P-X^3	86.3(1)	91	92.5(1)	85.0	86.2	88.4(2)	89.7(4)	84.6(2)	88.4(3)	80.0	80.7
X^2-P-X^4	90.5(5)	89	89.8(1)	78.1	78.3	87.8(2)	88.6(4)	92.8(2)	84.1(3)	79.0	91.0
X^1-P-X^4	92.6(4)	88	86.2(1)	90.3	90.1	90.8(2)	91.8(4)	85.0(2)	92.7(4)	100.2	92.8
X^3-P-X^2	77.5(4)	81	82.8(1)	92.7	92.3	91.6(2)	92.9(4)	87.6(2)	89.7(3)	98.1	93.6
Ref.	114)	115)	116)	117)	117)	118)	119)	120)	121)	122)	122)

[1]) Proton bonded to phosphorus was not refined.
[2]) No standard deviations given.

interconversion by Berry pseudorotation about the bond marked with an asterisk as a pivot. The aea configuration with equatorial annelation has been observed for all those bicyclics which have been studied by X-ray diffraction (Fig. 2.6.1.). However, the assumption of this configuration for derivatives (I) – (VI) [with the exception of the fused diphosphetidine (IV)] is already predestined by the aspiration of the most electronegative ring substituent (oxygen) to occupy the axial sites of the trigonal bipyramid. A further corroboratory factor, which is presumably decisive for deriva-

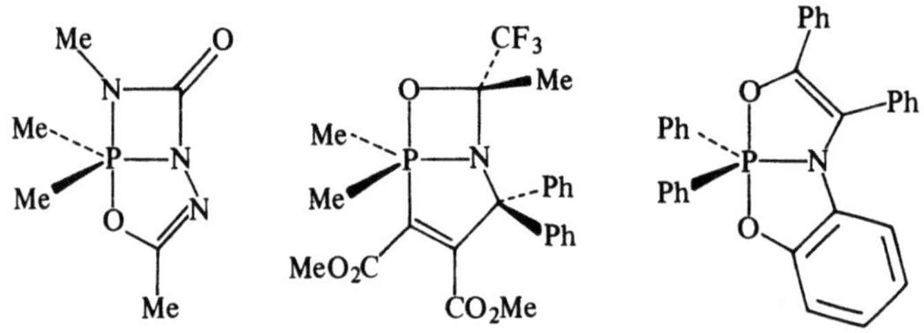

(I) 4:5 (II) 4:5 (III) 5:5

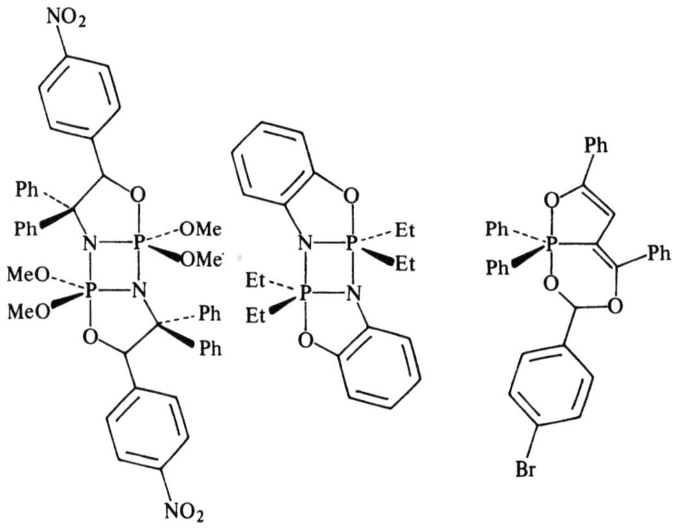

(IV) 5:4:5 (V) 5:4:5 (VI) 5:6

Fig. 2.6.1. Fused bicyclic pentacoordinate phosphorus derivatives which have been analysed by X-ray diffraction

tive (IV), is the fact that only an equatorially substituted nitrogen bridgehead is capable of adopting the preferred planar trigonal coordination. This geometry at nitrogen is always observed in the planar diphosphetidine $[PN]_2$ ring system. A trigonally coordinated equatorial nitrogen bridgehead is also observed in derivatives (I), (III) and (V), where extended π-delocalisation is possible over the atoms of the fused four- and five-membered rings, leading to planarity of one five-membered ring (that with a fused benzene ring) in (III), or of the whole bicyclic ring system in (I) and (V). In contrast, the equatorial nitrogen in (II) is non-planar, indicating that the stabilisation which may be achieved by the reduction of ring strain through puckering of the constituent rings of a bicyclic system, for which extended π-delocalisation is not possible, is more important than that which could be achieved by a potential π-bonding component in the P-N bond.

Recent studies of tricyclic derivatives [Fig. 2.7.1. (II) and (III)] have, however, demonstrated that an eae configuration is possible if the two exocyclic phosphorus substituents of the bicyclic become themselves members of a five-membered ring (*e.g.* molecule C). This suggests that this configuration would also be adopted if dictated by the ligand electronegativities *e.g.* for a bicyclic system with a nitrogen bridgehead, two endocyclic substituents of low electronegativity and two exocyclic substituents of high electronegativity as illustrated by B. Unfortunately structural studies on such potentially eae bicyclic systems have yet to be carried out, so that information concerning the relative importance of ligand electronegativity (favouring eae) and possible extended π-delocalisation over the members of the fused ring systems (favouring aea) is not available.

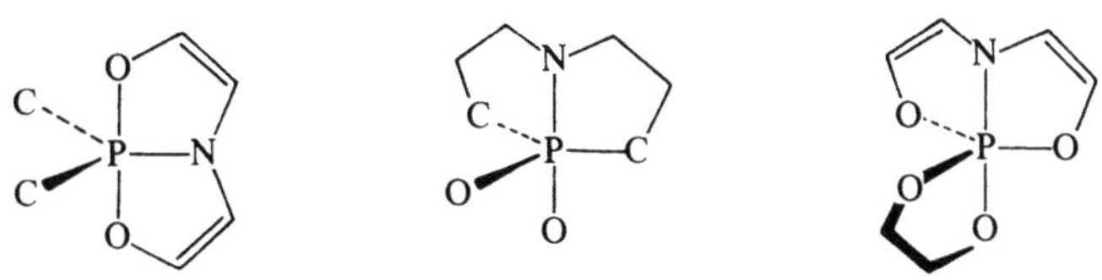

A aea *B* eae *C* eae

On the basis of the structural information for derivatives (I) – (VI) it is possible to recognise the following stereochemical trends in fused bicyclic derivatives:

1. The deviation from axial linearity increases on going from fused five/six- to five/five- to five/four-membered bicyclic systems, in accordance with increasing ring strain. The fused diazadiphosphetidines (5/4) (IV) and (V) assume an intermediate position between 5/5 and simple 5/4 bicyclic systems, which is to be expected on account of the lesser degree of strain in the centrosymmetric $[PN]_2$ ring.
(VI) $3.1^\circ <$ (III) $8.4^\circ <$ (IV) $15.5^\circ \approx$ (V) $15.6^\circ <$ (II) $20.1^\circ <$ (I) 25.3°
The increase in the deviation on going from (II) to (I) is also predictable, because of the planar bicyclic system and shorter axial bond in the five-membered ring which lead to an increased ring strain in the latter derivative.

2. The $P\text{-}O_{ax}$ distances are longer than those observed for acyclic bonds and increase with increasing ring strain *i.e.* on going from puckered to planar five-membered rings to four-membered rings. The former increase is exemplified by derivative (III) where $P\text{-}O^1$ in the puckered five-membered ring is 1.700(5), whereas $P\text{-}O^2$ in the planar five-membered ring (fused to a benzene ring) is 1.763(5) Å. The P-N axial and equatorial distances are somewhat longer for acyclic bonds though the differences are not so marked as for the P-O bond (Table 2.6.1. and 2.8.2).

The angular pattern observed for the 5/5 bicyclic (III) may be interpreted to a good approximation as an incipient pseudorotation (15% transition) to the eae con-

Table 2.6.1. Bond lengths and angles in fused bicyclic derivatives of pentacoordinate phosphorus (Fig. 2.6.1.)

	Bond lengths (Å					Bond angles (°)					
Compound	P-(1)	P-(2)	P-(3)	P-(4)	P-(5)	(1)-P-(3)	(2)-P-(3)	(3)-P-(4)	(3)-P-(5)	(4)-P-(5)	Ref.
(I)[1]	1.799(2)	1.798(1)	1.676(2)	1.709(7)	1.858(6)	71.9(1)	82.9(1)	122.2(5)	125.0(4)	112.8(1)	123)
(II)	1.811(3)	1.888(4)	1.684(3)	1.812(6)	1.805(8)	74.2(1)	86.0(2)	125.8(3)	122.8(2)	109.9(3)	124)
(III)	1.700(5)	1.763(5)	1.703(6)	1.818(5)	1.811(6)	86.6(3)	87.6(3)	130.6(4)	116.5(4)	112.8(5)	125)
(IV)	1.782(4)	1.694(4)	1.667(4)	1.591(4)	1.590(4)	77.4(2)	88.1(2)	118.8(2)	135.3(2)	105.7(2)	126)
(V)	1.812(4)	1.774(4)	1.681(4)	1.812(6)	1.819(6)	76.4(2)	88.0(2)	–	–	111.3(2)	127)
(VI)	1.786(6)	1.737(6)	1.815(6)	1.821(7)	1.824(7)	87.3(2)	92.7(2)	124.0(2)	119.1(2)	116.8(2)	128)

[1]) P-C^4 and P-C^5 in (I) are chemically meaningless and result from an inadequacy in the crystallographic model. Refinement in $P2_1/m$ instead of $P2_1$ leads to a sensible value of 1.80 Å for the now equivalent P-C bonds.

figuration with C^5 as pivot[118]. The O-P-C^5 angles are widened to 93.3(4) and 94.7(4)° (theoretical 94.5°) and the N-P-C^4 angle to 130.6(4)° (129°), whereas the O-P-O angle narrows to 171.6(3)° (171), the N-P-C^5 angle to 116.5(4)° (115.5) and the C-P-C angle to 112.8(5)° (115.5°). The shortness of P-C^4 and its striking difference to P-C^5 in (I) is chemically meaningless and results from the inadequacy of the crystallographic model for this molecule with approximate C_m-symmetry in the non-centrosymmetric space group $P\,2_1$.

2.7. Tricyclic Derivatives

Two possible configurations are available for tricyclic derivatives, which contain a fused bicyclic and a single ring system, namely aea/ee and eae/ae placements for these systems. If the single ring system adopts the energetically more favourable ae placement then the bicyclic system will be forced to adopt an eae configuration with axial

aea/ee eae/ae

annelation. This has been observed for derivatives (II) and (III) in Fig. 2.7.1. A reduction of the equatorial angle between the non-ring substituents at phosphorus to 106–117° has, however, been observed for fused bicyclic systems (Table 2.6.1.). The flexibility of this angle suggests that it might be capable of narrowing still further in order to accomodate a small ee placed ring. This has recently been confirmed for derivative (I) in which the four-membered phosphetan ring is diequatorially placed. An ae placement would force the two highly electronegative oxygens of the fused

(I) (II) (III)

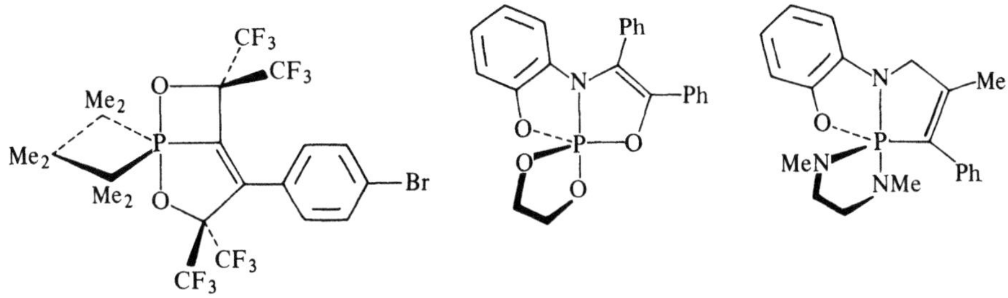

aea/ee eae/ae eae/ae

Fig. 2.7.1. Tricyclic pentacoordinate phosphorus derivatives which have been analysed by X-ray diffraction

bicyclic system to adopt the electronically unfavourable equatorial site. It is probable that ee placement will only occur in stable species for small rings with endocyclic P-C bonds. As discussed in Section 2.4. a heteroatom equatorially bonded to phosphorus will preferentially orientate itself so that its π-donor orbital is in the equatorial plane. This is not possible for ee placement making the ae placement relatively much more energetically favourable for such heterorings than for rings with two P-C bonds. Furthermore, in derivatives (II) and (III) an aea/ae configuration will also provide the most favourable ligand arrangement from the standpoint of ligand electronegativities.

It may be seen from Table 2.7.1. that very considerable distortion of the ideal trigonal bipyramidal skeleton is necessary to accomodate the ee ring placement. The distortion may also be interpreted as being towards a 0°-TR configuration with the pair being the O and C of the oxaphosphetan ring. This structure demonstrates that ee placement of small rings, which is required for high energy intermediates in the permutational isomerisation of a variety of spirophosphoranes if this takes place by a Berry pseudorotation[87], is energetically reasonable.

Although the axial nitrogen in derivatives (II) and (III) is now pyramidal, the benzoaxazaphospholin ring remains planar. The second five-membered ring in (II) is puckered as was observed for the same fused bicyclic system in (IV). It is instructive to compare the geometries of the bicyclic system in these two derivatives. As a result of the decrease in the P-O bond distances and the increase in the P-N distance on going from (IV) to (II), the two O-C bonds become 3% longer (1.347(8), 1.376(5) vs 1.395(4), 1.413(3) Å), the two N-C bonds 2% shorter (1.421(8), 1.427(5) vs 1.400(3), 1.406(3) Å).

2.8. Conclusion

Theoretical and experimental studies on acyclic and cyclic pentacoordinate phosphorus derivatives demonstrate that their stereochemistry is determined by a combination of factors:

1. Substituent electronegativities.
2. Steric interactions between substituents.
3. Strain reduction for small (4- or 5-membered) ring systems.
4. π-Interaction between donor lone pairs on (particularly equatorial) heteroatom substituents and the framework or the phosphorus 3*d*-orbitals; possible π-delocalisation over all members of a small unsaturated ring system. It is now possible to make the following generalisations about the geometry of such derivatives:

1. The trigonal bipyramid is the preferred configuration for acyclic, monocyclic and fused bicyclic derivatives. It has demonstrated itself as capable of very considerable distortion to allow for the accomodation of strained small ring systems.

2. Square (rectangular) pyramidal geometry is observed for spirobicyclic systems with very strained small ring systems *e.g.* for derivatives of the type $(C_6H_4X_2)_2PY$, (X highly electronegative, Y less electronegative than X), where ring strain, substituent electronegativity and ligand subset stabilisation will all combine to favour the rectangular pyramidal geometry. Introduction of X substituents of lower electronegativity (or Y substituents of higher electronegativity than X, or both), lowering of the ligand

Table 2.7.1. Bond lengths and angles in tricyclic derivatives of pentacoordinate phosphorus (Fig. 2.7.1.)

	Bond lengths (Å)					Bond angles (°)					
Compound	P-(1)	P-(2)	P-(3)	P-(4)	P-(5)	(1)-P-(3)	(2)-P-(3)	(3)-P-(4)	(3)-P-(5)	(4)-P-(5)	Ref.
(I) aea/ee	1.85(2)	1.68(2)	1.81(2)	1.86(2)	1.81(2)	71.6(9)	88.1(9)	141.1(9)	133.4(9)	82.9(9)	129)
(II) eae/ae	1.755(2)	1.654(2)	1.635(2)	1.637(2)	1.595(2)	88.4(1)	89.6(1)	134.0(1)	110.4(1)	115.6(1)	130)
(III) eae/ae	1.819(2)	1.735(2)	1.849(3)	1.693(2)	1.663(2)	86.6(1)	94.8(1)	124.4(1)	119.3(1)	116.2(1)	131)
(IV)[1]) aea	1.703(6)	–	1.700(5)	1.763(5)	–	86.6(3)	–	171.6(3)	–	–	125)

[1]) (IV) is compound (III) from Fig. 2.6.1. which contains the same fused bicyclic system as (II), and is included for comparison purposes. The numbering has been altered to correspond with that in (II).

subset symmetry (*e.g.* to $X_2X'_2PY$), or particularly a relaxation of ring strain will all lead to a relative stabilisation of the trigonal bipyramid. The fine balance between these factors and the relatively small energy difference between the trigonal bipyramidal and square pyramidal geometries may lead in particular examples to the establishment of a C_{2v} distorted intermediate configuration *e.g.* $(C_6H_4O_2)_2PF$.

3. In the absence of complicating small ring systems, the axial sites of a trigonal bipyramid will be occupied by the more electronegative substituents. An "apicophilicity" series, in which the electronegativity order is modified by the proclivity to π-interaction and the steric repulsion of a ligand, has not as yet been established by diffraction studies.

4. Four- and five-membered ring systems will preferentially span the axial-equatorial sites in a trigonal bipyramid, even at the expense of forcing more electronegative substituents to occupy the electronically less favourable sites. As a result of π-interaction, the axial-equatorial placement is relatively more favourable for ring systems with heteroatoms substituted at phosphorus than for those with P-C bonds. In certain cases such latter ring systems may be capable of taking up a diequatorial placement (*e.g.* when two or three of the other substituents are highly electronegative heteroatoms and members of a fused bicyclic ring system).

5. The energy stabilisation achieved by π-delocalisation over all members of unsaturated five-membered rings with heteroatoms bonded to phosphorus, with the associated ring planarity and relative lengthening and shortening of constituent bonds, is greater than that which may be achieved through the relaxation of strain through ring puckering. This leads to the relative stabilisation of the rectangular pyramid for $(C_6H_4O_2)_2PY$ derivatives and may be of significance in leading to aea placement of fused bicyclic systems, where extended π-delocalisation is feasible.

6. The axial and equatorial P-O distances in cyclic pentacoordinate phosphorus derivatives increase significantly with increasing ring strain. A similar, though less marked, trend is indicated for P-N bonds (Table 2.9.2., Fig. 2.8.1.). In contrast, the

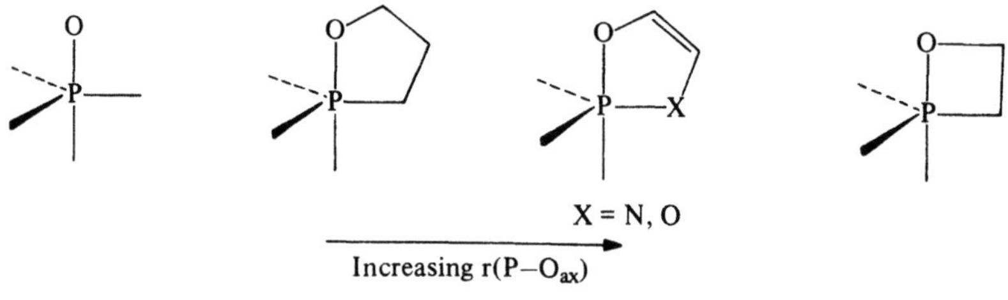

Fig. 2.8.1. The correlation of the lengthening of the P-O_{ax} bond distance in trigonal bipyramidal derivatives with the increase in ring strain

equatorial P-(sp^3)C distances display a wide scatter, indicating the flexibility of which this bond, which has of course no π-component, is capable of in order to produce a reduction in ring strain.

7. The orientation of the π-donor orbital of an equatorial substituent is dictated by a compromise between steric interactions and electronic energy and is not normally perpendicular to the axial plane of a trigonal bipyramid for bulky ligands.

8. The great flexibility of the trigonal bipyramidal skeleton displayed in the structures of many cyclic oxyphosphoranes suggests that the Berry pseudorotation mechanism may be an oversimplified representation for the permutational isomerisation of such species. Theoretical calculations and statistical considerations combined with these studies lead to the conclusion that this isomerisation may take place by a continuum of different routes (including turnstile rotation pathways) over a broad relatively flat potential surface, which have similar energy barriers. The Berry pseudorotation will, however, represent the energetically most favourable pathway for acyclic derivatives.

Table 2.8.1. Range limits for bond distances (Å) in pentacoordinate phosphorus derivatives[1])

	Acyclic [5]		Cyclic [5]			Sum of covalent radii	
Bond	Axial	Equatorial	Axial	Equatorial	Basal[2])	Pauling	Schomaker-Stevenson corrected
P-C	1.95–1.99	1.71–1.89	1.89	1.81–1.95	1.82–1.97	1.87	1.83
P-N	–	1.65–1.67	1.71–1.82	1.60–1.70	–	1.83	1.75
P-O	1.60–1.71	1.55–1.60	1.65–1.85	1.60–1.69	1.65–1.74	1.84	1.71
P-F	1.58–1.68	1.53–1.57	–	-	–	1.82	1.65
P-S	–	–	–	2.08	2.13–2.21	2.14	2.10
P-Cl	2.12–2.16	2.02–2.10	–	–	–	2.09	2.01

[1]) Not included in this stable are the monocyclic betaine type structures (I) and (II), Fig. 2.4.1., the intermediate square pyramidal/trigonal bipyramidal spirobicyclic derivatives 6, 10, Fig. 2.5.1., and the electron diffraction sutdies on Cl_2PF_3 and $(CF_3)_3PCl_2$ (see Section 2.3.).
[2]) Apical distances in these square/rectangular pyramidal derivatives have been included under "equatorial" acyclic bond distances.

Table 2.8.2. Specific bond distances in cyclic pentacoordinate phosphorus derivatives

Bond type	r(P-O$_{ax}$) Å	r(P-O$_{eq}$) Å
Acyclic	1.60–1.71	1.55–1.60
Puckered 5-membered rings	1.65–1.70	1.60–1.63
Planar 5-membered rings	1.68–1.80	1.64–1.69
4-membered rings	1.79–1.85	–
	r(P-N$_{ax}$) Å	r(P-N$_{eq}$) Å
Acyclic	–	1.65–1.67
Diazadiphosphetidines	1.74–1.78[1])	1.60–1.65
Other cyclic systems	1.76–1.82	1.67–1.70

[1]) A P-N$_{ax}$ bond distance of 1.71 Å was reported for the extended spirobicyclic diazadiphosphetidine 18 in Fig. 2.5.2. (see Table 2.5.2.).

The rapidly growing fund of structural data on pentacoordinate phosphorus derivatives has led to a better qualitative appreciation of those factors which determine the molecular geometry of such species. It is now essential to quantify this knowledge by systematic study of the interplay of ligand electronegativities, steric interactions, π-bonding and ring strain minimisation in the establishment of the equilibrium geometries for model compounds, in order to further develop our models for the mechanisms of those vital biochemical processes which involve transitory pentacoordinate phosphorus species.

3. Stereochemistry of Hexacoordinate Phosphorus Derivatives

Phosphorus is able to make use of its 3*d*-orbitals to some extent to increase its coordination number from five to six, as in the anions PF_6^-, PCl_6^-. The facility of PF_5 to form such stable non-ionic complexes with organic donor molecules such as amines, phosphines and ethers is also well established[132–136]. For tetrafluorophosphoranes, RPF_4 (R = Me, Et, Ph), however, relatively few such complexes have been reported[137]. A rapid decline in phosphorus acceptor strength with decreasing electronegativity of the attached ligands is observed in the series $PF_5 > ArPF_4 > AlkPF_4 \geqslant R_2PF_3$, (R = Ar, R = Alk). Acceptor strength is so severely depressed in R_2PF_3 and R_3PF_2 that no detectable interaction is observable by NMR between these derivatives and strong donor molecules[137]. The cationic hexacoordinate phosphorus species depicted here have also recently been reported[138].

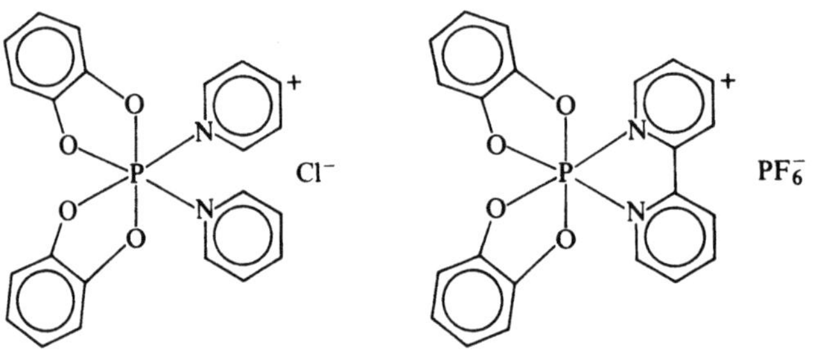

Regular octahedral geometry and equivalency of all bonds are the characteristic stereochemical features of the PF_6^- and PCl_6^- anions. Small angular distortions from 90° and differences in bond distances may be observed as a result of crystal packing effects. The configuration may be formally explained by assumption of *d*-orbital participation ($d_{x^2-y^2}, d_{z^2}$) in an sp^3d^2 hybridisation at phosphorus. Salts of the type $M^+PF_6^-$, ($M^+ = NH_4^+, K^+$), display an NaCl type of structure with crystallographically octahedral PF_6^- anions, for which a bond length of 1.58 Å has been determined[139], similar to that reported in other analyses[140, 141]. Below −15°, KPF_6 adopts a CsCl type of structure[142]. Anomalously long P-F bond distances of 1.73 Å, which are involved in F···H-O hydrogen bonding, have been observed in the hydrated salts $MPF_6 \cdot H_2O$ ($M = H^+, Na^+$)[143, 144]. The other five P-F bond distances in the octahedral are normal. A regular octahedral coordination with P-Cl bond distances of 2.04 and 2.08 Å have been found for the anion in $PCl_4^+ PCl_6^-$, contrasting with the shorter bond length of 1.97 Å in the tetrahedral cation[145].

Inter- and intramolecular complexes of fluorophosphoranes with nitrogen bases, which have been structurally characterised by X-ray diffraction, are displayed in Fig. 3.1. The N → P coordinate bond length of 1.980(3) in (III)[146] is much longer

(I) (II) (III)

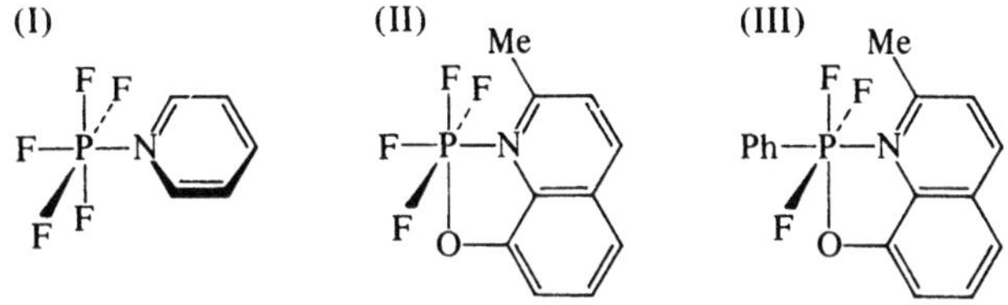

C_{2v} C_v C_v

than those of 1.898(4) (librationally corrected) and 1.911(4) Å observed for (I)[147] and (II)[148] respectively. These values contrast with 1.75 Å calculated for the sum of the covalent radii after application of the Schomaker-Stevenson correction and with endocyclic axial N-P bond lengths of 1.71–1.82 Å in trigonal bipyramidal species (Table 2.8.1.). The weakness of the N → P coordinate bond in (III) is in accordance with the much weaker Lewis acid acceptor properties exhibited by the PhF_3PO-grouping in comparison to PF_5 or the F_4PO-grouping, as evidenced by the ease of formation of hexacoordinate complexes ($F_4P(ox) > PhF_3P(ox) > MeF_3P(ox) \sim EtF_3P(ox)$, ox = 8-hydroxyquinoline)[148]. Considerable distortion from perfect octahedral coordination is observed at phosphorus in (III) with the P-O and P-F bonds being displaced away from the P-C bond at angles of 92.1(1), 95.5(1) and 93.5(1)°. A similar distortion of the coplanar fluorine atoms towards the nitrogen is observed in (I); the mean F-P-F angle is 91.8(2)°. In (III) although the coplanar fluorines are likewise bent away from the remaining fluorine at F-P-F angles of 92.6(1) and 92.0(1), the P-O bond is, in fact, displaced towards this P-F bond [O-P-F 88.5(2)°]. The extent of the distortion in (III) may be explained by the VSEPR model in terms of two complementary factors:

1. The electron density at phosphorus in the P-C bond of (III) is higher than that in the P-F bonds opposite N in (I) and (II) and will therefore exert a greater degree of repulsion upon the coplanar P-O and P-F bonds.
2. The N → P coordinate bond in (III) is much weaker than in these other complexes, and the subsequent electron density at phosphorus and hence the degree of repulsion upon the coplanar P-O and P-F bonds will therefore be smaller.

The (F)P-F and (O)P-F bond lengths of 1.610(2) and 1.593(3) Å in (III) are significantly longer than the equivalent values [1.598(3) and 1.572(3) Å] observed in (II), as would be predicted on account of the decrease in effective electronegativity of phosphorus resulting from the replacement of one fluorine by a phenyl group on going from (II) to (III). An interesting trend (F)P-F > (N →)P-F > (O)P-F is observed for the different P-F bond lengths in the derivatives (I) – (III). These differences, which are significant for all three complexes, may be associated with the electronegativity of the opposite partner in the octahedron and the nature of its bond to phosphorus. The values of the P-F bond lengths in (I) – (III) are close to that of 1.58 Å observed in the PF_6^- anion. The P-O bond lengths of 1.678(3) and 1.689(3) in (II) and (III) are significantly shorter than those of 1.723(7) and 1.711(8) Å in the hexacoordinate tris-(*o*-phenylenedioxy)phosphate anion, which has a crystallographically imposed C_3 symmetry[149]. This and the widening of the P-O-C bond angle at oxygen in (II) and (III) in comparison to that in the anion indicate that the degree of π-interaction in the P-O bond is greater in these complexes.

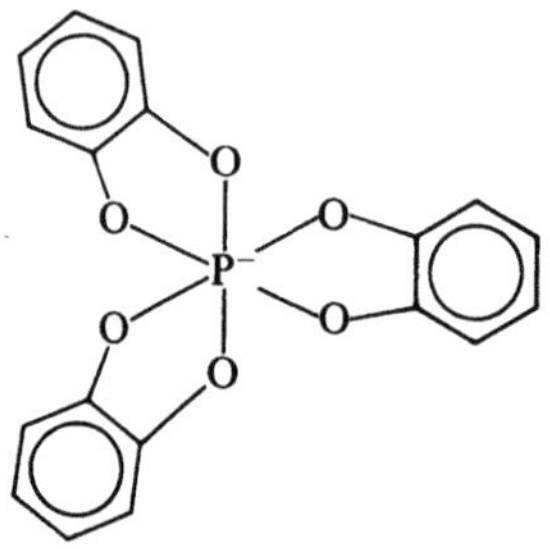

It is interesting that derivatives (II) and (III) both display a crystallographically imposed plane of symmetry. A similar phenomenon has been observed for $F_5P \cdot NMe_3$[150]. The plane of the pyridine ring in (I) is staggered at an angle of 40.8° to one of the atom planes of the octahedron, giving it an approximate C_{2v} symmetry. It appears that the adoption of a particular configuration by these adducts is dictated by steric interactions. In $F_5P \cdot NMe_3$ the methyl groups are staggered with respect to the coplanar fluorines, and likewise the pyridine ring in (II). In this way potential intramolecular interactions will be minimised. In (III) the phenyl group, which is perpendicular to the oxMe-grouping, occupies the most sterically favourable position, namely that opposite the N → P coordinate bond, thereby avoiding a potential interaction with the quinoline aromatic group.

Acknowledgements. This article is based on the author's Habilitationsschrift which has been accepted by the Technical University of Braunschweig (1976). I am most grateful to the Gesellschaft für Biotechnologische Forschung mbH for providing me with optimal facilities to pursue my work in the field described in this review, to my collaborators Prof. Dr. R. Schmutzler (Braunschweig) and Prof. Dr. A. Schmidpeter (München), and to my co-workers Frau A. Borkenstein and Dr. D. Schomburg. I am also indebted to Prof. Dr. R. R. Holmes (Massachusetts) for allowing me to quote his unpublished results.

Note Added in Proof

The following structural studies, several of which represent new structural classes, have appeared since the original submission of this article:

1. A monocyclic derivative 2,2-bis-(*p*-bromophenoxy)-2-dimethylamino-4,4,5,5-tetrakistrifluoromethyl-1,3,2-dioxaphosph(V)olan[151].

2. The spirobicyclic derivatives 5-*p*-bromophenoxy-10-methyl-2,2,3,3-tetrakis-trifluoromethyl-1,4,6-trioxa-10-aza-5-phosphaspiro[4.5]decane[151], 1,6-dioxa-4,9-diaza-5-λ^5-phosphaspiro[4.4]nonane[152] and 2-phenoxy-2,2-*o*-phenylenedioxy-4,5-bis(trifluoromethyl)-2,2-dihydro-1,3,2-dioxaphospholene[153].

3. A fused tricyclic phosphonium cation in $[HP(OCH_2CH_2)_3N]BF_4$, which contains an axial P-H bond and an axially annelating N → P coordinate bond of length 1.986 Å in a TBP framework[154].

4. The fused tetracyclic derivative cyclen fluorophosphorane, which displays a structure intermediate between TBP and SP with the lone fluorine pseudo-apicel[155].

5. A hexacoordinate phosphonium cation 2,2′-bispyridine-bis(*o*-phenylenedioxy)-phosphonium hexafluorophosphate, which contains two N → P coordinate bonds of length 1.898(4) and 1.901(3) Å[156].

The continued interest in spirobicyclic derivatives indicated by the structures listed above is also reflected in theoretical contributions on the quantitative description of structural distortions of cyclic phosphoranes in terms of the Berry exchange coordinate[157] and on computer simulation of phosphorane structures[158].

4. References

1) Vilkov, L., Khaikin, L. S.: Topics Curr. Chem. *53*, 25 (1975)

2) Davidson, G., Ebsworth, E. A. V., Sheldrick, G. M., Woodward, L. A.: Spectrochim. Acta *22*, 67 (1966)

3) Davidson, G., Woodward, L. A., Ebsworth, E. A. V., Sheldrick, G. M.: Spectrochim. Acta *23A*, 2609 (1967)

4) a) Beagley, B., Robiette, A. G., Sheldrick, G. M.: J. Chem. Soc. (A), *1968*, 3002
b) Beagley, B., Robiette, A. G., Sheldrick, G. M.: J. Chem. Soc. (A), *1968*, 3006

5) Rankin, D. W. H., Robiette, A. G., Sheldrick, G. M., Beagley, B., Hewitt, T. G.: J. Inorg. Nucl. Chem. *31*, 2351 (1969)

6) Cruickshank, D. W. J.: J. Chem. Soc. *1961*, 5486

7) Muetterties, E. L., Schunn, R. A.: Quart. Rev. (Chemical Society London) *20*, 245 (1966)

8) Sacconi, L.: Pure Appl. Chem. *17*, 95 (1968)

9) Wood, J. S.: Progr. Inorg. Chem. *16*, 227 (1972)

10) Westheimer, F. H.: Accounts Chem. Res. *1*, 70 (1968)

11) Mislow, K.: Accounts Chem. Res. *3*, 321 (1970)

12) Gutowsky, H. S., McCall, D. W., Slichter, C. P.: J. Chem. Phys. *21*, 279 (1953)

13) a) Ugi, I., Ramirez, F., Marquarding, D., Klusacek, H., Gokel, G., Gillespie, P.: Angew. Chem. *82*, 766 (1970)
b) Ugi, I., Marquarding, D., Klusacek, H., Gillespie, P., Ramirez, F.: Accounts Chem. Res. *4*, 288 (1971)

14) Berry, R. S.: J. Chem. Phys. *32*, 933 (1960)

15) Corbridge, D. E. C.: Structural chemistry of phosphorus compounds. In: Griffith, E. J., Grayson, M. (eds.). Topics in phosphorus chemistry, Vol. 3. New York – London: Interscience 1966, p. 57

16) Corbridge, D. E. C.: The structural chemistry of phosphorus. Amsterdam: Elsevier 1974

17) Daly, J. J.: Stereochemical aspects of organophosphorus compounds. In: Dunitz, J. D., Ibers, J. A. (eds.). Perspectives in structural chemistry. London – Sydney – Toronto: John Wiley & Sons *1970*, 165

18) Hansen, K. W., Bartell, L. S.: Inorg. Chem. *4*, 1775 (1965)
19) Bartell, L. S., Hansen, K. W.: Inorg. Chem. *4*, 1777 (1965)
20) Yow, H., Bartell, L. S.: J. Mol. Struct. *15*, 209 (1973)
21) Braune, H., Pinnow, P.: Zeit. physik. Chem. *B35*, 239 (1937)
22) Brockway, L. O., Beach, J. Y.: J. Amer. Chem. Soc. *60*, 1836 (1938)
23) Rouault, M.: Ann. Phys. *14*, 78 (1940)
24) Muetterties, E. L., Mahler, W., Schmutzler, R.: Inorg. Chem. *2*, 613 (1963)
25) Muetterties, E. L., Mahler, W., Packer, K. J., Schmutzler, R.: Inorg. Chem. *3*, 1298 (1964)
26) Holmes, R. R.: J. Chem. Phys. *46*, 3724 (1967)
27) Holmes, R. R.: J. Chem. Phys. *46*, 3730 (1967)
28) Holmes, R. R., Deiters, R. M., Golen, J. A.: Inorg. Chem. *8*, 2612 (1969)
29) Muetterties, E. L.: J. Amer. Chem. Soc. *91*, 1636 (1969)
30) Muetterties, E. L.: J. Amer. Chem. Soc. *91*, 4115 (1969)
31) Ugi, I., Marquarding, D., Klusacek, H., Gillespie, P., Ramirez, F.: Accounts Chem. Res. *4*, 288 (1971)
32) Gillespie, P., Hoffmann, P., Klusacek, H., Marquarding, D., Pfohl, S., Ramirez, F., Tsolis, E. A., Ugi, I.: Angew. Chem. *83*, 691 (1971)
33) Ramirez, F., Ugi, I.: Adv. Phys. Org. Chem. *9*, 256 (1971)
34) Hoffmann, R., Howell, J. M., Muetterties, E. L.: J. Amer. Chem. Soc. *94*, 3047 (1972)
35) Strich, A., Veillard, A.: J. Amer. Chem. Soc. *95*, 5574 (1973)
36) Whitesides, G. M., Mitchell, H. G.: J. Amer. Chem. Soc. *91*, 5384 (1969)
37) Ramirez, F., Ugi, I.: Bull. Soc. Chim. Fr. 453, 1974
38) Bartell, L. S.: Inorg. Chem. *5*, 1635 (1966)
39) Gillespie, R. J.: Inorg. Chem. *5*, 1634 (1966)
40) Zemann, J.: Z. anorg. allgem. Chem. *324*, 241 (1963)
41) Foppl, L.: J. reine angew. Math. *141*, 251 (1912)
42) Gillespie, R. J., Nyholm, R. S.: Quart. Rev. (Chemical Society London) *11*, 339 (1957)
43) Gillespie, R. J.: J. Chem. Educ. *40*, 295 (1963)
44) Gillespie, R. J.: J. Amer. Chem. Soc. *85*, 4671 (1963)
45) Gillespie, R. J.: Can. J. Chem. *38*, 818 (1960)
46) Gillespie, R. J.: J. Chem. Soc. *1963*, 4672
47) Gillespie, R. J.: Can. J. Chem. *39*, 318 (1961)
48) Sidgwick, N. V., Powell, H. M.: Proc. Roy. Soc. *A176*, 153 (1940)
49) Schutte, K., Van der Waerden, B. L.: Math. Ann. *123*, 96 (1951)
50) Cotton, F. A.: J. Chem. Phys. *35*, 228 (1961)
51) Mulliken, R. S., Riecke, C. A., Orloff, D., Orloff, H.: J. Chem. Phys. *16*, 1248 (1949)
52) Craig, D. P., Nyholm, R. S., Maccoll, A., Orgel, L. E., Sutton, L. E.: J. Chem. Soc. *1954*, 332
53) Rundle, R. E.: Acta Cryst. *14*, 585 (1961)
54) Rundle, R. E.: Rec. Chem. Progr. *23*, 195 (1962)
55) Rundle, R. E.: J. Amer. Chem. Soc. *85*, 112 (1963)
56) Rundle, R. E.: Surv. Progr. Chem. *1*, 81 (1963)
57) Van der Voorn, P. C., Drago, R. S.: J. Amer. Chem. Soc. *88*, 3255 (1966)
58) Issleib, K., Gründler, W.: Theor. Chim. Acta *8*, 70 (1967)
59) Boyd, D. B.: J. Amer. Chem. Soc. *91*, 1200 (1969)
60) Florey, J. B., Cusachs, L. C.: J. Amer. Chem. Soc. *94*, 3040 (1972)
61) Berry, R. S., Tamres, M., Ballhausen, C. J., Johansen, H.: Acta Chem. Scand. *22*, 231 (1968)
62) Santry, D. P., Segal, G. A.: J. Chem. Phys. *47*, 158 (1967)
63) Brown, R. D., Peel, J. B.: Aust. J. Chem. *21*, 2605 (1968)
64) Rauk, A., Allen, L. C., Mislow, K.: J. Amer. Chem. Soc. *94*, 3035 (1972)
65) Howell, J. M., Van Wazer, J. R., Rossi, A. R.: Inorg. Chem. *13*, 1747 (1974)
66) Boyd, D. B., Lipscomb, W. N.: J. Chem. Phys. *46*, 910 (1967)
67) Hewson, M. J. C., Schmutzler, R., Sheldrick, W. S.: Chem. Commun. *1973*, 190; Sheldrick, W. S.: J. Chem. Soc. (Dalton) *1973*, 2301
68) Pierce, S. B., Cornwall, C. D.: J. Chem. Phys. *48*, 2118 (1968)
69) Sheldrick, W. S.: Acta Cryst. *B31*, 1776 (1975)
70) Oberhammer, H., Schmutzler, R.: J. Chem. Soc. (Dalton) *1976*, 1454

71) Adams, W. J., Bartell, L. S.: J. Mol. Struct. *8*, 23 (1971)
72) Oberhammer, H., Grobe, J.: Z. Naturforsch. *30B*, 506 (1975)
73) Sarma, R., Ramirez, F., McKeever, B., Maracek, J. F., Lee, S.: J. Amer. Chem. Soc. *98*, 581 (1976); erratum, ibid *98*, 4691 (1976)
74) Wheatley, P. J.: J. Chem. Soc. *1964*, 2206
75) Holmes, R. R.: Accounts Chem. Res. *5*, 296 (1972)
76) Griffiths, J. E., Carter, R. P. jr., Holmes, R. R.: J. Chem. Phys. *41*, 863 (1964)
77) Cavell, R. G., Poulin, D. D., The, K. I., Tomlinson, A. J.: Chem. Commun. *1974*, 19
78) Poulin, D. D., Cavell, R. G.: Inorg. Chem. *13*, 2324 (1974)
79) Poulin, D. D., Cavell, R. G.: Inorg. Chem. *13*, 3012 (1974)
80) Lagowski, J. J.: Quart. Rev. (Chemical Society London) *13*, 233 (1959)
81) Griffiths, J. E.: J. Chem. Phys. *41*, 3510 (1964)
82) Griffiths, J. E., Beach, A. L.: J. Chem. Phys. *44*, 2686 (1966)
83) Peake, S. C., Schmutzler, R.: J. Chem. Soc. (A) *1970*, 1049
84) Landau, M. A., Sheluchenko, V. V., Drozd, G. I., Dubov, S. S., Ivin, S. Z.: Zh. Strukt. Khim. *8*, 1097 (1967)
85) Harris, J. J., Rudner, B.: J. Org. Chem. *33*, 1392 (1968)
86) Harman, J. S., Sharp, D. W. A.: Inorg. Chem. *10*, 1538 (1971)
87) Muetterties, E. L., Meakin, P., Hoffmann, R.: J. Amer. Chem. Soc. *94*, 5674 (1972)
88) Bone, S. A., Trippett, S., White, M. W., Whittle, P. J.: Tetrahedron Letters *20*, 1795 (1974)
89) Marquarding, D., Ramirez, F., Ugi, I., Gillespie, P.: Angew. Chem. 85, 99 (1973)
90) Chioccola, G., Daly, J. J.: J. Chem. Soc. (A) *1968*, 568
91) Kawamoto, T., Hata, T., Kishida, Y., Tamura, C.: Tetrahedron Letters *26*, 2417 (1971)
92) Carrell, H. L., Berman, H. M., Ricci jr., J. S., Hamilton, W. C. Ramirez, F., Maracek, J. F., Kramer, L., Ugi, I.: J. Amer. Chem. Soc. *97*, 38 (1975)
93) Hamilton, W. C., Ricci, J. S. Jr., Ramirez, F., Kramer, L., Stern, P.: J. Amer. Chem. Soc. *95*, 6335 (1973)
94) Hamilton, W. C., La Placa, S. J., Ramirez, F., Smith, C. P.: J. Amer. Chem. Soc. *89*, 2268 (1967)
95) Spratley, D. D., Hamilton, W. C., Ladell, J.: J. Amer. Chem. Soc. *89*, 2272 (1967)
96) Cadogan, J. I. G., Gould, R. O., Gould, S. E. B., Sadler, P. A., Swire, S. J., Tait, B. S.: J. Chem. Soc. (Perkin I) *1975*, 2392
97) Mazhar-Ul-Haque, Caughlan, C. N., Ramirez, F., Pilot, J. F., Smith, C. P.: J. Amer. Chem. Soc. *93*, 5229 (1973)
98) Pauling, L.: The nature of the chemical bond. 3 rd. edit. Ithaca, New York, Cornell University Press 1960
99) Holmes, R. R.: J. Amer. Chem. Soc. *97*, 5379 (1975)
100) Hoard, L. G., Jacobson, R. A.: J. Chem. Soc. (A) *1966*, 1203
101) Hess, H., Forst, D.: Z. anorg. allgem. Chemie *342*, 240 (1966)
102) Almenningen, A., Andersen, B., Astrup, E. E.: Acta Chem. Scand. *23*, 2179 (1969)
103) Sheldrick, W. S., Hewson, M. J. C.: Acta Cryst. *B31*, 1209 (1975)
104) Fild, M., Sheldrick, W. S., Stankiewicz, T.: Z. anorg. allgem. Chemie *415*, 43 (1975)
105) Cox, J. W., Corey, E. R.: Chem. Commun. *1967*, 123
106) Harris, R. K., Wazeer, M. I. M., Schlak, O., Schmutzler, R., Sheldrick, W. S.: J. Chem. Soc. (Dalton), *1977*, 517
107) Wunderlich, H., Mootz, D., Schmutzler, R., Wieber, M.: Z. Naturforsch. *29b*, 32 (1974)
108) Wunderlich, H.: Acta Cryst *B30*, 939 (1974)
109) Schomburg, D., Sheldrick, W. S.: unpublished results
110) Brown, R. K., Holmes, R. R.: J. Amer. Chem. Soc. *99*, 3326 (1977)
111) Sarma, R., Ramirez, F., Maracek, J. F.: J. Org. Chem. *41*, 473 (1976)
112) Wunderlich, H., Mootz, D.: Acta Cryst. *B30*, 935 (1974)
113) Eisenhut, M., Schmutzler, R., Sheldrick, W. S.: Chem. Commun. *1973*, 144
114) Devillers, J. R., Holmes, R. R.: J. Amer. Chem. Soc. *99*, 3332 (1977)
115) Cadogan, J. I. G., Gould, R. O., Tweddle, N. J.: Chem. Commun. *1975*, 773
116) Zobota, J. S., Holmes, R. R.: unpublished results

117) Howard, J. A., Russell, D. R., Trippett, S.: Chem. Commun. *1973*, 856
118) Hellwinkel, D., Krapp, W., Schomburg, D., Sheldrick, W. S.: Z. Naturforsch. *31b*, 948 (1976)
119) Newton, M. G., Collier, J. E., Wolf, R.: J. Amer. Chem. Soc. *96*, 6888 (1974)
120) Duff, E., Russell, D. R., Trippett, S.: Phosphorus *4*, 203 (1974)
121) Gibson, J. A., Röschenthaler, G.-V., Schomburg, D., Sheldrick, W. S.: Chem. Ber., *110*, 1887 (1977)
122) Weiss, J., Hartmann, G.: Z. anorg. allgem. Chemie *351*, 152 (1967)
123) Schmidpeter, A., Luber, J., Schomburg, D., Sheldrick, W. S.: Chem. Ber. *109*, 3581 (1976)
124) Schmidpeter, A., Schomburg, D., Sheldrick, W. S.: unpublished results
125) Sheldrick, W. S., Schmidpeter, A., Weinmaier, J.-H.: Angew. Chem. *87*, 519 (1975); Sheldrick, W. S.: Acta Cryst. *B32*, 925 (1976)
126) Schmidpeter, A., Schomburg, D., Sheldrick, W. S.: unpublished results
127) Kabachnik, M. I., Gilyarov, V. A., Tikhonina, N. A., Kalinin, A. E., Andrianov, V. G., Struchkov, Yu. T., Timofeeva, G. I.: Phosphorus *5*, 65 (1974)
128) Swank, D. D., Caughlan, C. N., Ramirez, F., Pilot, J. F.: J. Amer. Chem. Soc. *93*, 5236 (1971)
129) Aly, H. A. E., Barlow, J. H., Russell, D. R., Smith, D. J. H., Swindles, M., Trippett, S.: Chem. Commun. *1976*, 449
130) Schmidpeter, A., Schomburg, D., Sheldrick, W. S., Weinmaier, J. H.: Angew. Chemie *88*, 851 (1976)
131) Schmidpeter, A., Schomburg, D., Sheldrick, W. S.: unpublished results
132) Muetterties, E. L., Bither, T. A., Farlow, M. W., Coffman, D. D.: J. Inorg. Nucl. Chem. *16*, 52 (1960)
133) Webster, M.: Chem. Rev. *66*, 87 (1966)
134) Tebbe, F. N., Muetterties, E. L.: Inorg. Chem. *6*, 129 (1967)
135) Brown, D. H., Crosbie, K. D., Fraser, G. W., Sharp, D. W. A.: J. Chem. Soc. (A) *1969*, 551
136) Schulz, C. W., Rudolph, R. W.: J. Amer. Chem. Soc. *93*, 1898 (1971)
137) Muetterties, E. L., Mahler, W.: Inorg. Chem. *4*, 119 (1965)
138) Schmidpeter, A., Von Criegern, T., Blanck, K.: Z. Naturforsch. *31b*, 1058 (1976)
139) Bode, H., Clausen, H.: Z. anorg. allgem. Chemie *265*, 229 (1951)
140) Field, J. S., Wheatley, P. J.: J. Chem. Soc. (Dalton) *1972*, 2269
141) Hardy, A. D. U., Sim, G. A.: J. Chem. Soc. (Dalton) *1972*, 2305
142) Bode, H., Teufer, G.: Z. anorg. allgem. Chemie *268*, 20 (1952)
143) Bode, H., Teufer, G.: Acta Cryst. *8*, 611 (1955)
144) Bode, H., Teufer, G.: Acta Cryst. *9*, 825 (1956)
145) Clark, D., Powell, H. M., Wells, A. F.: J. Chem. Soc. *1942*, 642
146) John, K.-P., Schmutzler, R., Sheldrick, W. S.: J. Chem. Soc. (Dalton) *1974*, 2466
147) Sheldrick, W. S.: J. Chem. Soc. (Dalton) *1974*, 1402
148) John, K.-P., Schmutzler, R., Sheldrick, W. S.: J. Chem. Soc. (Dalton) *1974*, 1841
149) Allcock, H. R., Bissell, E. C.: J. Amer. Chem. Soc. *95*, 3154 (1973)
150) Sheldrick, W. S.: unpublished results
151) Barlow, J. H., Bone, S. A., Russell, D. R., Trippett, S., Whittle, P. J.: J. C. S. Chem. Comm. *1976*, 1031
152) Meunier, P. F., Deiters, J. A., Holmes, R. R.: Inorg. Chem. *15*, 2572 (1976)
153) Narayanan, P., Berman, H. M., Ramirez, F., Maracek, J. F., YuFen Chaw, Prasad, V. A. V.: J. Amer. Chem. Soc. *99*, 3336 (1977)
154) Clardy, J. C., Milbrath, D. S., Springer, J. P., Verkade, J. G.: J. Amer. Chem. Soc. *98*, 623 (1976)
155) Richman, J. E.: Tetrahedron Lett. *1977*, 562
156) Schmidpeter, A., Sheldrick, W. S.: unpublished results
157) Holmes, R. R., Deiters, J. A.: J. Amer. Chem. Soc. *99*, 3318 (1977)
158) Deiters, J. A., Gailucci, J. C., Clark, T. E., Holmes, R. R.: J. Amer. Chem. Soc. *99*, 5461 (1977)

Received January 1, 1977, October 1, 1977

Complex Bases and Complex Reducing Agents
New Tools in Organic Synthesis

I. Activation of Sodamide: Sodamide Containing Complex Bases

Paul Caubère

Université de Nancy I, Laboratoire de Chimie Organique I (Associated to CNRS), Case Officielle 140, 54037 Nancy Cedex, France

Table of Contents

I. Introduction

I.1. Generalities

Reactions between solid and liquid substrates are often slow. However rates can be increased by using an appropriate solvent able to dissolve the insoluble reagent. During the dissolution, molecular or ionic architectures of the starting products are destroyed. New solvated entities more or less aggregated and eventually more or less ionised are formed.

Thus, even without taking into account the change in molecular arrangements of the liquid reagent when a solvent is added to the above defined system, after such structure modifications of the insoluble reagent we have to expect a change in reaction rates as well as appearance of new reactions.

On the other hand, if we look at reactions where at least one reagent is ionic, or pseudoionic, it is well known that use of dipolar aprotic solvent instead of a nonpolar one generally increases rates considerably[1)]. In those cases, very often, addition of a definite amount of polar solvent to the nonpolar is sufficient to obtain the wanted effect. The role of polar solvent is to reduce or destroy the aggregates of ionic reagent, to separate ionic particles and specifically to solvate cations[1)]. In other words it dramatically changes the degree of aggregation and the state of aggregation[2)].

The reactivity increase in reactions with "ionic" reagents can also be obtained with the help of complexing cations substrates such as crowns[3)] or cryptands[4)].

However we can also imagine that two ionic compounds having in a given solvent their own degree and state of aggregation will give, when mixed, new "complex aggregates" with properties entirely different from the ones observed for each reagent aggregate. Note that this concept is very similar to the well-known concept of solvation of a reagent by a solvent, each reagent playing here both parts of "solute" and "solvating agent".

By generalising this idea, we can also expect that an ionic reagent, usually more or less aggregated in a given solvent, should be able to have some interactions with a very poorly insoluble reagent leading to "complex aggregates". The consequence will be a "solubility" increase of the insoluble compound.

In fact, this phenomenon directs the formation of "complex bases" which constitute the subject of part one of our paper.

We shall see in Part two that this concept can also be applied to reducing properties of sodium hydride and that it leads to new reagents we called "complex reducing agents".

I.2. Complex Bases Story

In Scheme 1 is pictured the general reaction all our work started from.

It had been well known since the pioneering works of Wittig[5)] and Roberts[6)] that nucleophilic substitutions of haloaromatic compounds, nonactivated by electro withdrawing group, are feasible by means of elimination-additions.

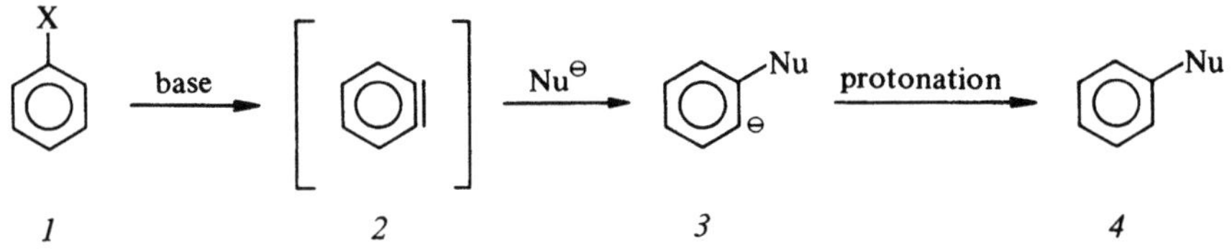

$Nu^{\ominus}$ = Nucleophile which may be uncharged.

Scheme 1

In these reactions (Scheme 1), a strong base eliminates H–X from *1* to give benzyne *2*[7,8]. Benzyne is the parent of a family of very reactive intermediates known as arynes or 1,2-dehydroarenes.

Benzyne is able to add nucleophiles very easily[7,8] to give anions *3* which after quenching by a proton leads to substituted products *4*. It is very important to note that in these condensations *nucleophiles operate only after the benzyne formation.*

During a study of several arynic reactions in aprotic solvents[9,10], we made some curious observations summarized in Scheme 2.

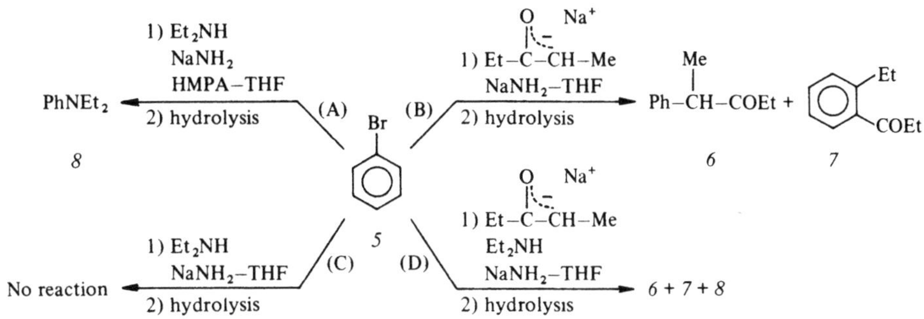

HMPA = Hexamethylphosphotriamide; THF = Tetrahydrofuran

Scheme 2

Comparison of Reactions (A) and (C) indicates that in the conditions used, sodamide is not able to eliminate H–Br in THF and HMPA is needed to perform the reaction. However in Reaction (B), benzyne is the reaction intermediary (we shall discuss later the mechanism of formation of ketones *6* and *7*) although THF is the solvent of the reaction. Also, if diethylamine is present with enolate ion [Reaction (D)] diethylaniline is formed although in THF and in the absence of enolate this product was not obtained [Reaction (C)]. From all these observations it appears that *generation of benzyne in THF depends upon the nature of the nucleophile to be condensed.*

This is in contradiction to what is known about elimination-addition condensations. Having verified that under our conditions, the ketone enolate and $NaNH_2$ taken separately do not react with bromobenzene, we concluded that *in Reactions (B) and (D) the very base is not $NaNH_2$ but the mixture $NaNH_2$-enolate.* In other words, we thought that in the presence of enolate ions, $NaNH_2$ properties are modified and that diethylamine, an uncharged nucleophile, does not have the same effect.

Now, if we compare Reactions (C) and (D) we can say that diethylamine may be condensed on 5 in THF if $NaNH_2$-enolate is used as a base. However, enolate presents the disadvantage of being a better nucleophile than diethylamine and competes with it. Then we thought that if all these observations were more than fortuitous, it should be possible to find a low nucleophilic sodium salt of an organic compound, able to modify (like the enolate ion) the basic properties of $NaNH_2$ but allowing the condensation of diethylamine. After some preliminary experiments, sodium *tert*-butoxide (*t*-BuONa) seemed to fit those requirements.

As a matter of fact, condensation of diethylamine on bromobenzene is very easily performed in THF in the presence of $NaNH_2$-*t*-BuONa (Scheme 3).

$$\underset{5}{\text{Ph–Br}} + \text{Et}_2\text{NH} \xrightarrow[\text{2) hydrolysis}]{\text{1) NaNH}_2\text{-}t\text{-BuONa, THF}} \underset{8}{\text{Ph–NEt}_2} \quad 100\%$$

Scheme 3

We call "complex bases"[11] such mixtures of two bases and an activating agent the alcoholate (now every salt of organic or possibly inorganic compound used in the preparation of complex bases is also called activating agent).

Remark: In complex bases, cations of the activated base and cations of the activating agent are identical. We never activated, for instance, a base having a sodium cation with potassium alcoholates.

Moreover, it is very important to note that a complex base is very easily prepared by addition of the precursor (alcohol, ketone, etc...) of the activating agent to the calculated excess of commercial $NaNH_2$ in suspension in the desired solvent at about 40 to 50 °C. It is prejudicial to prepare the pure activating agent (as alcoholate for instance) and to add it to the $NaNH_2$ suspension; the mixtures thus obtained are generally of much poorer reactivity. In other words, the activating agent has to be prepared *in situ*. Thus complex bases combine simplicity with efficacy. Finally, it must be noted that those reagents are heterogeneous and look like "milks" with some $NaNH_2$ suspension.

If we compare Reaction (A) (Scheme 2) with the reaction of Scheme 3 we can say that *t*-BuONa communicates to $NaNH_2$ in THF similar properties to those that this amide has in the presence of HMPA. We thought that if this supposition was true, it should be possible to perform in THF reactions which usually need HMPA as a solvent.

This was readily verified (Scheme 4). Thus Normant and Cuvigny[12] prepared trityl sodium in HMPA by reaction of triphenylmethane with $NaNH_2$. With similar conditions but in THF, this reaction is not obtained. On the contrary in this latter solvent, if the base used is $NaNH_2$-*t*-BuONa, the deep red color of the anion immediately appears at room temperature[10, 11]. However, the transformation is not quantitative (*vide infra*).

Later on we used this reaction in testing several potential activating agents (Chapter III).

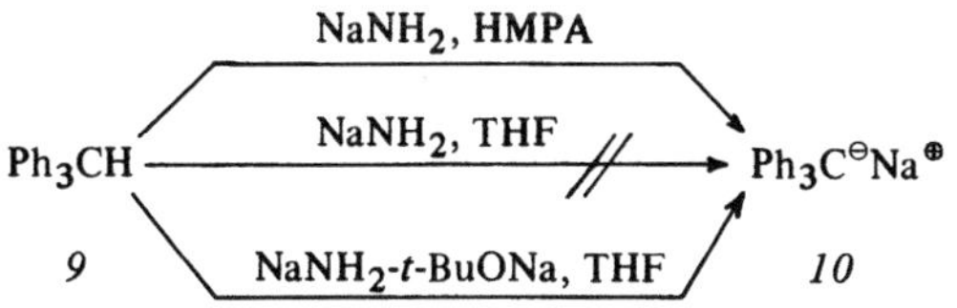

Scheme 4

II. Factors Influencing the Complex Bases Properties

II.1. Reactions Used

From what we said, it is clear that numerous factors intervene in the properties of complex bases. Thus we have to take in account the ratio $NaNH_2$/activating agent, the nature of the solvent and, of course, the nature of the activating agent. These parameters were studied empirically and we shall give the main results.

First we had to choose test reactions. We chose the ones described in Scheme 5[11, 13, 14].

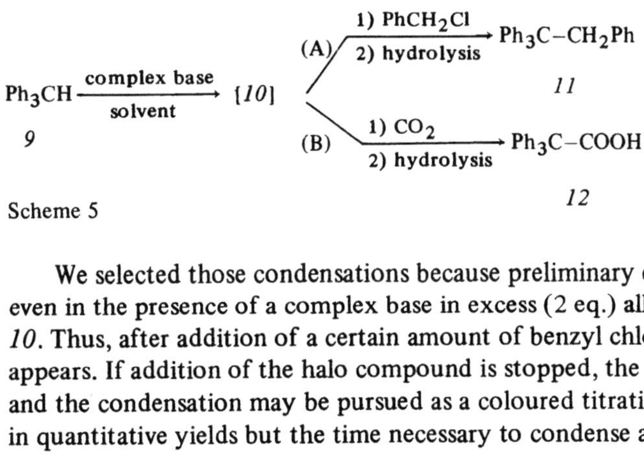

Scheme 5

We selected those condensations because preliminary experiments showed that even in the presence of a complex base in excess (2 eq.) all *9* is not transformed into *10*. Thus, after addition of a certain amount of benzyl chloride the red colour disappears. If addition of the halo compound is stopped, the coloration appears again and the condensation may be pursued as a coloured titration. Generally *11* is obtained in quantitative yields but the time necessary to condense a determined quantity of benzyl chloride to a determined quantity of *9* depends upon the parameters cited above. From those times, we can obtain data concerning the influence of a given factor.

Other information is obtained by means of carbonatation of the anion. This reaction constitutes a satisfactory estimation of the amount of *10* formed in the medium.

II.2. Ratio $NaNH_2$-Activating Agent

We carefully studied[13, 14] two activating agents: *t*-BuONa and $(Et{-}CO{-}CH{-}CH_3)^-Na^+$. The reaction test was Reaction (A) Scheme 5.

For the complex base $NaNH_2$-*t*-BuONa, the maximum activation is reached when $NaNH_2$/*t*-BuONa is worth about 2 and between 5 and 7. However in this latter case, experiments are poorly reproducible and usually ratio 2 is only used.

With the enolate as activating agent, maximum of reactivity is only obtained when $NaNH_2/(Et{-}CO{-}CH{-}CH_3)^-Na^+$ is worth 2. In this case it is interesting to note that $Ph{-}CH_2{-}Cl$ condenses only on the trityl anion, although enolates were in the reactional medium. We have here the first indication that in a complex base the specific properties of the activating agent are strongly masked.

In summary we can say that, for abstraction of a proton from a substrate, the highest efficiency of a complex base is obtained when the starting ratio $NaNH_2$/ Activating agent is worth 2. Of course, this result gives no data on the stoechiometry of the complex base since the medium is heterogeneous (*vide supra*) and since the ratio evolves during the reaction.

II.3. The Solvent

The rôle of the solvent was studied with complex base $NaNH_2$-*t*-BuONa with the help of both reactions described in Scheme 5. From numerous experiments[13, 14] we can conclude that with these test reactions, polar aprotic solvents lead to the most efficient bases. Thus we found the following decreasing order:

$$\text{HMPA} > \text{DME} > \text{THF} > \text{Diglyme} > \text{Triglyme} \gg C_6H_6, C_6H_{12}$$

DME = 1,2-Dimethoxyethane; Di and Triglyme = $CH_3O(CH_2CH_2O)_n{-}CH_3$ with $n = 2$ and 3 respectively. This classification deserves some comments:

With a few of the activating agents that we shall give hereafter, (Chapter III) the efficiency order of THF and DME is reversed.

HMPA must be used with care because complex bases react with this solvent as soon as the temperature reaches 40 °C[10]. Condensation of benzyl chloride on the degradation products thus obtained (see Scheme 6) leads to *17*. The same compound has been isolated by Normant and collaborators after condensation of $PhCH_2Cl$ on the reaction product of HMPA with alkali metals[15].

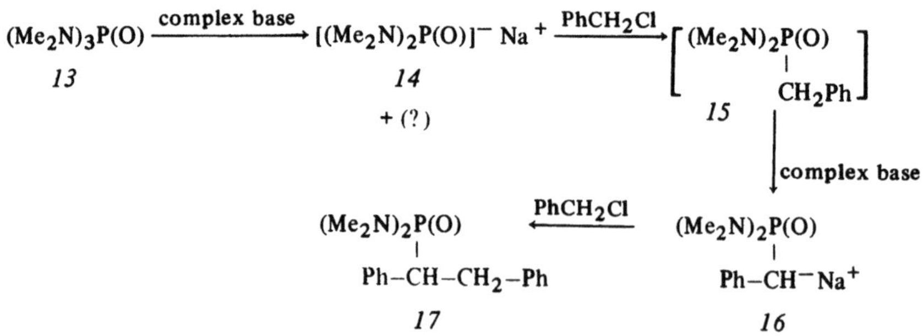

Scheme 6

We never could identify the other part of HMPA degradation.

In the same way, some others dipolar aprotic solvents are inconsistent with the use of complex bases. Thus dimethyl formamide is rapidly destroyed and condensation of Ph–CH_2–Cl leads to a substantial amount of N,N-dimethyl benzylamine. N-Methyl pyrrolidone and, of course, dimethyl sulfoxyde react very easily with complex bases. Pyridine turns brown very rapidly.

Finally, in nonpolar aprotic solvents such as benzene or cyclohexane trityl anion formation is possible only with activating agents more powerful than *t*-BuONa (*vide infra*).

II.4. The Activating Agent

The chief part in modification of complex bases basic properties is devoted to the nature of the activating agent. Several problems arise when we want to study this parameter.

First we have to recognize whether a reagent is activating. Thus we meet the first difficulty. We have to define a reaction allowing for detection of an activating agent and, of course, a complex base may be a good base for a given reaction but not for another. Thus choosing a reaction leads to a defective list which will need to be further perfected.

Secondly we have to give a classification following the activating power of the activating agents. Such a classification also depends upon the studied reaction and the solvent.

Thirdly we have to try to obtain rules allowing us to define the molecular design of the right activating agent needed to perform a given reaction.

Thus, all we shall describe below will be important in practice but is still incomplete and a lot of further work is needed.

To define activating agent we chose reactions pictured in Scheme 5 performed in THF and DME. We said earlier that Reaction (A) (Scheme 5) is in fact a coloured titration. Thus appearance of the red colour of $Ph_3C^-Na^+$ shows that the organic salt studied is an activating agent.

The time needed to condense with quantitative yield a definite amount of $PhCH_2Cl$ to a known quantity of Ph_3CH gives a good qualitative idea of what we can call the "basic renewal power" (b.r.p.; the shorter the condensation time will be better the b.r.p.).

Remark: In our opinion it is important to note that sometimes basic systems are described with the help of halogeno compounds condensing on carbanions generated from more or less acidic organic substrates. When yields are quantitative it is sometimes concluded that the basic system is able to transform quantitatively the substrate to the corresponding anion. This conclusion can be erroneous if the condensation of the halogeno compound is slower than the formation of the carbanion.

For instance we met complex bases in some solvents which, at definite temperature, allow condensation of $PhCH_2Cl$ on Ph_3CH without discoloration. However, lowering the temperature forced us to perform the condensation like a coloured titration. Moreover, carbonatation of $Ph_3C^-Na^+$ prepared in conditions where discoloration is not observed led to acid formation with yields far from quantitative.

This is one of the reasons for which we also used Reaction (B) (Scheme 5) with the aim to define activating agents.

Yields of isolated acid give a good quantitative picture of the generated carbanions. Thus, this reaction gives a very good idea of the "carbanionic generation power" (c.g.p.).

Now, from our numerous experiments [11, 13, 14, 16] we can say:

Among strong nucleophilic organic salts, most ketone enolates are good activating agents (numerous examples are given in Chapter IV.2.). Compounds like PhSNa, $NCCH_2^-Na^+$ are also rather good activating agents.

Among much less nucleophilic substrates, alcoholates are very interesting. They are often very inexpensive, they allow the formation of carbanions, they generally do not compete with carbanions in condensation reactions, they lead to complex bases able to give elimination reactions (*vide infra*) and finally they allow a large "modulation" of the basic properties. For these reasons we studied and we are continuing to study a lot of them.

We have briefly summarized in Scheme 7 the formula of alcohols of which corresponding alcoholates have activating properties.

$R-CH_2OH$ R = *t*-Bu, *n*-Bu, *n*-$C_{11}H_{23}$, *o*-$MeOC_6H_4$, *p*-$MeOC_6H_4$

R^1R^2>CHOH $R^1 = R^2$ = Me, *i*-Pr, *p*-$MeOC_6H_4$; $[R^1, R^2] = (CH_2)_5$, (*i*-Pr)$CH-(CH_2)_2-CH(Me)-CH_2$

$R^1R^2R^3$>C–OH $R^1 = R^2 = R^3$ = Me, *n*-Bu; $R^1 = R^2$ = Me R^3 = Et, *n*-Bu, *n*-C_8H_{17};
R^1 = Me, Et $[R^2, R^3] = (CH_2)_5$; $[R^1, R^2, R^3]$ = Adamantyl

$R-(OCH_2CH_2)_n-OH$ $n = 1$ R = Me, *n*-Bu, Ph; $n = 2$ R = Me, Et, *n*-Bu, *n*-$C_{10}H_{21}$;
$n = 3$ R = Et, *n*-Bu; $n = 4$ R = Et; $n = 5$ R = Et;
$n = 6$ R = Et, *o*-$HOCH_2-C_6H_4$

p-$R-C_6H_4-OH$ R = H, MeO; Ph–CH(OH)–CH(Me)–NH(Me); $Me_2C(OH)-(CH_2)_n-CMe_2(OH)$ (n = 0, 2);

$MeN(CH_2-CH(Me)OH)_2$;

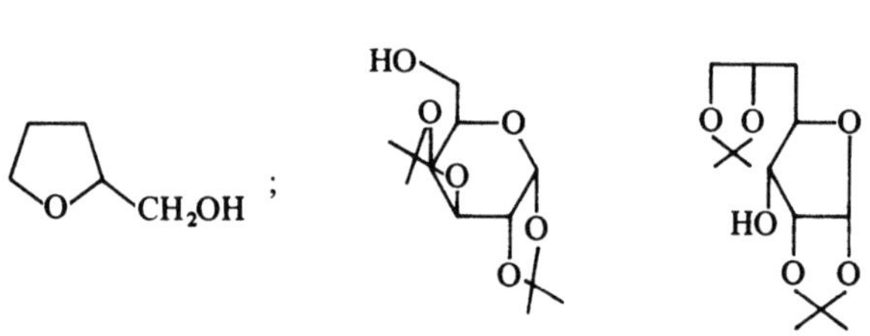

Scheme 7

It is clear that several kinds of alcoholates are able to activate sodamide and, of course, this is not a limiting list. Among those compounds, a few are poor activating agents.

On the other hand some alcohols with structures looking like the ones described in Scheme 7 have sodium salts which are not at all activating agents for the test reactions. This is the case, for example, with MeOH, EtOH, $HO(CH_2)_nOH$ (n = 2, 3, 4, 6), $HO(CH_2CH_2O)_2H$. It is very surprising that this last compound has no activating properties when it is among its monoethers that we found the strongest activation!

From all our results it is very difficult to obtain other than very general rules. However we can say that:

1. Glycol monoethers often have the best b.r.p. in THF except for $R-(OCH_2CH_2)_nOH$ with $n \geqslant 4$; as a matter of fact the best results are obtained for $n = 2$. Moreover the activity seems to decrease when R, being linear, the carbon condensation increases.
2. After glycols monoethers (for b.r.p. in THF) are secondary and primary alcohols with rather light ramified chain and tertiary alcohols having not too heavy chains.
3. Other alcohols have not very good b.r.p. in THF.
4. In DME the b.r.p. order of class 1 and 2 is rather reversed.
5. Concerning the c.g.p. we find in THF and DME an order similar to the one found in THF for b.r.p.
6. Generally speaking, DME diminishes the differences between activating powers comparatively with what is found in THF. In other words DME has a levelling effect.

Obviously those rules are only general indications and, at the present time, some results seem very strange. For example, 1-ethyl cyclohexanol has a better b.r.p. than 1-methyl cyclohexanol in THF as well as in DME. This order is just reversed for c.g.p.!

Any way, these very simple rules help in the choice of activating agents and we are working with the aim of refining them. Moreover, we shall see later (Chapter IV. 1.2.2.) that we have started to define rules for elimination reactions.

In passing, we note that Biehl and co-workers[17] have subsequently observed enhancement of the "basic power" of $NaNH_2$ in the presence of certain sodium salts in dimethylamine. They called their observation a "Novel Catalytic Salt Effect in Base-induced Aryne Reactions". The structure of activating agents used led these authors to erroneously conclude that only salts with linear and resonating anions are good activating compounds.

III. Attempted Interpretation of the Phenomena

At first we must give complementary data. During our first work on activation of sodamide[11], we found that complex bases react with benzyl chloride giving chiefly *trans*-stilbene even when the activating agent is a ketone enolate.

These results strengthen the concept that specific properties of a complex base constituant are strongly masked. However, they do not disappear completely. Thus reactions of benzylbromide with a complex base having a ketone enolate as activating agent give substantial amounts of benzylated ketone.

On the other hand during the preparation of a complex base by addition of an alcohol on sodamide, NH_3 formed may be accountable for the observed activation. In fact this hypothesis is easily eliminated because it does not explain why the properties of a complex base vary from one base to another. Moreover, we verified[11] that a suspension of $NaNH_2$ in THF solution of NH_3 has no special properties.

Finally, some potentiometric measurements[18, 19] show that, for example, in THF, $NaNH_2$ is dissolved very slowly. The highest value of the pH is obtained after a few hours and the concentration of base is very low. On the contrary the sodium salt of diethylen glycol monoethylether (prepared *in situ*) enables the maximum basicity to be reached almost instantaneously. Analogous observations were made with others alcoholates. Moreover, the pH obtained with complex bases are, of course, much higher than the ones obtained from alcoholates alone.

We think that these experiments complete the demonstration of activation of $NaNH_2$ by alcoholates. Now, what is our interpretation of these observations?

Owing to the complexity of the reagents, our interpretation cannot be more than qualitative. We think that the rôle played by alcoholate is to complex $NaNH_2$ on the crystal surface (Scheme 8) and to bring it into the solvent in a complexed form.

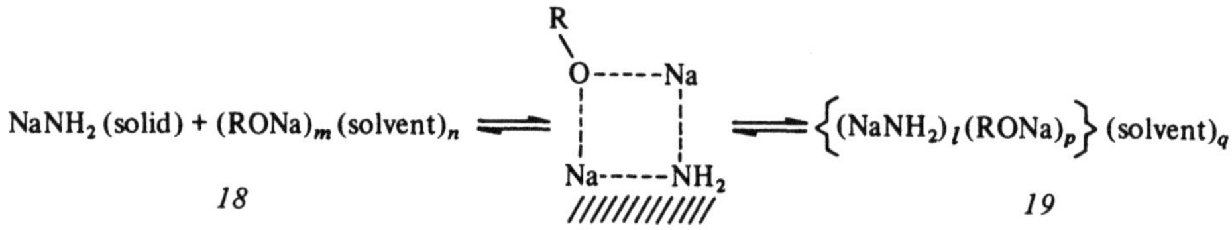

Scheme 8

By accepting this hypothesis, it is easy to understand why the complexation is much more difficult, if not impossible, when an activating agent is added crystallized to the $NaNH_2$ suspension instead of being generated *in situ*.

Our supposition is well supported by the following observations. When a large amount of divided $NaNH_2$ is kept in contact with THF for a long time, addition of Ph_3CH leads to the appearance of the red colour. If this coloration is destroyed by means of a condensation, we have to wait a very long time before the coloration appears again. Moreover from the amount of the reagent added we can say that the concentration of dissolved $NaNH_2$ is very low. In other words, $NaNH_2$ dissolves very slowly in THF and in very little concentration.

In the presence of an alcoholate the time of dissolution is much shorter and the base concentration much higher. Thus the first rôle of activating agent is to enhance the rate of "dissolution" of $NaNH_2$ as well as its "solubility".

Complex bases thus formed must be under aggregated form (*19*, Scheme 8) and several different aggregates (with different values of l and p in *19*) are certainly in equilibrium in the medium. An observation, which will be very important for further applications of complex bases, supports this affirmation. We said before that complex bases are a heterogeneous medium looking like a "milk" with some $NaNH_2$ in suspension. However after decantation, we can observe three phases: a clear "solution", a white colloidal compound and solid $NaNH_2$. The "solution" as well as the colloidal compound react easily with Ph_3CH to give the carbanion.

It is clear that the nature of the aggregates *19* depends upon the nature of the activating agent and of the solvent. Particularly the ionisation state of a complex base must be a function of the nature of the alcoholate chain (aliphatic or bearing heteroatom, linear or ramified, etc...) as well as of the "solvating properties" of the solvent.

In our opinion, another very important characteristic of complex bases is the surface structure of aggregates. By surface structure we mean the external geometric repartition of the charged atoms. Of course, this charge distribution is a function of the solvent and, for a determined solvent, of the structure of the activating agent. We think that surfaces of aggregates are the predominant factors in reactions needing two ionic sites separated by a determined distance (for example *syn* eliminations with Sicher's transition state[20]). Note that the rôle of the aggregation state of bases is now well evidenced[21]; in all the reactions studied the surface of the aggregates must be very important.

Anyway, we thought that it would be possible to find activating agents giving complex bases with surfaces adaptable to a wanted reaction. We shall see arguments in Chapter IV. 1.2. which support this hypothesis.

In summary, sodamide containing complex bases are the result of complexation of $NaNH_2$ by more or less soluble sodium salts with the formation of aggregates. These latter are more or less ionized and are able to abstract a proton in unusual conditions. Moreover they can react by means of two reaction sites with the help of ions distributed on their surface.

From a general view-point, it is not surprising that complex bases properties are different from those of each constituant. Indeed if two compounds, more or less aggregated in a solvent are mixed, the equilibrium symbolized in Scheme 9 must be realized.

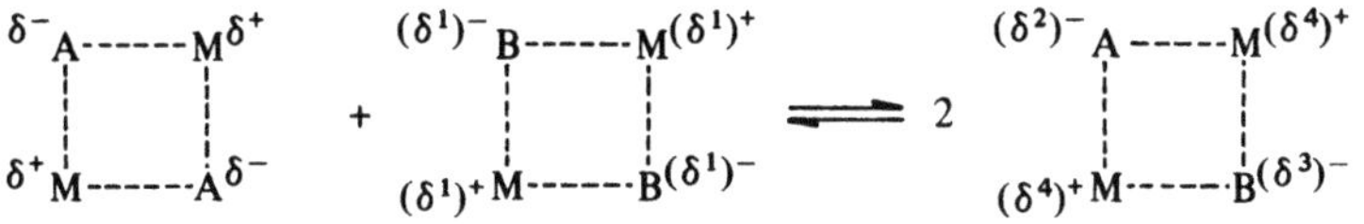

20 *21* *22*

Scheme 9

Of course ionization state and interionic distances in *22* must be different from those in *20* and *21*. Thus the chemical properties must also be different. As we said in the introduction, there is a certain analogy between the activation of, for example $(A^-M^+)_n$, by formation of mixed complexes with $(B^-M^+)_m$ and the activation of $(A^-M^+)_n$ by solvatation with dipolar aprotic solvents.

In the particular case of sodamide containing complex bases where the base to be activated is insoluble, alcoholates can complex sodamide with help of two ionic sites ($RO^{\delta-}$ and $Na^{\delta+}$). This must be more favorable for "dissolution" of $NaNH_2$ than simple solvation of the cation by dipolar aprotic solvent (solvation of anions being low). However, in "solution" because of the "double complexation" the basic power of complex bases must be lower than the basic power of $NaNH_2$ in a dipolar aprotic solvent *if this latter was able to dissolve it substantially.*

Last but not least following the activating agent used, complex bases have to be divided in two classes:

1. Nonnucleophilic complex bases with poor nucleophiles as activating agents. This class has to be divided in two subgroups. One containing bases able to abstract a proton and the other containing bases able to give two site reactions. Of course some activating agents belong to both classes.

2. Nucleophilic complex bases with good nucleophiles as activating agents. In this group, nucleophilic properties are masked but they do not completely disappear.

Complex bases of the first class will be used only as bases while those of the second class will be used as bases and nucleophiles.

We shall conclude by saying that the "activation" phenomenon discussed here is not limited to the activation of an insoluble compound by a more soluble one, but also to the mixture of two or more ionic compounds. As a matter of fact, such property modifications were cited in the literature with organosodium[22)] and organolithium[23)] compounds for example. What is important, in our opinion, is that in a reaction medium formed by, for example, two basic or nucleophilic entities (A^-, M^+) and (B^-, M^+) the observed reactions must be the result of mixed bases or nucleophiles. It is easy to understand what misinterpretation of reaction mechanisms can happen if this possibility is neglected.

IV. Use of Sodamide Complex Bases

IV.1. Nonnucleophilic Complex Bases

IV.1.1. Preparation and Alkylation of Carbanions

We described earlier the formation and benzylation of tritylsodium. This kind of condensation is rather general. Thus in THF or DME and in the presence of the complex base $NaNH_2$-*t*-BuONa it is possible to alkylate Ph_3CH not only by $PhCH_2Cl$ but also by EtBr, *n*-BuBr, Br–$(CH_2)_4$–X (X = Br, Cl). Yields in isolated alkyl triphenylmethanes are always between 85 to 95%[13, 14)].

In the same conditions, less acidic Ph_2CH_2 is also alkylated by the same halogeno compounds and even by *i*-Pr–Cl. Yields are in the same range than for tritane[13, 14)].

In benzene and cyclohexane, which are much less polar than THF, use of complex base $NaNH_2$-*t*-BuONa was unsuccessful. However, with the help of the very strong base $NaNH_2$–$Et(OCH_2CH_2)_2ONa$ alkylations of both hydrocarbons are possible. Yields are also about 90%[16, 19)].

We think that the success obtained with sodium salt of ethyleneglycol monoethylether is the consequence of some internal complexation by oxygen atoms (Scheme 10).

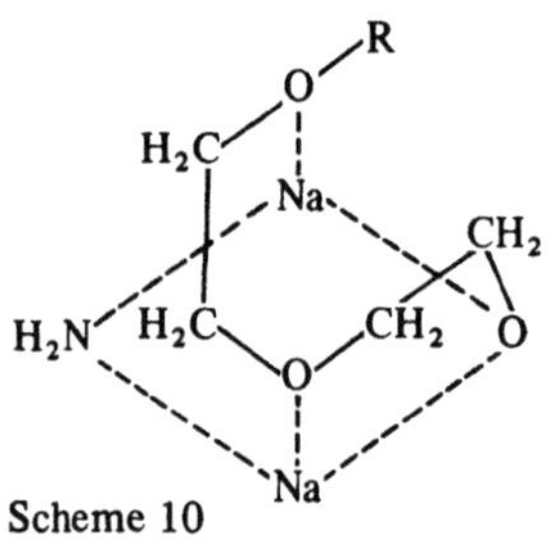

Scheme 10

On the other hand, recent work[16] shows that it is also possible to alkylate imines in THF in the presence of the complex base $NaNH_2$–$Et(OCH_2CH_2)_2ONa$. Thus cyclohexylimines of butyraldehyde and cyclohexanone are monobenzylated by Ph–CH_2–Cl with 40% and 80% yields respectively. Dialkylation is never more than 5%.

Until now we did not develop very much of our investigations on alkylations but at the present time we are working on this subject with the aim of learning if it would be possible to replace lithium dialkylamides by sodamide containing complex bases, and above all, to perform anionic chemistry in a nonpolar medium.

IV.1.2. Elimination Reactions

IV.1.2.1. Synthesis of Methylenecyclopropane

It is well known that KNH_2 is a stronger base than $NaNH_2$. We thought that if all we said before was true, the properties of sodamide complex bases should look like more those of KNH_2 than those of $NaNH_2$.

This hypothesis is verified with help of the now well-known preparation of methylenecyclopropane of Koster, Arora and Binger[24]. Thus, methallyl chloride *23* (Scheme 11) reacts with $NaNH_2$ in THF to give methylcyclopropene *24*. On the other hand, in the presence of KNH_2 the reaction leads to methylenecyclopropane *25*. Moreover KNH_2 isomerises *24* to *25* in THF.

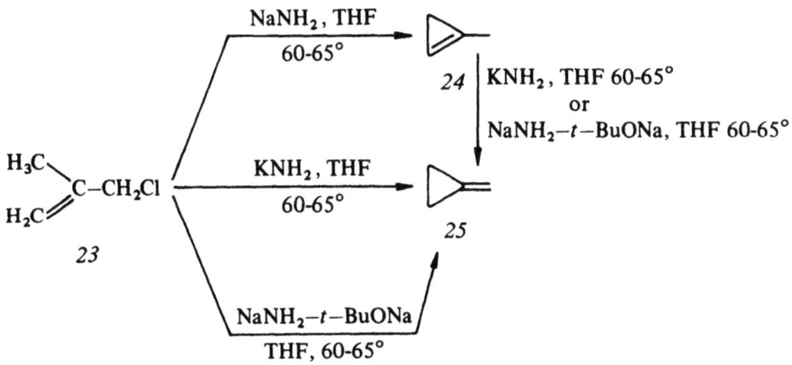

Scheme 11

As expected we have shown[13] that complex base $NaNH_2$-*t*-BuONa advantageously replaces KNH_2 in this synthesis. The chief advantage of the complex base is its simplicity of use which readily allows work on a large scale. Moreover the reagents are neither dangerous nor expensive. This reaction has been used by Conia and Salaün in their elegant synthesis of cyclobutanone[25].

IV.1.2.2. Syn Eliminations on trans Dibromo Cycloalkane Derivatives

Reaction pictured in Scheme 12 may be performed by way of *syn* elimination.

26 *27*

X = halogen, tosylate
Scheme 12

It is well known that in cyclohexan series (*26*, $n = 6$) this elimination is very difficult to perform with classical bases[26]. For example, to our knowledge, no satisfactory method has so far been found for preparation of 1-bromo cyclohexene from *trans* 1,2-dibromo cyclohexane. This apparently very simple reaction, leads only with great difficulties to the desired compound (yields are low and tedious purifications are required to obtain a pure product[27]).

We thought that a good way to realize this reaction would be to find a complex base allowing a Sicher's transition state[20] as pictured in Scheme 13.

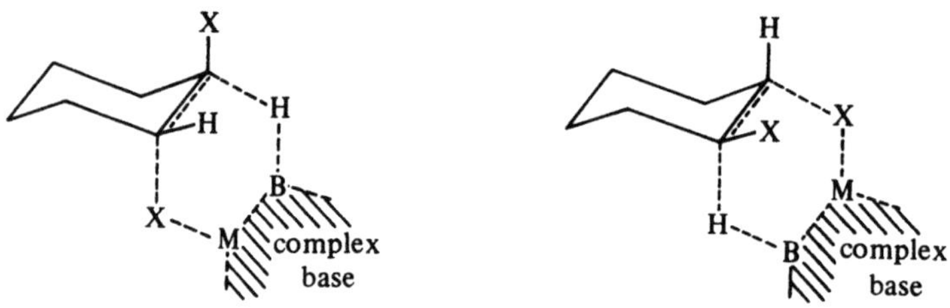

28 *29*

Scheme 13

If the hypothesis given in Chapter III were true, the problem was to find the good activating agent leading to aggregates with ionic surfaces able to give the desired two-sites reaction. Of course an efficient order of activating agents for this elimination reaction would not have anything to do with that found for b.r.p. or c.g.p. (Chapter II.4.). As a matter of fact, as it may be seen from Scheme 14, we found[16, 28, 29] a new classification of the reagents.

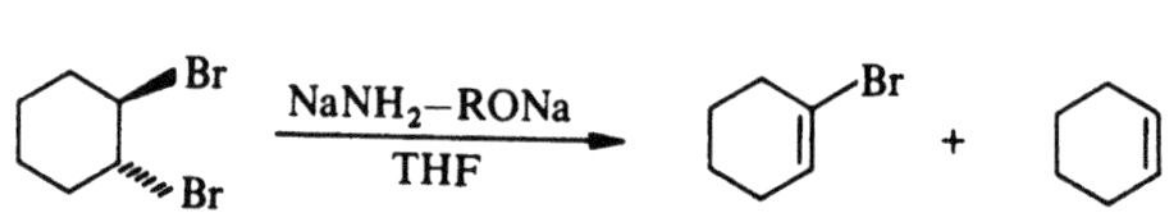

30 *31* *32*

		31 yield[1]	32 yield
R =	Me_3CCH_2, $Me_2(C_8H_{17})C$, $(n\text{-}C_4H_9)_3C$, Me_3C	60–65%	∿30–35%
	$PhOCH_2CH_2$, 1-MeC_6H_{10}, Me_2CH, $Me_2(nC_4H_9)C$	50–55%	∿30%
	$C_{11}H_{23}CH_2$, p$MeOC_6H_4CH_2$, C_6H_{11}, $Me_2(Et)C$	25–40%	∿25%
	$Et(OCH_2CH_2)_2$	15%	60%

[1]) Yield measured by cpv. Isolated yields are a little lower.
Scheme 14

The best activating agents are those having very ramified chains behind the oxygen atom. As expected, results obtained are not in relation with the proton abstracting power of the complex base and the observed elimination must not be of the E_2H type[30].

Formation of cyclohexene is the consequence of the well-known debrominations by bases[26, 30–33]. This elimination is obviously favored by the *trans* diaxial position of the halogens. Thus it was of interest to study *trans* dibromocyclohexanes derivatives with blocked conformations.

Taking in account the accessibility of starting materials, reactions described in Scheme 15 were performed[29].

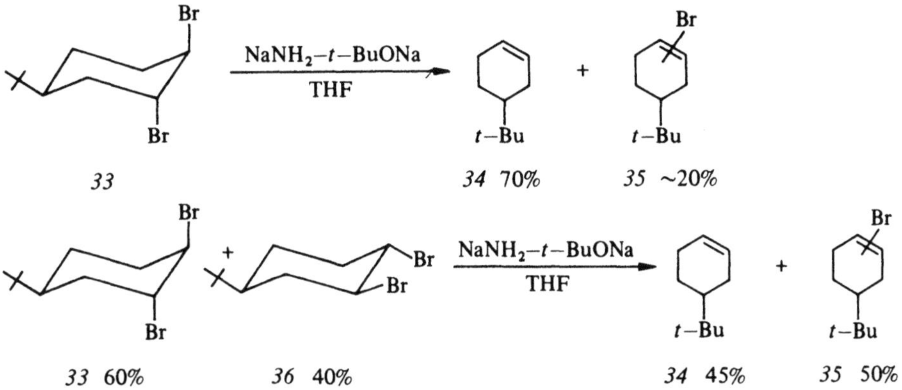

Scheme 15

It is noteworthy that even with *33*, HBr *syn* elimination still happens with 20% yield. Comparison of the two reactions of Scheme 15 easily shows that *36* only gives HBr *syn* elimination. Moreover, we have shown that HBr *syn* elimination on *36* is faster than *trans* debromination on *33*.

Finally, if we look at the difference of complexing properties between oxygen of a tosyl group and halogens, for a given cation, and if *28* and *29* (Scheme 13) were operating in our eliminations, we would have to observe a large behaviour difference between *trans* dibromocyclohexane and the corresponding *bis* tosylate. Indeed, this latter does not react with complex base $NaNH_2$-*t*-BuONa[29].

In our opinion all those observations support the hypothesis of the main rôle played by the geometry of the base and the intervention of Sicher's transition states *28* and *29* in these reactions. Thus, complex bases resolve partially the problem of *syn* eliminations in cyclohexan series. However, at the present time, their effectiveness seems to be limited to substrates closely related to *trans* dibromides.

Remark: In those reactions the $NaNH_2$ quality is important, particularly on the reaction time and possibly on the required temperature. It is recommended that the elimination by gas-liquid chromatography be followed.

IV.1.2.3. Preparation of Linear and Cyclic Acetylenic Hydrocarbons

Linear Acetylenic Hydrocarbons. It is well known that several of the eliminations of the type described in Scheme 16 are difficult or impossible to perform.

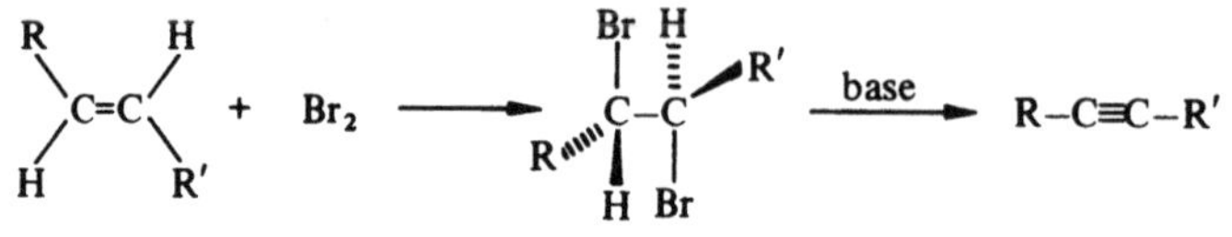

37 38 39

Scheme 16

For example Davis and Ansari[33] proposed a synthesis of optically active *39* (R = Me(Et)CH, R′ = Ph) by way of this kind of reaction. However, these authors never succeeded in performing the last step with classical bases. Moreover, with $NaNH_2$, they obtained a mixture of ethylenic and saturated hydrocarbons (Scheme 17).

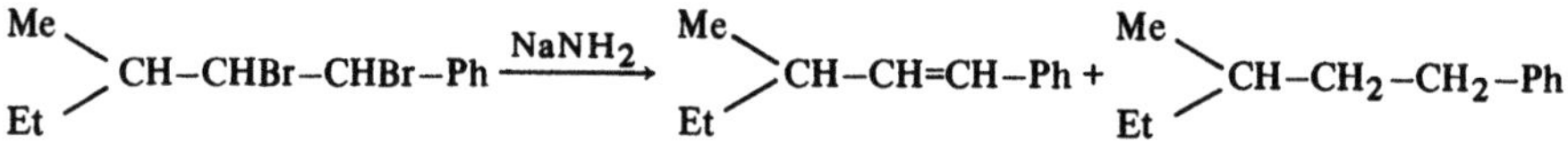

Scheme 17

Saturated hydrocarbon formation is a consequence of debromination by $NaNH_2$[33]. We thought that complex base $NaNH_2$-*t*-BuONa might been a solution to this problem; results obtained are briefly summarized in Scheme 18[34].

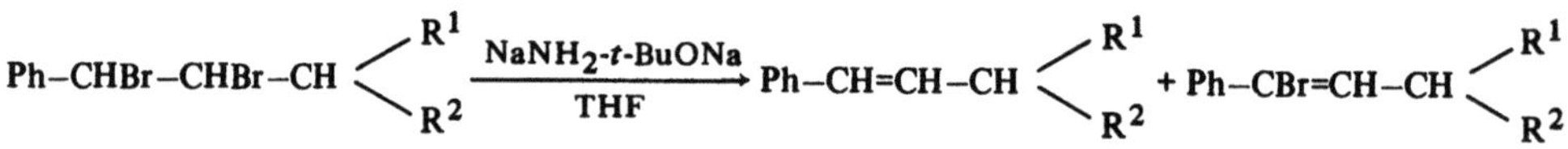

	41	*42*
40a $R^1 = R^2 = Me$	20%	10%
40b $R^1 = Me$ $R^2 = Et$	25%	20%

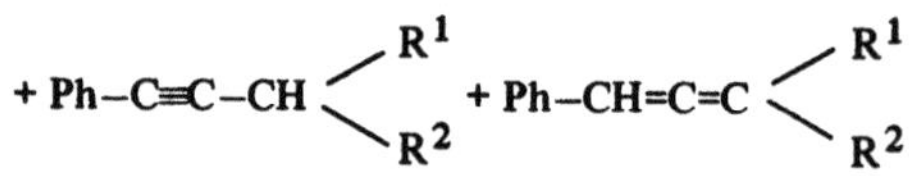

	43	*44*
40a	45%	25%
40b	42%	10%

Scheme 18

The main difficulty in these reactions is the isomerization of *43* to *44*. If conditions are too drastic, the only products isolated with good yields are *44*. However, with the help of $NaNH_2$-*t*-BuONa we synthesized[34] optically active *43b* of which the synthesis was proposed in Scheme 16[33].

Cyclic Acetylenic Hydrocarbons. This work was essentially undertaken for the purpose of confirming the special properties of complex bases in *syn* eliminations. Thus, we studied the formation of cyclododecyne from cyclododecenes by means of reactions described in Scheme 19[35].

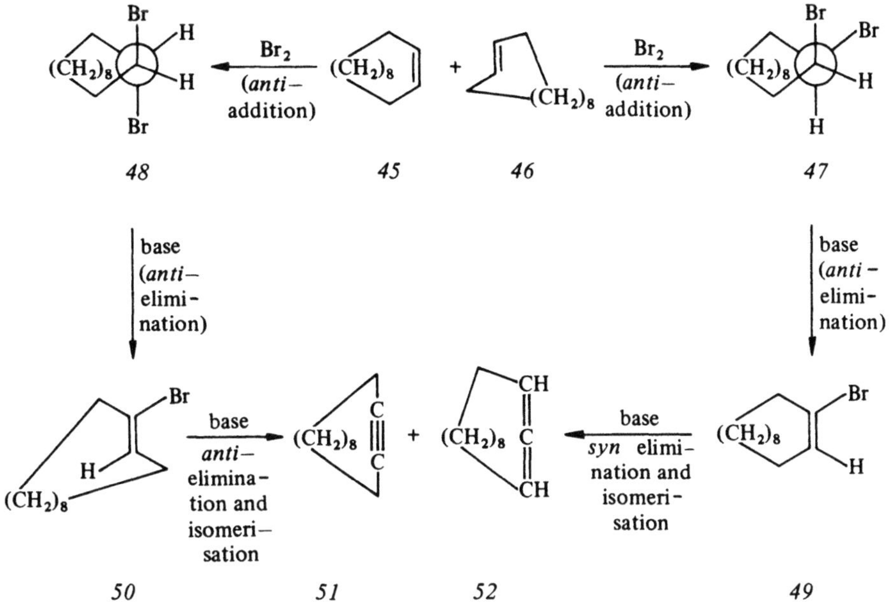

Scheme 19

Taking in account the composition of the commercial mixture *45* + *46* (35% *cis*, 65% *trans*), of the *anti* addition of Br_2, of the structure of *50* and *49* as well as their formation ratio, we can say that the first eliminations of H–Br are *anti*. Thus the second eliminations of H–Br have to be *anti* from *50* and *syn* from *49*. Comparative experiments showed that *t*-BuONa alone and less efficiently $NaNH_2$ alone in THF are able to give the *anti*-eliminations but not the *syn* ones. On the contrary, $NaNH_2$-*t*-BuONa in the same solvent is able to perform all the reactions. The side reaction with complex base is still isomerization of *51* to *52*. An application of this study was the preparation of cyclooctyne with 30–35% yield[35].

We can conclude this chapter by saying that complex bases may be good reagents for performing H–X *syn* eliminations. Their limitation is their isomerizing power; but we think that it should be possible to eliminate this disadvantage in the near future.

IV.2. Nucleophilic or Nonnucleophilic Complex Bases

IV.2.1. Aromatic Elimination-Addition Reactions (Arynic Condensations)

IV.2.1.1. Generalities

Several reviews have been written on arynic condensations (see for example[7, 8, 36]) and we shall recall only briefly the basic principle of these reactions (Scheme 20).

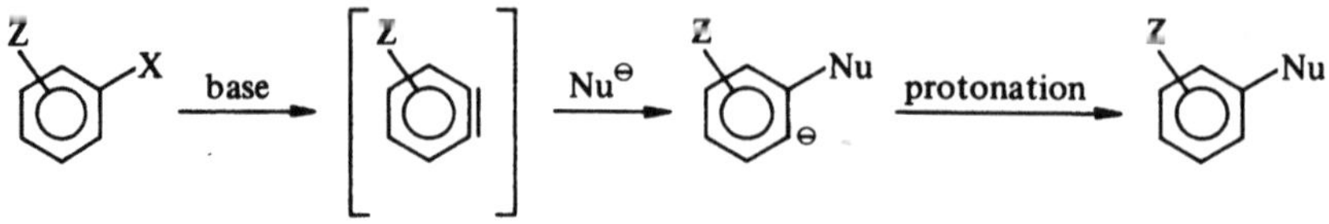

53 *54* *55* *56*

Nu = nucleophile

Scheme 20

We said before that the first step is the elimination leading to a dehydrobenzene *54*. This latter can be attacked on two carbon atoms which are equivalent if Z = H.

If Z ≠ H, two isomers may be formed. The isomer ratio depends upon electronic and steric properties of Z and its position on the ring.

Note that, of course, the directing effects of Z have nothing to do with the ones encountered in electrophilic substitutions. For instance, MeO has a very strong *meta*-directing effect in 1-methoxy 2,3-dihydrobenzene. Indeed the strong inductive electron withdrawing effect of the methoxy group made the negative charge of *55* more stable on *ortho* than on *meta* position. Thus *56 meta* isomer is only formed.

IV.2.1.2. Thiolates and Amines Condensations on Halogeno Aromatic Compounds

IV.2.1.2.1. Monohalogenobenzenes. From the literature[7)] it appears that arynic condensations of thiolates in the presence of $NaNH_2$ or KNH_2 in liquid ammonia give poor to fair yields in desired products. On the other hand, amines generally do not condense in liquid ammonia. They nearly always need the use of the corresponding lithium amide and/or they constitute the solvent.

On the contrary (Scheme 21) complex base $NaNH_2$-*t*-BuONa enables those condensations on C_6H_5Br in THF at 30–45 °C with good to very good yields[10, 37)]. Moreover, with amines, these latter are used in only slight excess (two moles per mole of bromo compounds), and unreacted amines may be recovered.

$$\underset{57}{\text{R–SNa}} + \text{Ph–Br} \xrightarrow[\text{THF}]{NaNH_2\text{-}t\text{-BuONa}} \underset{58}{\text{Ph–S–R}} \qquad 80–100\%$$

R = Ph, Et

$$\underset{59}{R^1R^2NH} + \text{Ph–Br} \xrightarrow[\text{THF}]{NaNH_2\text{-}t\text{-BuONa}} \underset{60}{\text{Ph–NR}^1R^2} \qquad 60–100\%$$

$R^1 = R^2$ = Me, Et, Pr, *i*-Pr, *n*-Bu, *i*-Bu

$[R^1, R^2] = (CH_2)_5, (CH_2)_2–O–(CH_2)_2$

R^1 = Ph; R^2 = Me

Scheme 21. Examples of condensations of thiolates and amines on PhBr in the presence of a complex base

Those reactions may be performed in the presence of $NaNH_2$ alone but HMPA is needed as a solvent or cosolvent[38, 39] and yields are nearly always lower.

Complex bases also enable condensations of primary amines on Ph–Br in good yields but we always obtain a mixture of mono- and diphenylamines[37]. Of course all the condensations are feasible with substituted halogeno compounds.

It is of interest to mention results obtained with monofluorobenzenes. These derivatives have highly acidic hydrogens at *ortho*-positions of fluorine and it is possible to perform hydrogen-lithium exchange with the help of organolithium compounds. *Ortho* fluoro phenyllithium thus obtained easily lose LiF leading to benzyne[7, 40, 41]. However generation of the same arynic species by mean of $NaNH_2$ in liquid NH_3 is generally not possible except in a few very special cases[42]. On the contrary, arynic condensation of sodium thiolates and amines on fluoro-aromatic derivatives is feasible with the help of $NaNH_2$ in the mixture HMPA–THF or in THF alone, but in this latter case complex base $NaNH_2$-*t*-AmONa is needed[43]. Of course fluoro compounds are less reactive than corresponding bromo or chloro derivatives.

With thiolates there is no particular problem. Both condensation methods lead to the same qualitative results. Generally speaking, complex base method gives better overall yields and shorter reaction times. We have illustrated these reactions in Scheme 22.

$$\underset{57}{R\text{–}SNa} + \underset{61}{x\text{-}Z\text{-}C_6H_4\text{–}F} \xrightarrow[\text{THF}]{NaNH_2\text{-}t\text{-}AmONa} \underset{62}{y\text{-}Z\text{-}C_6H_4\text{–}SR} + \underset{63}{z\text{-}Z\text{-}C_6H_4\text{–}SR}$$

R = Et, Ph

x-Z =		
H	y-Z = z-Z = H	55–65%
o-Me	y-Z = *o*-Me; z-Z = *m*-Me (*62*/*63* = 46/54)	50%
p-Me	y-Z = *m*-Me; z-Z = *p*-Me (*62*/*63* = 55/45)	60%
p-MeO	y-Z = *m*-MeO; z-Z = *p*-MeO (*62*/*63* = 60/40)	78%
o-MeO	y-Z = z-Z = *m*-MeO	30%

Scheme 22

Condensation of amines on fluorobenzenes is nothing less than curious. Indeed, for the first time we met a strong difference in the nature of formed products between $NaNH_2$ in HMPA–THF and a complex base (here $NaNH_2$-*t*-AmONa) in THF. Our observations are illustrated in Scheme 23.

$$\underset{61}{x\text{-}ZC_6H_4\text{–}F} + Et_2NH \xrightarrow[\text{HMPA–THF}]{NaNH_2} x\text{-}Z\text{-}C_6H_4\text{–}NEt_2 \quad (30\text{–}78\%)$$

$$\underset{61}{x\text{-}ZC_6H_4\text{–}F} + Et_2NH \xrightarrow[\text{THF}]{NaNH_2\text{-}t\text{-}AmONa} \underset{64}{y\text{-}Z\text{-}C_6H_4\text{–}NEt_2} + \underset{65}{z\text{-}Z\text{-}C_6H_4\text{–}NEt_2}$$

x-Z =		
H	y-Z = z-Z	75%
o-Me	y-Z = *o*-Me; z-Z = *m*-Me (*64*/*65* = 50/50)	55%
p-Me	y-Z = *m*-Me; z-Z = *p*-Me (*64*/*65* = 52/48)	55%
o-MeO	y-Z = z-Z = *m*-MeO	90%
p-MeO	y-Z = *m*-MeO; z-Z = *p*-MeO (*64*/*65* = 53/47)	35%

Scheme 23

As for the other reactions described above, complex base gives more satisfactory results (better yields and shorter reaction times). However, the most important result is that $NaNH_2$ in HMPA–THF leads to direct substitutions when $NaNH_2$-*t*-AmONa in THF leads to arynic condensations.

It is noteworthy that direct substitutions need the presence of $NaNH_2$. In other words, there is no condensation of amine on fluorobenzenes in HMPA–THF without sodamide.

We interpreted these observations[43)] by the intervention of an "arynoid" (Scheme 24).

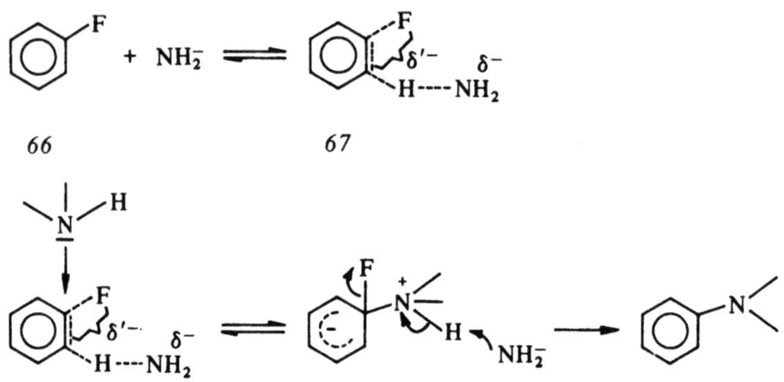

66 67

67 68

Scheme 24

In HMPA–THF, $NaNH_2$ reacts with hydrogen giving a slight ionisation of C–H bond. The consequence is a slight ionisation of C–F bond, insufficient to lead to aryne but making easy the amine attack in spite of the appearance of negative charges near the attacked carbon. With thiolates, charge repulsion would be unfavorable to this mechanism. Moreover, thiolates often give complex bases with $NaNH_2$ and those reagents, as we can see, are particularly efficient in aryne generations.

IV.2.1.2.2. Polyhalogenobenzenes. We were interested in the reactivity of polyhalogenobenzenes because these compounds are potential starting material for synthesis of numerous substituted benzenes.

Among the numerous polyhalogeno derivatives, we chose 1,2,4- and 1,2,3-trichlorobenzenes. Indeed several nucleophilic substitutions had already been studied with the first compound and results published[44)] did not give a good explanation of its reactivity. In fact, basic systems used by the authors chiefly lead to trisubstitutions and hardly give a good idea of the successive products formed. Moreover both trichlorobenzenes are side products of industrial benzene chlorination and are easily obtained in mixture.

Thus it was interesting to know the potential synthetic applications of each compound as well as of their mixture and complex bases might be good reagents to perform elimination-additions on derivatives of this type.

In this chapter we shall briefly describe results obtained for amine condensations[45–47)].

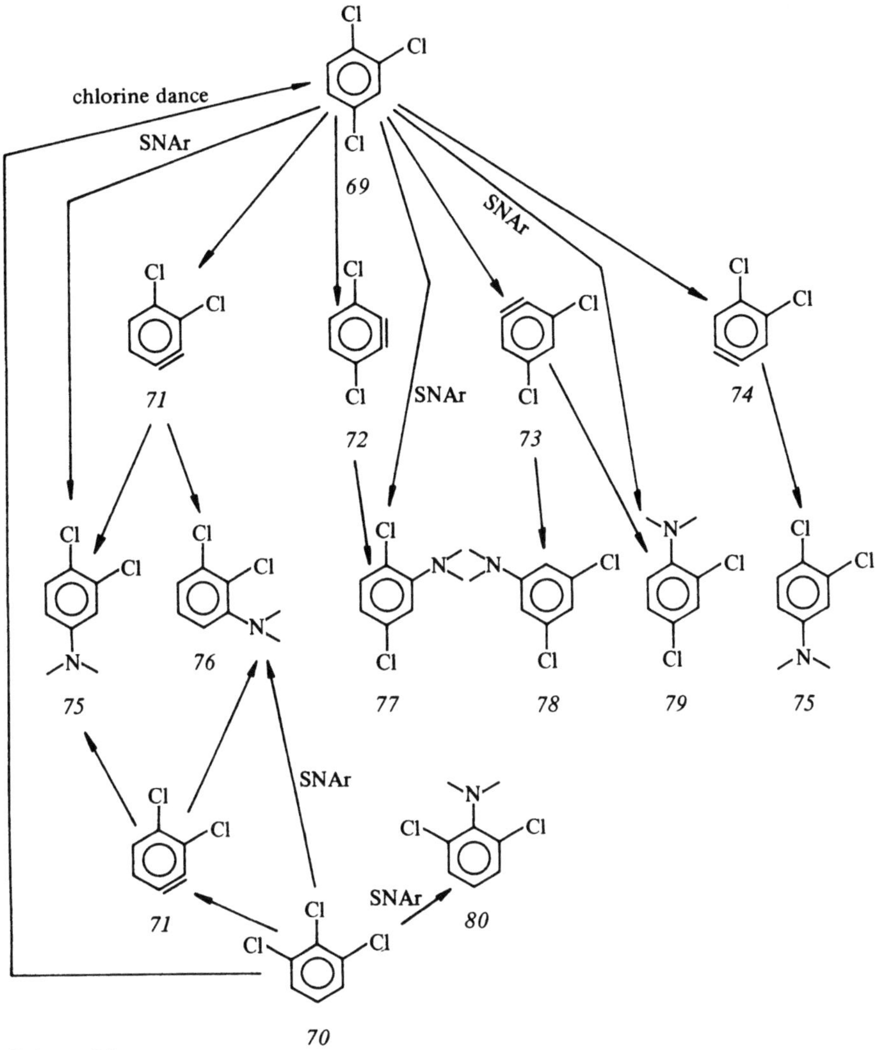

Scheme 25

We have pictured in Scheme 25 the whole mechanism we could expect for monosubstitutions on *69* and *70*. Fortunately all those reactions have not the same probability. Indeed, according to hydrogen acidities of halogenobenzenes[7a, 48], and chlorine elimination probabilities[49] arynes are formed following the decreasing order: *71* > *72* > *73* > *74*. Taking into account the chlorine *meta*-directing effect in arynic condensations[7, 50] we can see that *75* and *77* should be the main products formed by arynic condensation on *69*. On the other hand *70* must essentially lead to *75*.

The ways taken by SNAr condensations are not easily predictable. However, we can say that probability formation of *80* is very low for steric reasons. Finally the famous Bunnett's "Halogen Dance"[51] had to be considered and in fact it certainly happens with *70* (*vide infra*).

Condensations of NH_3 (in liquid NH_3–THF) and primary amines (in THF) on *69* and *70* in the presence of $NaNH_2$ or $NaNH_2$-*t*-BuONa only lead to dichloro aminobenzene mixtures. Products of the *75* and *77* types are very strongly dominant[45] and are formed by arynic paths. *77* from *70* can be only explained by means of the chlorine dance preceding the elimination-addition.

In contradiction with the results expressed before, we found that here the nature of the base is of little importance. However, to obtain good yields we were forced to use an excess of amines and it is known[52] that $NaNH_2$ is rather reactive in primary amines. Thus excess of primary amine must have a levelling effect on the basic power of $NaNH_2$ and complex base.

Anyway, with those reactions we were able to synthesize from *69* and *70* several amino dichlorobenzenes *75* and *77* with yields from 30 to 70%[45]. Moreover, the mixture *69* + *70* can be used in those syntheses.

With secondary amines[46] general Scheme 25 still fits well but results are more complicated than with primary amines. Indeed, sometimes SNAr condensations appear. Moreover, it is not possible to avoid completely the formation of disubstituted products (*i.e.* diamino chlorobenzenes).

Finally, paths taken by the reactions depend on the nature of base, amine and halogeno compound. We can briefly schematize our conclusions by the following table:

Table. Secondary amine condensations on *69* and *70*

Base	$NaNH_2$		$NaNH_2$-*t*-BuONa	
Halogeno compound \ Amine	R_2NH	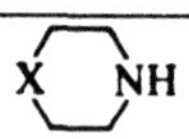	R_2NH	
	R = Et, i.Pr.	X = O, CH_2	R = Et, i.Pr.	X = O, CH_2
69	E A[1])	SNAr + E A[1])(?)[2])	E A[1])	E A[1]) + SNAr(?)[2])
70	E A[1]) + SNAr	SNAr + E A[1])(?)[2])	E A[1])	E A[1]) + SNAr

[1]) E A (Elimination Addition) = Arynic Condensation.
[2]) Very low part in the reaction, if happens.

It is clear that complex bases strongly favor arynic reactions. Moreover, $NaNH_2$-*t*-BuONa always gives better overall yields and higher ratios monocondensation/dicondensation than $NaNH_2$. Note that the chlorine dance is nearly always encountered with *70*.

As for primary amines, we were able to synthesize several N,N-dialkyl amino dichlorobenzenes *75, 77* and *76* with about 25% (and exceptionally 90%) isolated yields. These latter are not very good but must be improvable. Moreover, this very simple method is easily handled on a large scale, and very often the mixture *69* + *70* may be used.

In the aim of performing di- and trisubstitutions on *69* and *70*, we first studied condensations of amines on *75, 77,* and *76,* where >N– is a dialkylamino group[47]. Reactivities observed are summarized on Scheme 26.

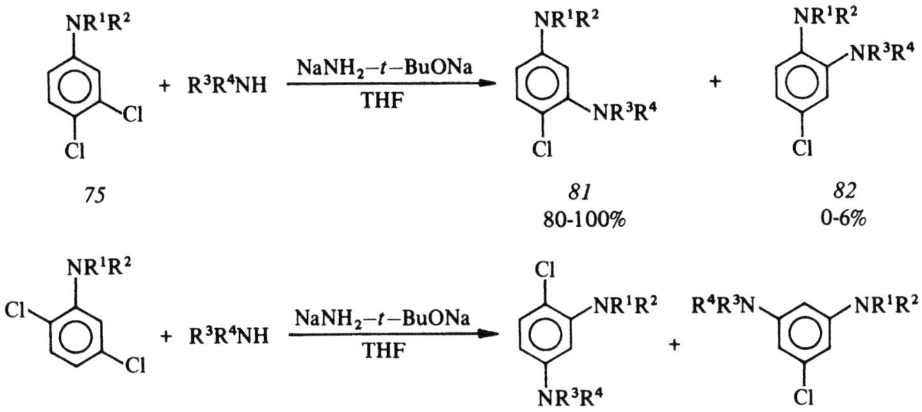

75 — 81 80-100% — 82 0-6%

77 — 83 50-65% — 84 3-15%

Scheme 26

Here mechanisms are essentially arynic with a very strong amino *meta*-directing effect. Taking into account what we saw about monosubstitutions and the two reactions of Scheme 26, it is easy to understand the results described in Scheme 27[47].

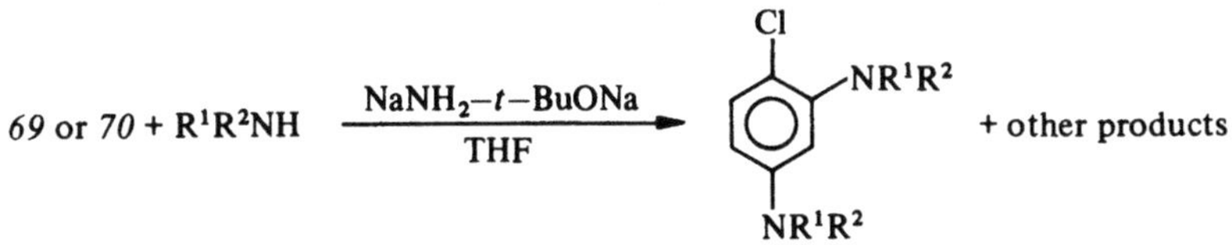

83 ($R^1 = R^3$; $R^2 = R^4$)
main product 50-70%

Scheme 27

Direct trisubstitutions are still simpler[47]. As a matter of fact, amines condense on a mixture of *83* and *84* to give chiefly one triaminobenzene (Scheme 28).

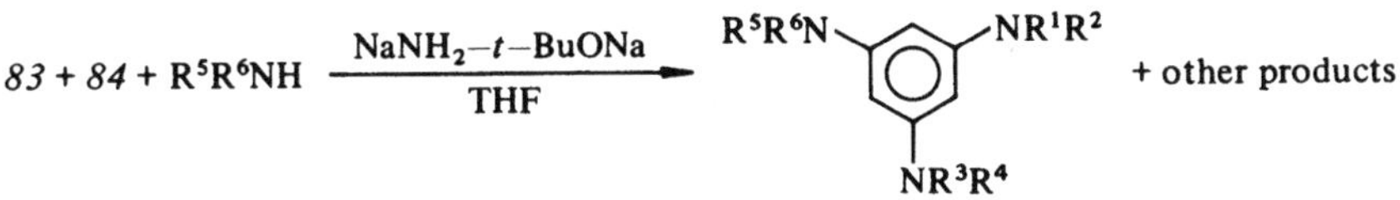

85
main product 70-100%

Scheme 28

Thus direct trisubstitutions allow to synthesize symmetrical triaminobenzenes in rather good yields (Scheme 29).

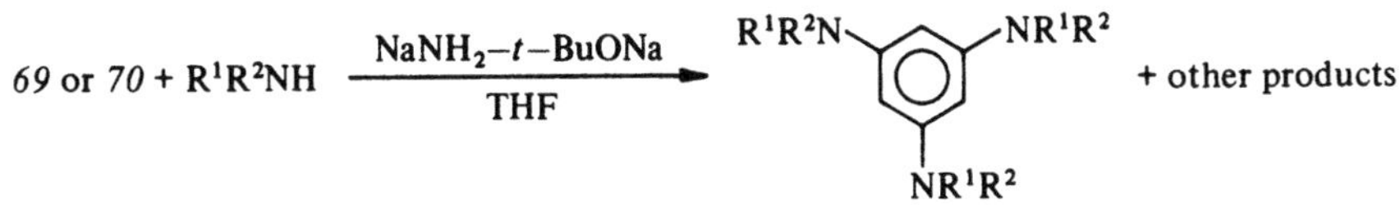

85 ($R^1 = R^3 = R^5$; $R^2 = R^4 = R^6$)
main product 50-60%

Scheme 29

Finally, application of the methods thus developed for the condensations of aliphatic diamines on polychlorobenzenes leads to a new synthesis of several 1,2,3,4-tetrahydroquinoxalines and 2,3,4,5-tetrahydro-1H-1,5-benzodiazapines[53].

Remark: Study of difluorobenzenes[54] shows that reactivities of these compounds are altogether what we saw with monofluorobenzenes (Chapter IV.2.1.2.1.) and with polychlorobenzenes. Those halogeno derivatives allow the synthesis of diamino benzenes and, more interestingly, of mono-aminofluorobenzenes.

In conclusion, this study gives some new light on nucleophilic substitutions of polyhalogenobenzenes and shows that complex bases are particularly useful when arynic condensations are wanted. Moreover, the described methods are of general interest in the synthesis of a lot of aminobenzenes and nitrogen heterocycles from several commercially cheap starting materials.

IV.2.1.3. Ketone Enolates Condensations on Halogeno Aromatic Compounds

IV.2.1.3.1. Introduction. As we said before, ketone enolates generally are good activating agents for $NaNH_2$[10, 11] and lead to nucleophilic complex bases. In other words, it is possible to use the "base" part to generate a dehydrobenzene and the "nucleophile" part to condense on the reactive intermediate.

From the literature[7, 8, 41, 55, 56] it appears that ketones were phenylated in liquid NH_3 in the presence of $NaNH_2$ by means of arynic condensations. Note that it has been shown for a few years that it is possible to synthesize phenyl ketones starting from halogenobenzenes and ketone enolates by means of SR_{N_1} reactions[57].

If we look at the first intermediary *86* of the condensation of a ketone enolate on dehydrobenzene, it appears that it bears altogether a negative charge and a ketonic function. In liquid NH_3 there is no further reaction but in aprotic solvent, *86* can evolve in different ways leading to new condensations (*vide infra*).

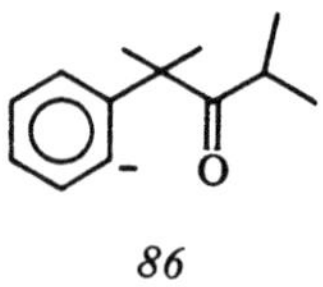

86

Concerning our own work on arynic ketone phenylations, we first performed experiments in the mixture HMPA–THF[9, 39]. They led us to the concept of complex bases. After that we generally used THF or DME, reserving HMPA to only a few special cases.

IV.2.1.3.2. Monohalogenobenzenes. [Condensation of Aliphatic Ketone Enolates: ketones $R{-}CH_2{-}CO{-}CH_2{-}R$ *(Two Hydrogens in* α *and* α' *Positions)].* Generally[9, 39, 58] two ketones (instead of one in liquid NH_3) are formed (Scheme 30). We called *89* normal ketones and *93* transposed ketones.

Ratio *93/89* increases with increasing reaction temperature and decreasing solvent polarity (C_6H_{12}, C_6H_6 > THF > HMPA). It is noteworthy that those reactions are feasible in solvents as little polar as cyclohexane!

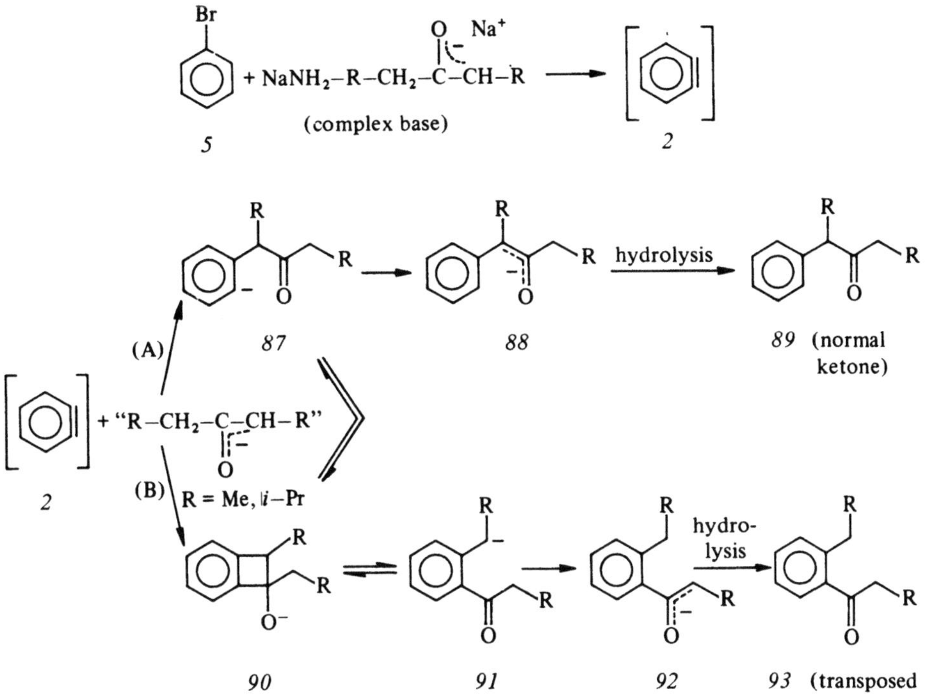

Scheme 30

Alcohols corresponding to *90* were never obtained with aliphatic ketone enolates. However, we shall see later with cyclic ketone enolates that *90* must be intermediates of these reactions and that their formation following way (B) (Scheme 30) is disfavored by polar aprotic solvents. Moreover, it is well known that polar aprotic solvents favor prototropies[59], as here in the *88* formation.

From a synthetic view-point, we have to say that these reactions are performed with a ratio ketone enolate/bromobenzene which is worth 2. However, excess of starting ketone is generally easily recovered. Yields being given relatively to bromobenzene, we can say that overall yields are between 60 and 80%. Moreover it is possible to obtain mixtures with *93*/*89* varying roughly from 60/40 to 90/10.

Thus these very simple reactions are easily handled on a large scale from current starting materials and constitute a good way for the synthesis of aromatic ketones, particularly of the *93* type.

Ketones $R^1R^2CH–CO–CHR^1R^2$ and $R^1R^2CH–CO–CR^3R^4R^5$ (*One Hydrogen in α and α′ or in α only*). In these cases normal ketones of the *89* type (see Scheme 30) are not detected. On the other hand, some new reactions may intervene[58, 60]. All the possible reactions are collected in Scheme 31 starting from *94* equivalent anion of *91*.

99 and *100* are typically Haller Bauer's degradation derivatives[61] and generally formed in rather drastic conditions. Unexpected "oxydoreduction" leading to *98* may take an important part in certain conditions. Finally, overall yields are from fair to good (55–75%).

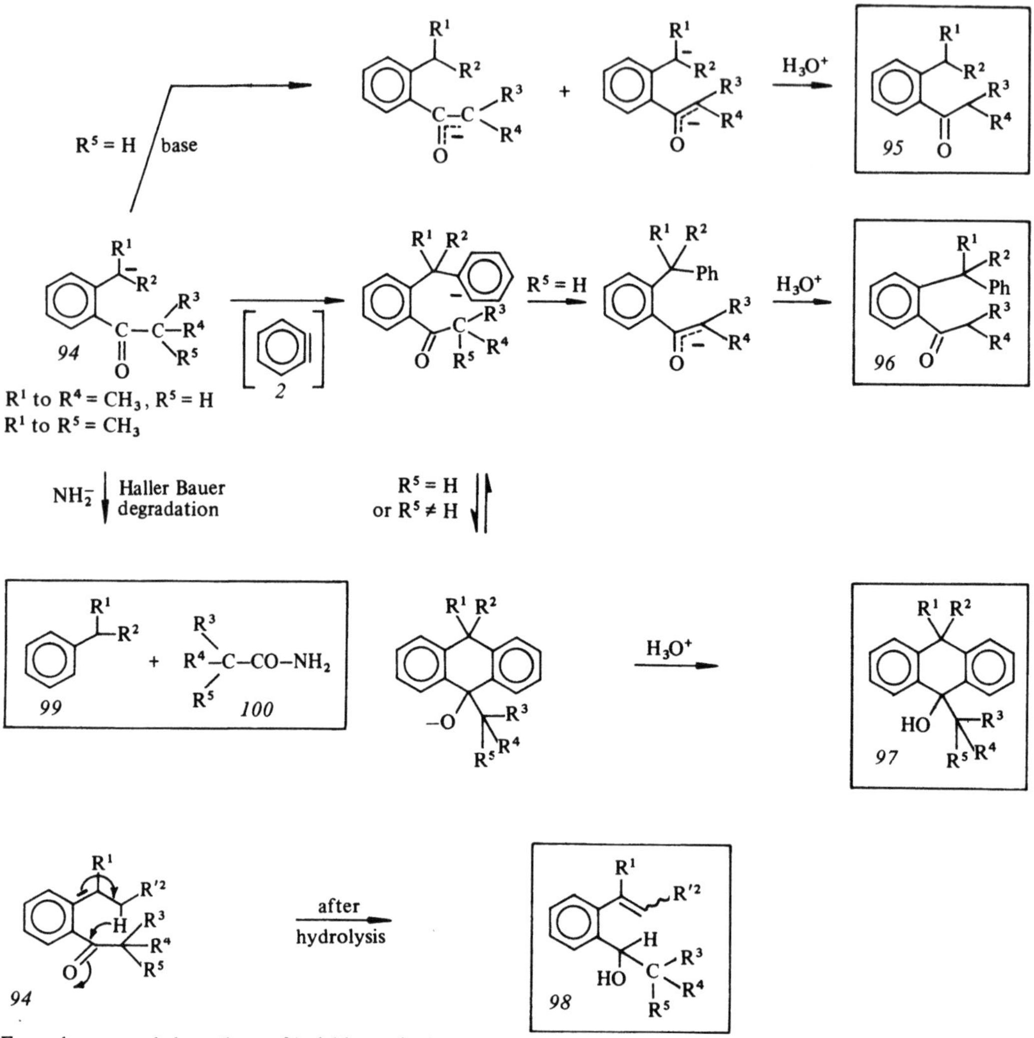

Formula surrounded are those of isolable products.

Scheme 31

Here too, ratio of formed products depends upon temperature, solvent and reaction time and it is possible to direct more or less selectively the condensations towards a determined product.

Thus taking into account the practical simplicity of these condensations and the elaborated structures they can led to, we can say that they constitute good ways in the synthesis of several structures.

Aromatic ketones Ph–C(=O)–CHR^{1}R^2. With these ketones, the only products isolated[60] are alcohols of the *98* type, anthranols of the *97* type (its formation depends on the structure of the starting ketone) and anthranones *102* which are formed following the mechanism pictured in Scheme 32.

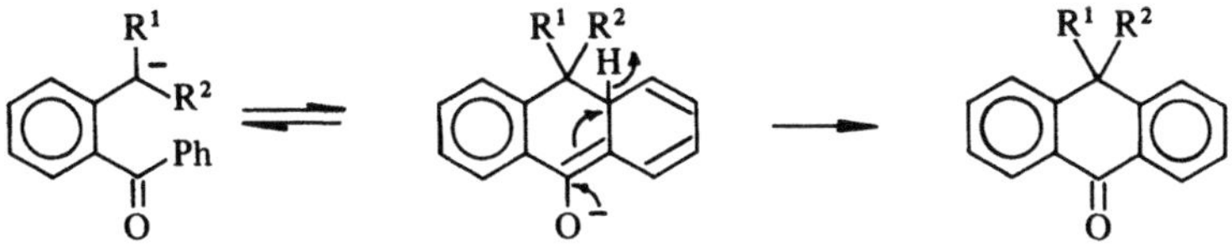

101 $R^1 = R^2 = CH_3$
$[R^1, R^2] = (CH_2)_5$

102

Scheme 32

Analogous cyclisation is given in the literature[62]. From a practical view-point overall yields are rather good (45 to 70%) and reactions are very simple to perform.

In conclusion, this first part concerning reactions of halogenobenzenes with complex bases $NaNH_2$-ketone enolates shows that these condensations constitute a new tool in the syntheses of a large variety of chemical structures. Moreover, even when yields are not very good, we are reminded that those reactions are generally easily performed, and starting materials are simple or commercial, when the obtained compounds very often need much more sophisticated methods to be synthesized by another way.

Condensation of alicyclic ketone enolates. Alicyclic ketone enolates are generally good $NaNH_2$ activating agents and give complex bases able to generate benzynes. A few exceptions are encountered. In those cases a complex base $NaNH_2$-alcoholate must be used together with the ketone enolate to be condensed.

Generally speaking, mechanism of these condensations looks like the one written for aliphatic ketones in Scheme 30. However, alicyclic ketones enolates allowed us to obtain many precise confirmations of it.

Thus in the most complicated case, we can observe the overall reactions pictured in Scheme 33.

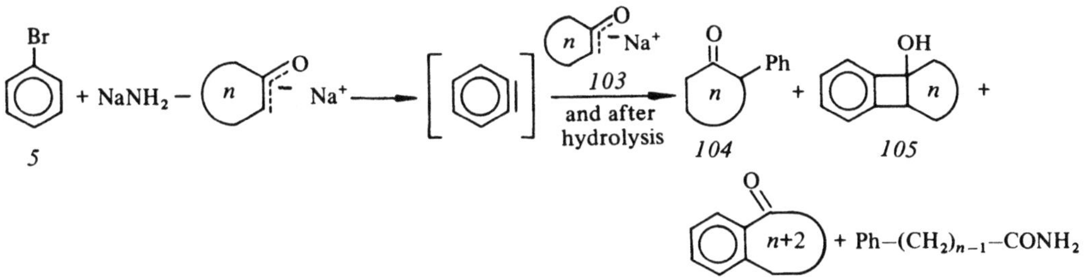

Scheme 33

Of course amide proceeds from Haller Bauer's degradation (*vide supra*) and may be generally avoided.

Ketones *104* and *106* respectively are normal and transposed ketones, corresponding to *89* and *93* (Scheme 30).

Alcohols *105* (only isolable for $n = 5, 6, 7$) are of interest. Indeed, their isolation brings a confirmation to our hypothesis concerning the participation of sodium benzocyclobutenolates as intermediaries. Moreover, we isolated them for the first time[63] and showed that they have rather special chemical properties[64–66].

Concerning arynic condensations, we made an intriguing observation. All the reactions are performed in a strongly basic medium. Alcohols *105* are isolated after hydrolysis although they are unstable in the presence of a base[64]! Thus they are spontaneously opened following the general path given in Scheme 34. As a matter of fact this is a well-known property of benzocyclobutenols[67].

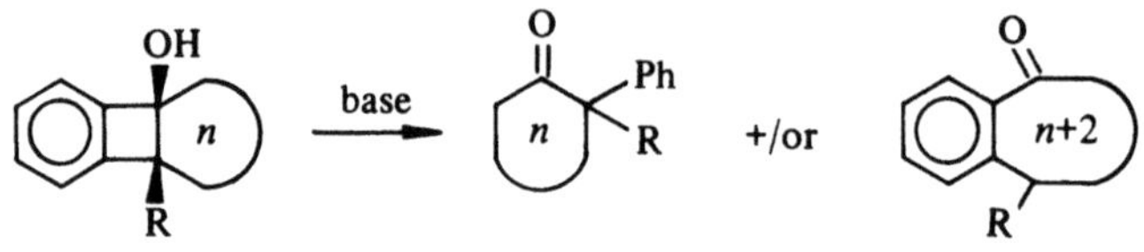

105 *104* *106*

Scheme 34

However, there were some contradictions between the results of our experiments. First lights were given when we compared the reactions described in Scheme 35.

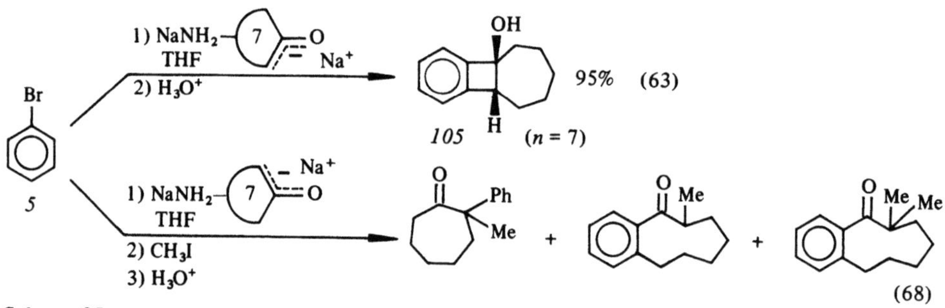

Scheme 35

In both cases, arynic condensations were performed in the same conditions. In the second reaction, hydrolysis was precedented by methyl iodide addition. As a consequence it was not possible to isolate *105* (n = 7) and ketonic compounds only were formed.

Our explanation is that arynic condensation leads to alcoholate corresponding to *105* (n = 7) which is stabilized by complexation of the oxygen by Na^+. Addition of CH_3I leads to a competition for Na^+ between the oxygen alcoholate and I^- or even CH_3I. The result of such a competition is a releasing of electronic charges allowing opening of the alcoholate (Scheme 36).

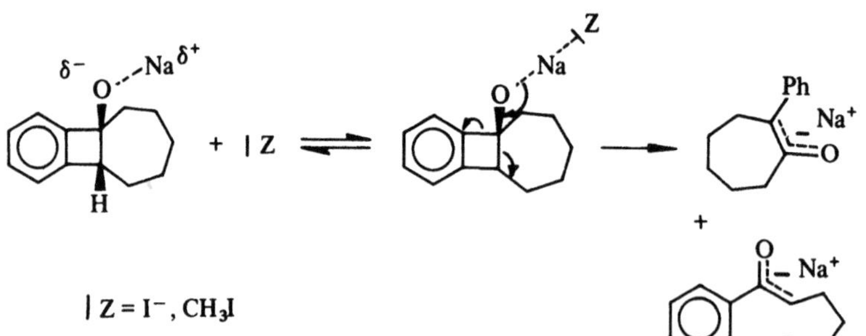

Scheme 36

This hypothesis was verified by adding NaI instead of CH_3I just before hydrolysis. In these conditions *104* ($n = 7$) and *106* ($n = 7$) were the only products isolated.

Thus from these experiments it must be concluded that if alcohols *105* are isolated after an arynic condensation, this means that during all the reaction, Na^+ remains near the oxygen of the starting enolate and that of course, all other things being the same, the alcoholate is stable in the reaction conditions. This hypothesis is confirmed by reaction pictured in Scheme 37[68].

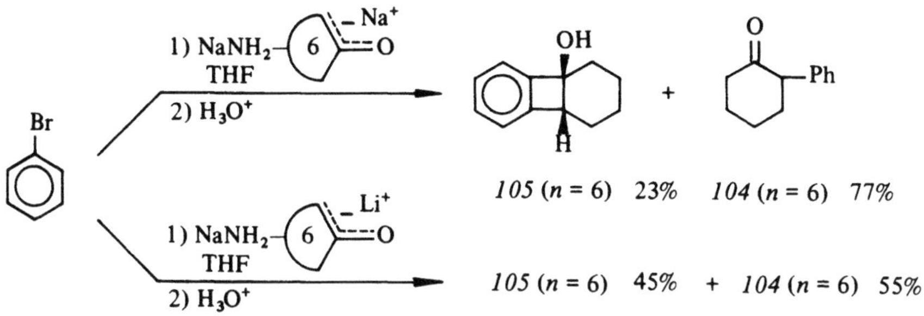

Scheme 37

Li^+ which is a more complexing cation than Na^+ leads to better yields in *105* ($n = 6$).

Taking into account all results, we proposed[68] the overall mechanism given in Scheme 38, (see page 78).

This mechanism looks strongly like the one given in Scheme 30 for aliphatic ketone enolate condensations. Paths (A) and (B) may intervene simultaneously. However, we think that (B) is the chief way[68].

We condensed a lot of cyclanone enolates with various ring size and numerous cyclohexanone enolates substituted in various positions by alkyl groups[63, 68, 69, 70]. From all this work we can make the following practical remarks:

Alcohols *105* are directly obtained from arynic condensations only when $n = 5, 6, 7$. We shall see briefly later that we have now another way to obtain several of them.

Cyclohexanone enolates never lead to ketones *106* except when an alkyl group (Me, *t*-Bu) is at position four in the starting compound. However, we shall see that it is possible to easily synthesize *106* from *105*.

THF and DME can be used as reactional solvent. DME allows performing condensations at lower temperature but, of course, favors the opening of sodium benzocyclobutenolates.

Haller Bauer's degradation is favored by increasing solvent polarity, temperature and steric hindrance in the neighbourhood of the carbonyl group.

Nonnucleophilic complex base $NaNH_2$-*t*-BuONa, used with low reactive ketones, favor the formation of ketonic compounds to the prejudice of benzocyclobutenols.

As for aliphatic ketones, two moles of starting carbonyl compound per mole of bromobenzene must be used. However, unreacted starting material is also generally easily recovered.

From a preparative chemistry view-point, these reactions may be useful and we shall give now a few examples of synthesized structures.

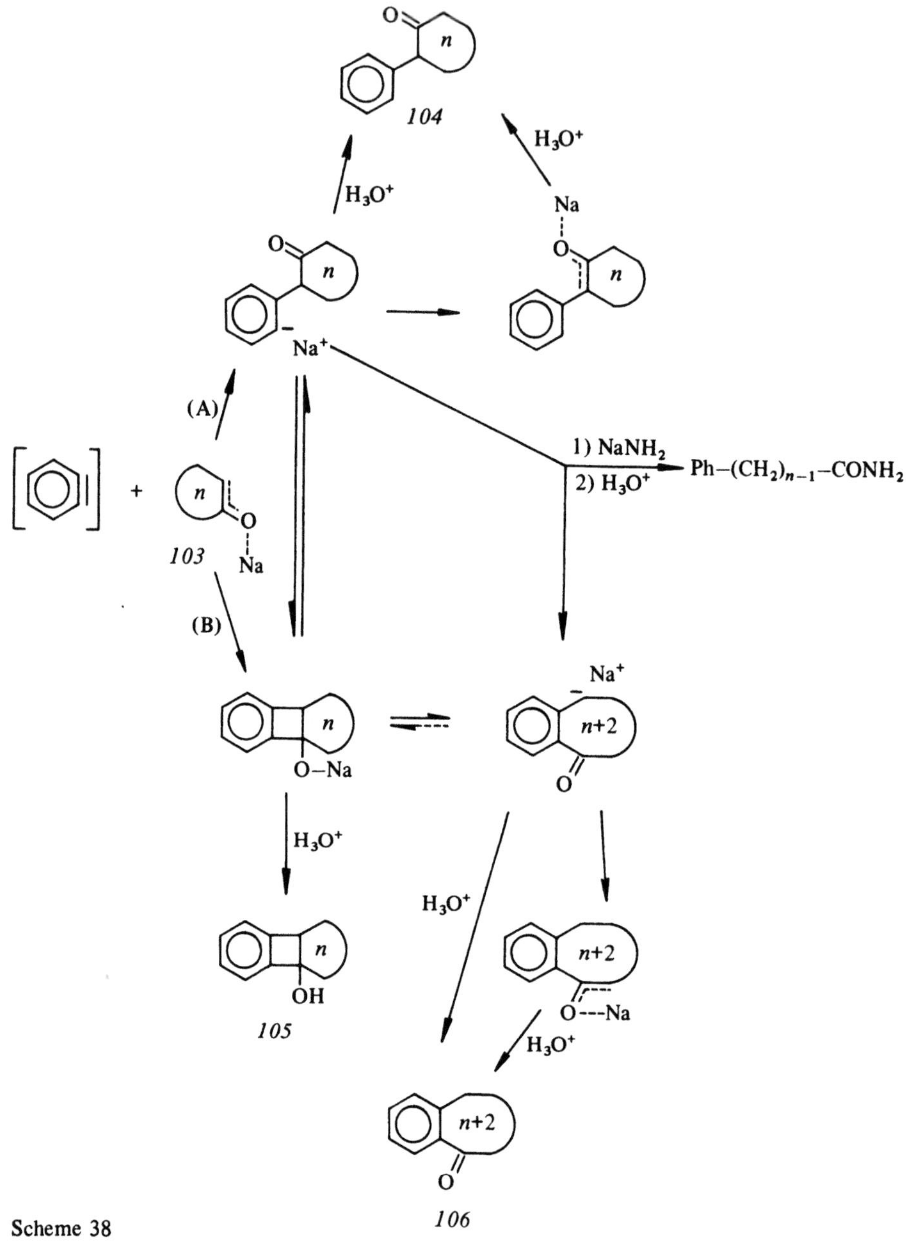

Scheme 38

It has been claimed that arynic phenylation of ketones is not a good reaction to obtain phenyl ketones[71]. Observation of our results show that arynic phenylation is one of the simplest syntheses (compare to methods given in Ref.[71]) and that many phenylcyclanones may be obtained in rather good yields as exemplified in Scheme 39.

Some examples of phenylcyclanones synthesized by reaction of Ph–Br with $NaNH_2$-ketone enolate or $NaNH_2$-*t*-BuONa-ketone enolate are shown:

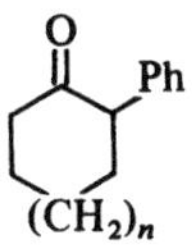
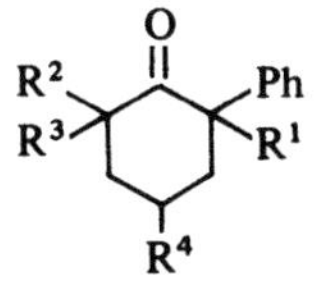
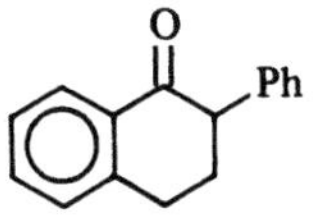
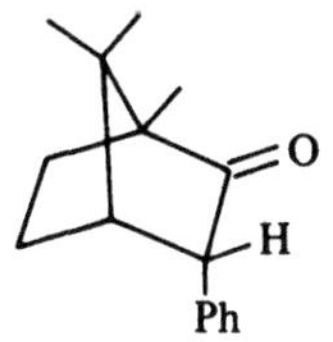

$n = 1, 2, 3, 7$

yields: 50–80%[63, 68, 69]

$R^1 = Me, R^2 = R^3 = R^4 = H$
$R^1 = R^2 = Me, R^3 = R^4 = H$
$R^1 = R^2 = R^3 = Me, R^4 = H$
$R^1 = R^2 = R^3 = H, R^4 = Me, t\text{-Bu}$
yields: 40–75%[69, 70]

yields: 75%[70]

yields: 55–60%[72]

Scheme 39

Interesting synthetic applications of arynic condensations of ketone enolates were also obtained in the synthesis of benzocyclenones. We shall comment them a bit.

Benzocyclenones may be often obtained directly after arynic condensations. However, as we saw before, benzocyclobutenols *105* are also obtained in several reactions. In those cases, *105* may be opened by reaction with a base in aprotic solvent to give the benzocyclenone *106* with very good yields[64].

On the other hand, arynic reactions are also of interest in the synthesis of benzocyclenones substituted on the aromatic ring by an alkyl group in the *ortho* position of the carbonyl group. In these syntheses, *ortho* alkyl bromobenzenes must be used and arynic condensation intermediaries are of the *108* type:

108

Of course *108* may be attacked in the *ortho* and *meta* positions leading to very complicated mixtures of compounds. Fortunately, ketone enolates are very sensitive to steric hindrance and condensations in *meta* position are strongly dominant[73].

Thus with the help of those reactions, we are able to prepare benzocyclenones as exemplified in Scheme 40. (Many other benzocyclenones substituted on the ketonic ring were also obtained during our studies[64, 72, 75].)

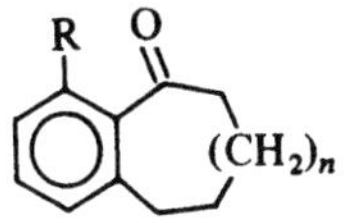

R = H	$n = 1–4, 6–9, 11, 12$	yields: 40–60%
$R = CH_3$	$n = 1–6, 8$	yields: 30–50%
R = Et, *i*-Pr	$n = 2–6, 8$	yields: 20–55%
[68, 73, 74]		

Scheme 40

We have to note that sometimes yields are not excellent and obtaining pure ketones needs some purifications. However, all the operations are very simple and may be performed on a large scale. Thus, taking into account that these reactions lead to benzocyclenones (without limitation of ring size) which are very often difficult to obtain by classical ways, we can say that they constitute interesting synthetic tools for these kind of derivatives.

Finally, we said before that our studies on arynic condensations of ketone enolates led us to synthesize for the first time benzocyclobutenols *105*. However, those compounds are directly obtainable by arynic reactions only when the starting ketone is a five, six or seven membered ring. Fortunately, benzocyclenones photochemically cyclize to give *105*. Moreover, when the aromatic ring bears an alkyl group in the *ortho* position of the carbonyl function[19, 66, 75], another photocyclization may take place leading to another class of benzocyclobutenols *109*. One of these latter has been independently prepared by Sammes and his colleagues[76].

In summary, numerous derivatives *105* and *109* may be now synthesized by one of the ways pictured in Scheme 41.

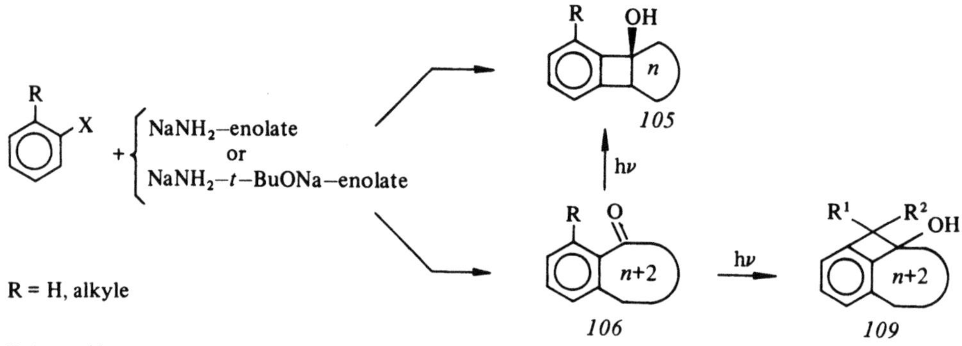

Scheme 41

A few illustrations of the structures thus obtainable are given in Scheme 42.

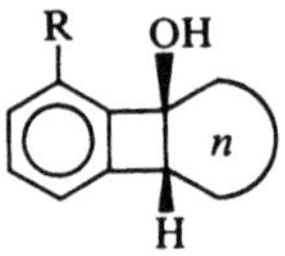

19, 63, 66, 69, 75)
R = H, n = 5, 6, 7, 8, 9
R = Me, n = 8, 9, 10, 12
R = Et, *i*-Pr, n = 8, 9

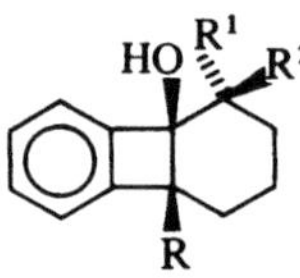

69, 70, 77)
R = Me, $R^1 = R^2$ = H
R = Me, R^1 = Me, R^2 = H
R = $R^1 = R^2$ = Me

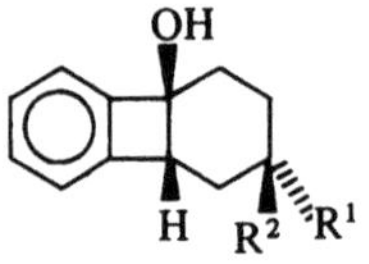

70)
R^1 = H, R^2 = *t*-Bu
R^1 = *t*-Bu, R^2 = H
$R^1 = R^2$ = Me

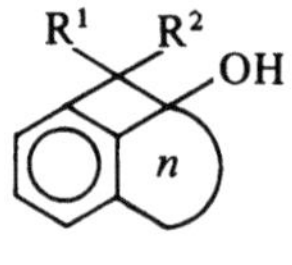

19, 66, 75)
n = 8, 9 $R^1 = R^2$ = H; R^1 = H R^2 = Me
n = 10 R^1 = H R^2 = Me; $R^1 = R^2$ = Me
n = 11 $R^1 = R^2$ = H, Me; R^1 = H R^2 = Me

Scheme 42

Closing this section we have to point out that arynic condensations of ketone enolates may incidentally take a curious way[78]. Thus $NaNH_2$-norcamphor lithium enolate condenses on bromobenzene to give directly the polycyclic alcohol *110!* (Scheme 43).

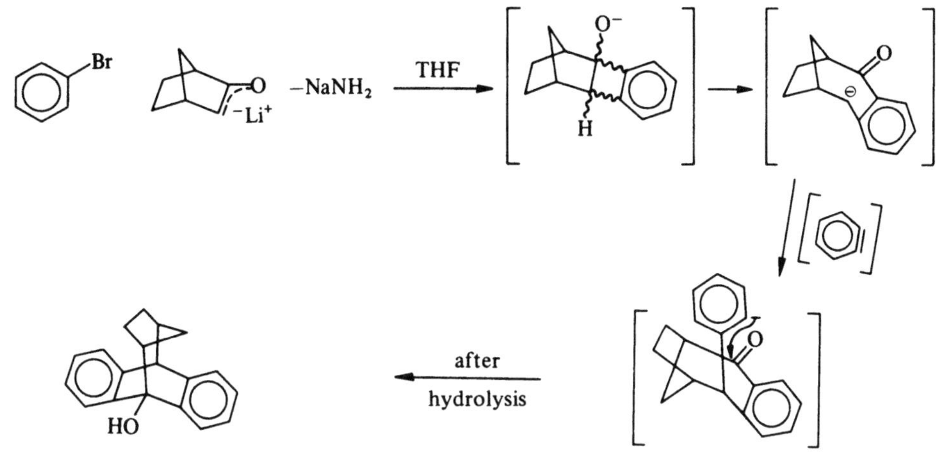

110 85%

Scheme 43

Condensation of α-β ethylenic ketone enolates. We started this work[79] independently and simultaneously with Sammes and Wallace[80] on ketones enolisable on the side, relative to the ketonic group, where the insaturation is. In those cases, use of complex base $NaNH_2$-*t*-BuONa is needed. Mechanisms are rather complicated as may be seen in Scheme 44 where we have collected our own results[79] and those of Sammes and Wallace[81].

Compounds *111, 112, 113* and *114* were effectively isolated. These reactions are of interest for the synthesis of naphthalenic compounds.

On the other hand, when the ketone is enolisable only at the saturated side, the reactions take another way as pictured in Scheme 45, where benzyne is also generated from bromobenzene and complex base $NaNH_2$-*t*-BuONa[82].

Structural effects on R^2 destabilizing the negative charge at the benzylic position favour attack (B). Thus, making R^2 = alkyl instead of R^2 = H leads to the formation of *116.* On the other hand, structural effects on R which promote delocalization of the ethylenic unsaturation towards the carbonyl group favour attack (A). From recent experiments we think that it should be possible to manage, at least partially, the reactions towards *115* or *116* by means of solvent and cation effects.

IV.2.1.3.3. Polyhalogenobenzenes. We essentially studied the reactions of $NaNH_2$-enolate or $NaNH_2$-*t*-BuONa-enolates with dialkylamino dichlorobenzenes *117* and *118* for which the corresponding generated arynes are *119* and *120* respectively.

119 is essentially attacked on position three and *120* on position four. Products formed[83] are of the same kind as the ones encountered with bromobenzenes. However, besides the usual aromatic ketones and benzocyclobutenols we observed the

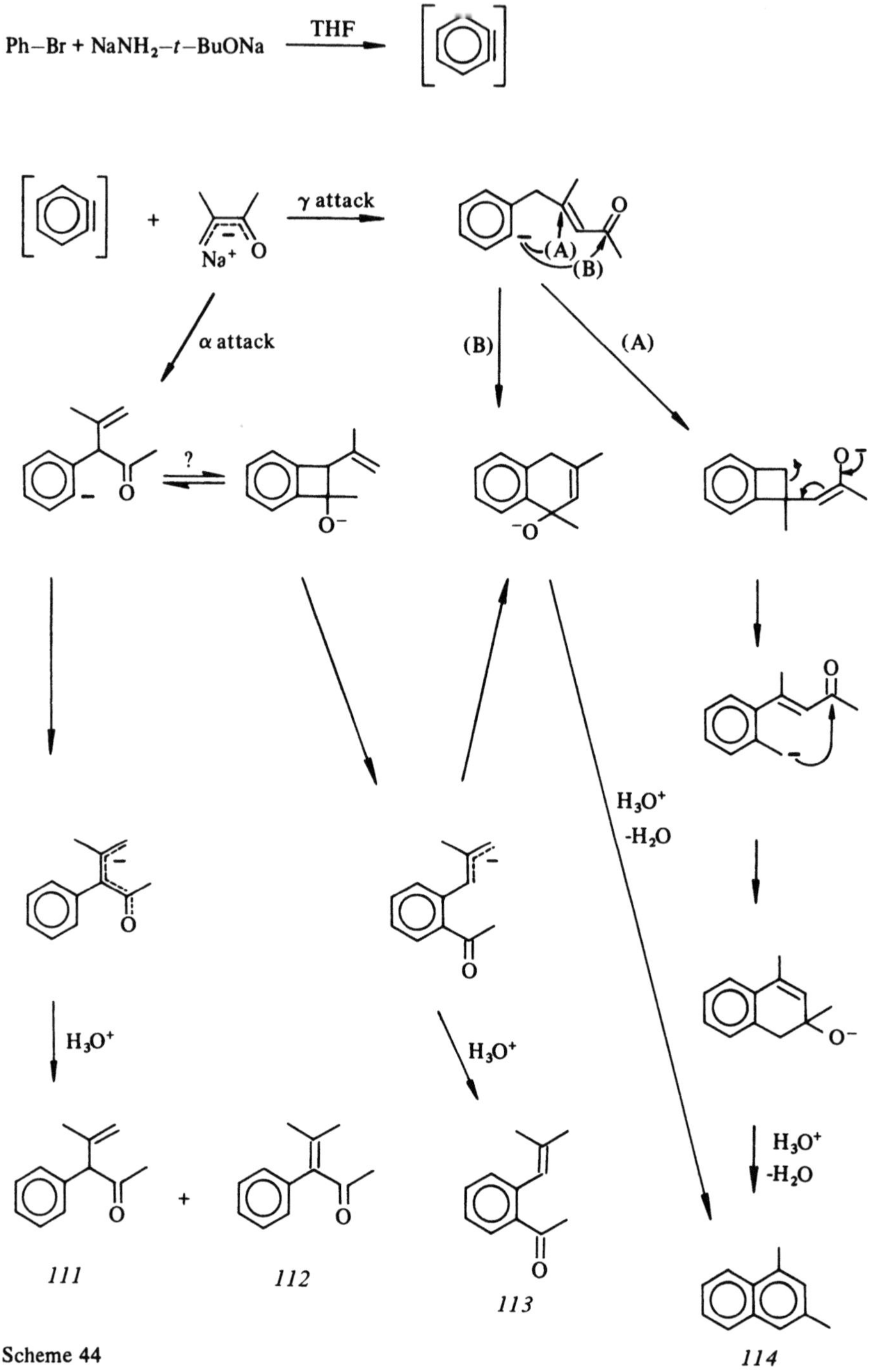

111

112

113

114

Scheme 44

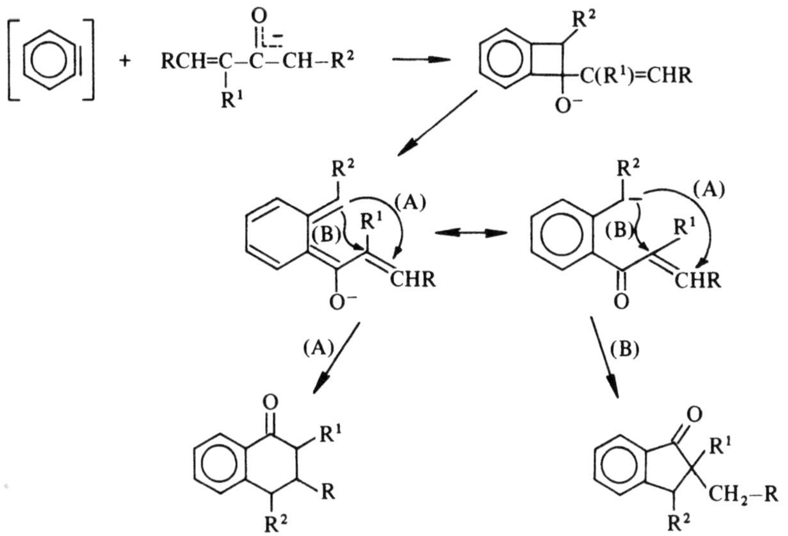

	115	*116*
R = C_6H_5 $R^1 = R^2 = H$	60%	10%
R = 4-$MeOC_6H_4$; 3,4-$(MeO)_2C_6H_3$ 2-Furyl, i-C_3H_7 $R^1 = R^2 = H$	45–60%	–
R = 4-$MeOC_6H_4$ $R^1 = H$ $R^2 = CH_3$	20%	40%
R = C_6H_5; 3,4-$(MeO)_2C_6H_3$ $R^1 = H$ $R^2 = CH_3$	–	50–60%
R = C_6H_5[R^1, R^2] = $(CH_2)_n$ (n = 3,4)	–	80–85%

Scheme 45

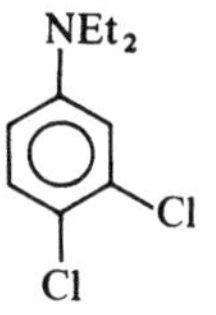

117

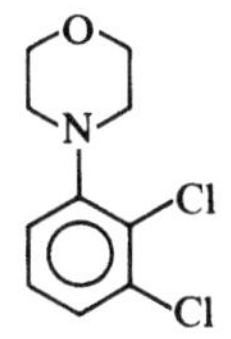

118

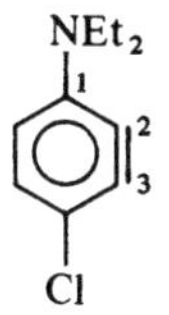

119

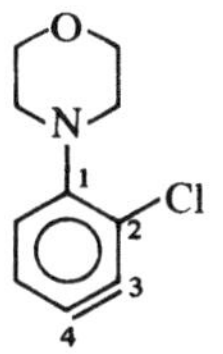

120

formation of benzofurans. Studying carefully this new reaction, we found[84] that heterocycles are formed following the mechanism pictured in Scheme 46 from *119*.

Starting from *117*, and aliphatic or alicyclic ketones, yields in 5-diethylamino benzofurans are between 25 and 50%. Reactions of the same type lead, from *118*, to 6-morpholinobenzofurans with about 25% yields. Finally benzofurans may be also obtained from 1,2 or 1,3-dichlorobenzenes[84].

An interesting point of these syntheses is the simplicity, compensating the yields which are not always very good.

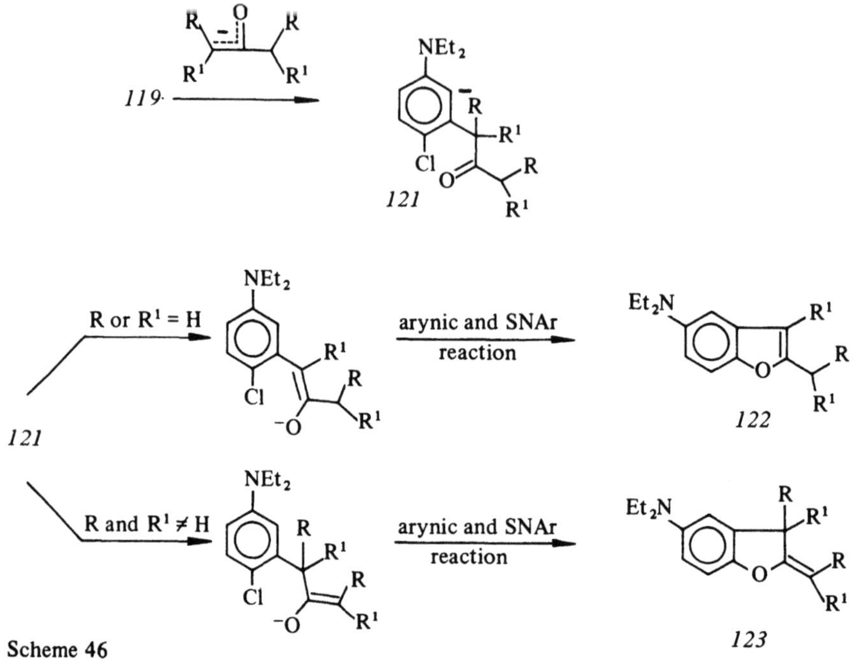

Scheme 46

IV.2.1.3.4. 1-Bromonaphthalene. Nucleophilic complex base $NaNH_2$-enolate very easily generate 1-naphthyne *124* from 1-bromonaphthalene[85].

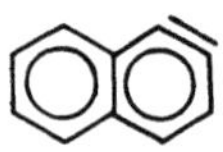

124

Of course ketone enolates condense very easily on *124* following a mechanism very similar to the one described with dehydrobenzenes. First nucleophilic attack must take place on two positions. In fact, as we said before, ketone enolates are very sensitive to steric hindrance and reactions on position two are strongly dominant. Thus numerous structures of the type exemplified in Scheme 47 may be prepared.

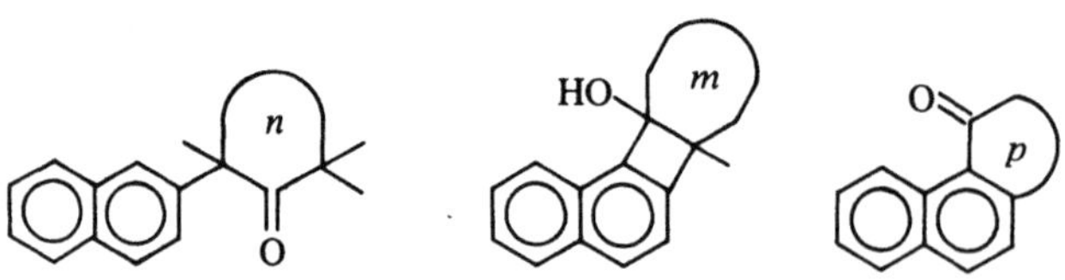

Scheme 47

IV.2.2. Cyclenic Eliminations-Additions: Cycloalkynes and 1,2-Cycloalkadienes Chemistry

IV.2.2.1. Generalities

The intermediacy of strained cycloalkynes and 1,2-cycloalkadienes[7] in certain reactions was proven long ago[86]. Thus reactions of 1-halocycloalkenes often occur by the elimination-addition mechanism pictured in Scheme 48[87].

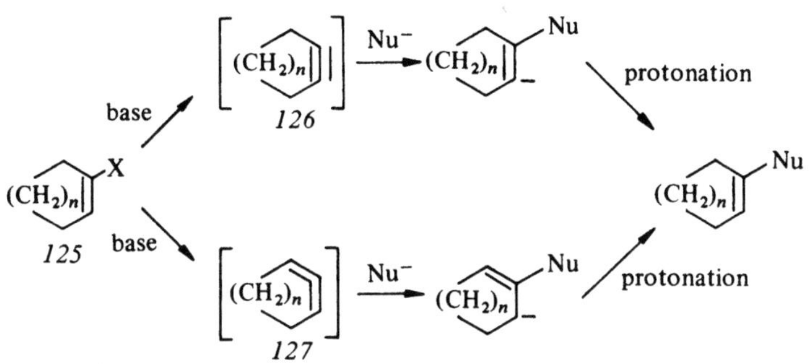

Scheme 48

The quantity *n* must be small enough for intermediaries *126* and *127* to be highly reactive with nucleophiles. Note that this field of chemistry still creates much interest[88].

We gave attention to cases where *n* is worth two and three because no general method allowing the reactions pictured in Scheme 48 had been pointed out. As a matter of fact in nearly all the preceding works cited, the base plays at the same time the rôle of nucleophile. Moreover, conditions used are often drastic and yields are rather low. Note that $NaNH_2$ in liquid ammonia never gives good results. Generally the major products formed are aminated polymers from 1-chlorocyclohexene and dimer derived from 1,2-cycloheptadiene (with 30% yields) when 1-chlorocycloheptene is used[89].

We thought that, as for arynic reactions, complex bases would be able to give the desired elimination step without condensation on the reactive intermediate, thus enabling the desired condensations feasible in good conditions. Of course we had no idea about the intermediate which could be formed with complex bases. Indeed it had been well shown that the nature of the base, of the halogen, of the solvent, and operating conditions all play a determining rôle on the nature of the generated intermediate[86–88].

IV.2.2.2. Amines and Thiolates Condensations on Vinylic Halogeno Cyclohexenes and Cycloheptenes

Vinylic halogeno cyclohexenes. With the help of complex bases $NaNH_2$-*t*-BuONa or certain $NaNH_2$–RSNa, we showed[90, 91] that it is possible to perform easily at low temperature (30–40 °C) the reactions described in Scheme 49.

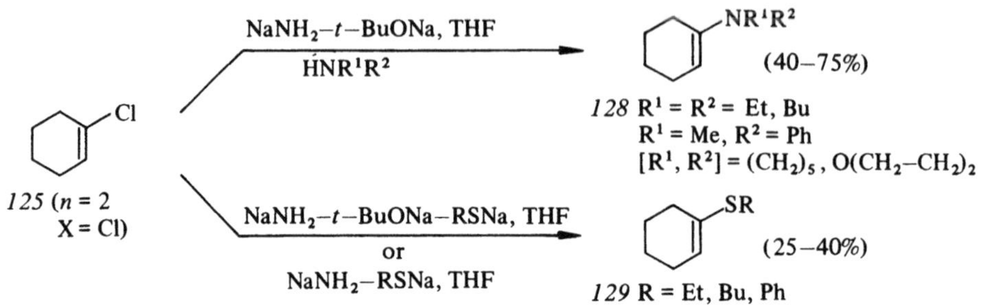

Scheme 49

Those results were encouraging, even for the thiolates, taking into account the yields generally observed in the reactions of bases with 1-chloro cyclohexenes[88]. Moreover, we had realized the first condensation of the kind pictured in Scheme 48 with a base different from the nucleophile to be condensed. However, nothing was known about the mechanism of those condensations. We have illustrated on 1-halogeno 4-alkyl cyclohexene (Scheme 50) all the hypothesis which have to be considered. In the worst case, we have to take into account that several mechanisms together intervene. To clarify the situation, we studied the condensation of amines on several chlorocyclohexenes[94]. We have summarized the results obtained in Scheme 51.

Without other details, it appears that enamines are formed only with chloro compounds able to give a cyclohexyne. This also explains the formation of two isomers corresponding to nucleophilic condensation on both carbons of the intermediate acetylenic bond. Of course, steric effects may eventually play a rôle on the relative ratios of the two products.

Comparison of Schemes 50 and 51 shows that elimination-addition by means of a cyclohexyne is certainly the mechanism leading to enamines from *130, 131* and *132.*

If cyclohexyne formation is not possible (*133* and *134*), the chloro compound reacts very slowly with complex base which degrades it without formation of enamines. Maybe there is the formation of a 1,2-diene but it does not condense with amines.

1-Chloro Cycloheptene. This chloro compound reacts in THF with $NaNH_2$ as well as with the complex base $NaNH_2$-*t*-BuONa at 35 °C. It does not react with *t*-BuONa in the conditions used[95, 96].

The product formed is the cycloheptadiene dimer (Scheme 52).

Thus we can say that elimination step leads essentially to 1,2-cycloheptadiene which easily dimerises.

In the presence of amines, the products obtained are mixtures of the expected enamine and *136* (Scheme 53).

Concerning the first condensations, total yields are in accord with the yield formation of *136* in Scheme 52. Thus enamine is formed to the prejudice of the dimer.

In the last condensations, yields of enamines are low but yields of dimer are not the expected complement. Thus we have to conclude that *135* disappears by another way.

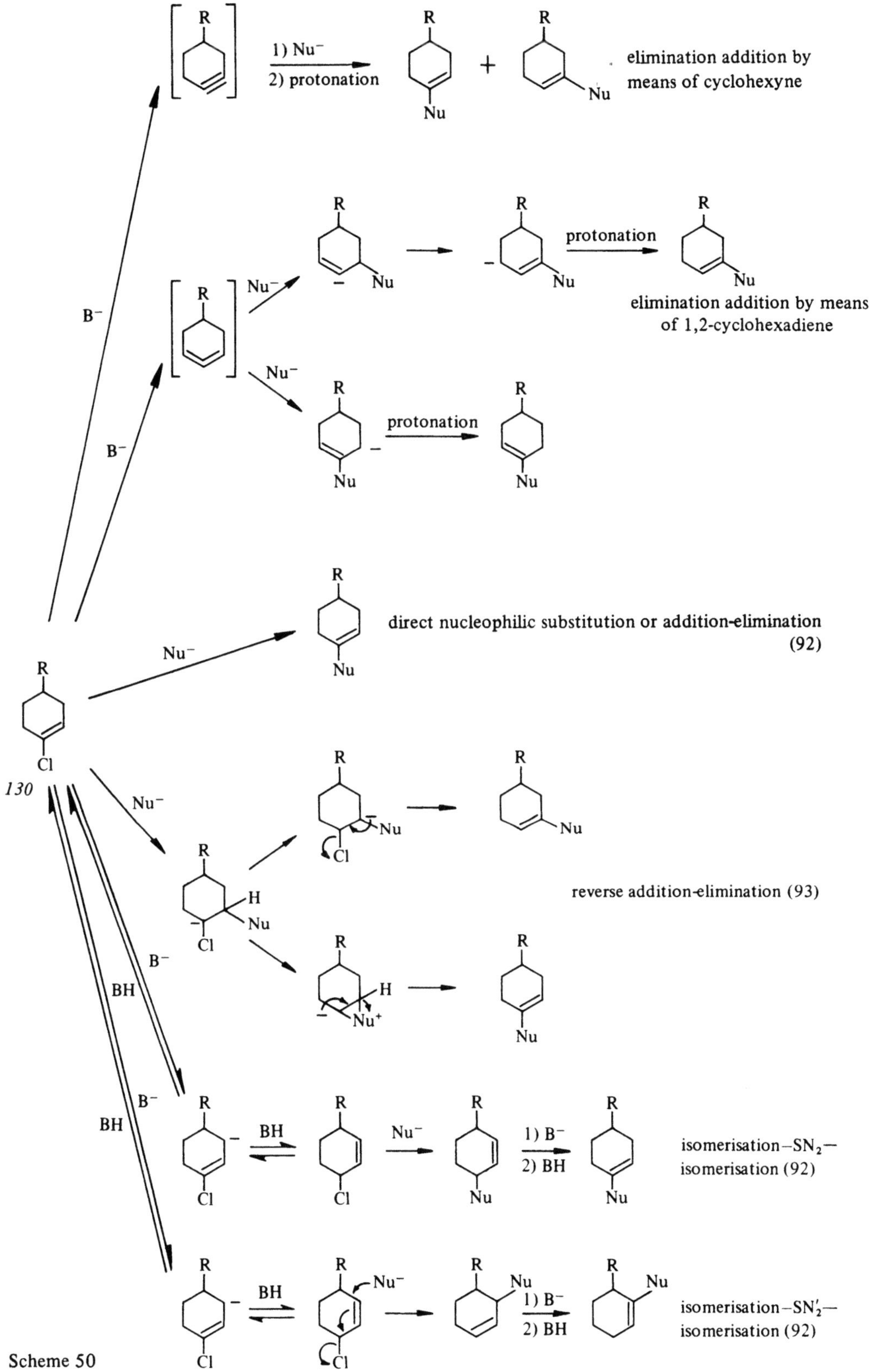
1) Nu⁻
2) protonation
elimination addition by means of cyclohexyne
protonation
elimination addition by means of 1,2-cyclohexadiene
protonation
direct nucleophilic substitution or addition-elimination (92)
130
reverse addition-elimination (93)
isomerisation–SN₂–isomerisation (92)
isomerisation–SN′₂–isomerisation (92)

Scheme 50

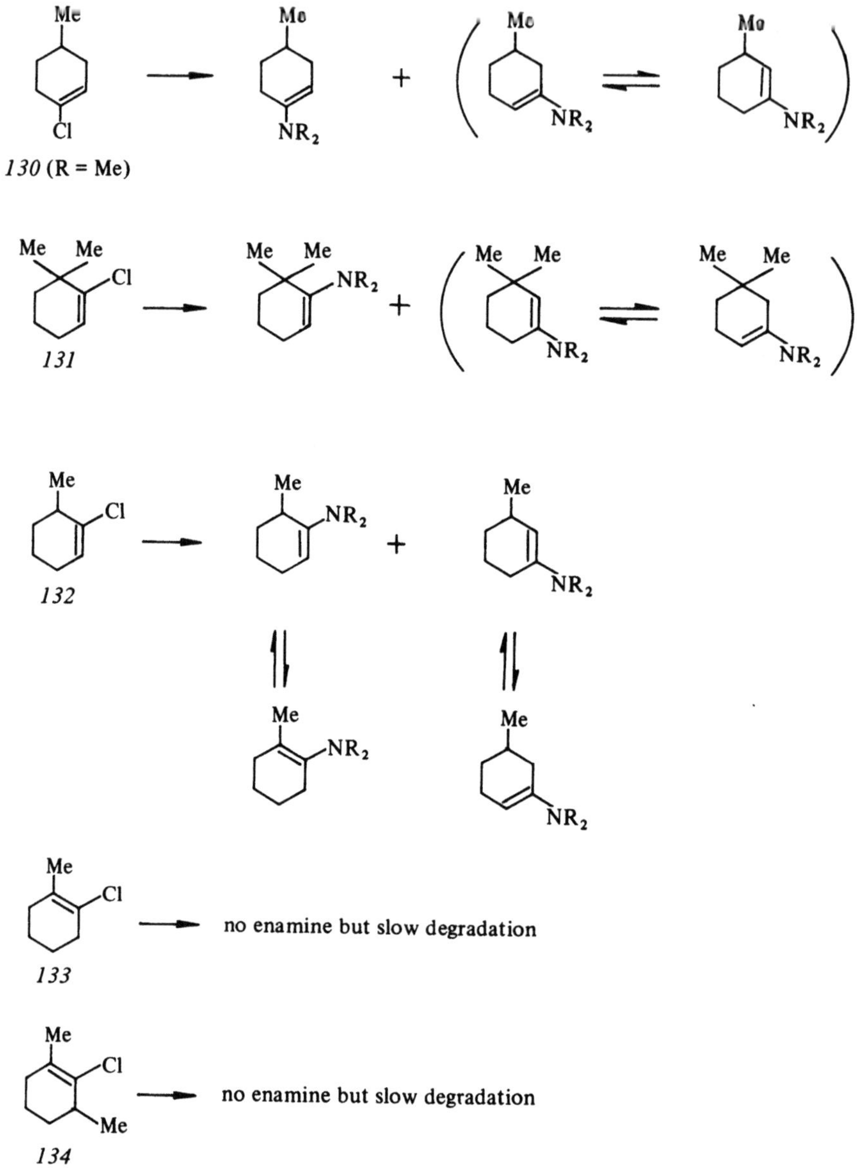

Scheme 51

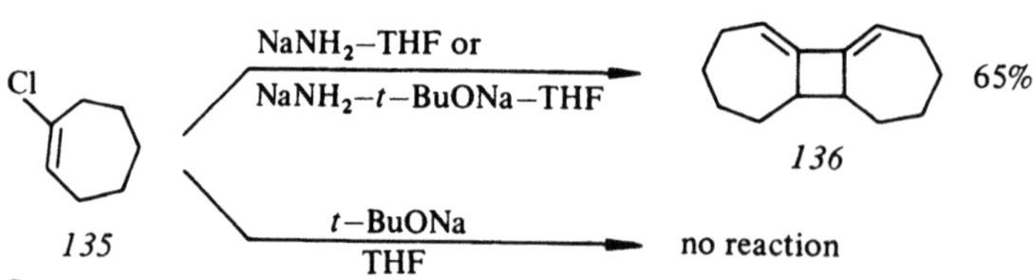

Scheme 52

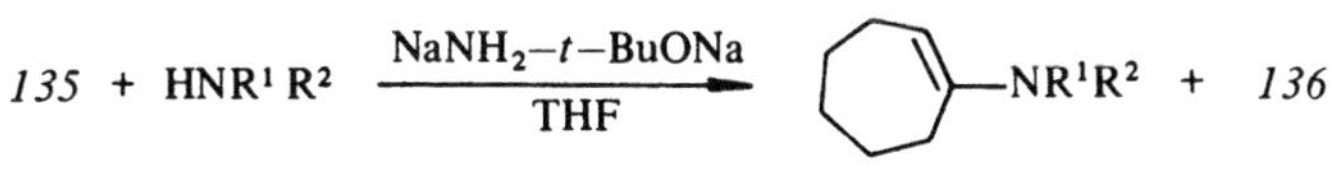

137

$[R^1, R^2] = O(CH_2CH_2)_2, (CH_2)_5$	30%	36% (66% total yield)
$R^1 = R^2 = Et$, *i*-Pr	0–6%	37–38% (38–43% total yield)

Scheme 53

This means that the elimination step depends upon the nature of the amine to be condensed. As it may be seen, we have to be careful in elimination-additions forecasting.

In our opinion it is not possible with only a few experiments to predict the path followed by an elimination-addition even when the halogeno compound and the base are determined. As a matter of fact, with these kinds of condensations, the nature of the nucleophile has also to be considered. Thus many experiments varying the nucleophilic reagent are needed to obtain enough data to allow some previsions. This conclusion is confirmed with what we shall now explain.

IV.2.2.3. Ketone Enolates Condensations on Vinylic Halogeno Cyclohexenes and Cycloheptenes

Vinylic Halogeno Cyclohexenes. Encouraging results described above led us to think that, as in the aromatic series, it would be possible to condense ketone enolates on cyclohexyne generated by reaction of nucleophilic complex bases $NaNH_2$-ketone enolates on 1-halogeno cyclohexene.

As a matter of fact, these kinds of bases easily react with 1-chloro cyclohexene at about 30–35 °C. Products formed are mixtures of ketones and alcohols in variable ratios. We have collected in Scheme 54, general formulas of all the products which can be formed from aliphatic or alicyclic ketones[19, 97, 98].

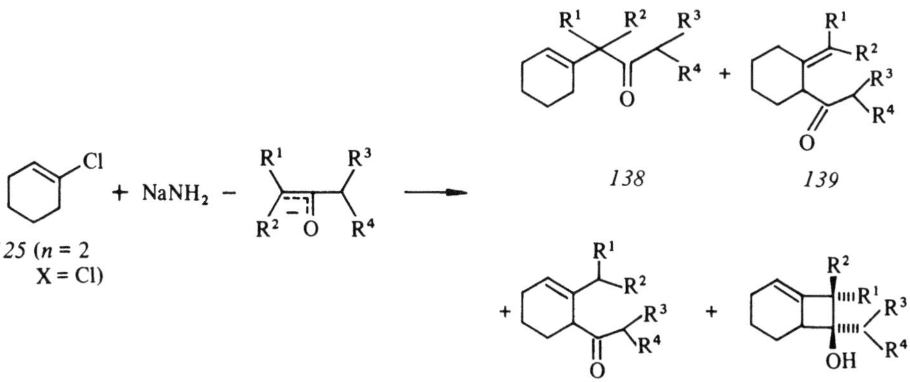

Scheme 54

Of course, as for reactions described in the aromatic series, relative ratios and total yields depend upon the nature of the starting ketone and of experimental conditions.

For example: With methylcyclopropylketone, *i.e.* $R^1 = R^2 = H$, $[R^3, R^4] = (CH_2)_2$, products formed in THF or DME are *139* and *140* in relative ratios varying from 80/20 to 70/30. Total yields are about 45%.

With cyclohexanone, *i.e.* $[R^1, R^3] = (CH_2)_3$, $R^2 = R^4 = H$, main product (98–100%) is *141* in THF, *138* (∿100%) in HMPA. Both compounds (*141*/*138* ≃ 1,7) are obtained in DME. Total yields are between 45 and 65%[97].

Observation of Scheme 54 shows a close analogy between these results and the ones obtained in arynic condensations. Thus we observe the formation of "normal ketones" *138*, "transposed ketones" *139* or *140*, and methylene cyclobutanols *141*. However, we met large difficulties with the mechanism. Indeed, starting from the hypothesis of cyclohexyne being the intermediary (what is consistent with our works on amine condensations) and by analogy with arynic condensations, *the reaction should be written as in Scheme 55.*

It is clear that this scheme is hardly consistent with our results:

We never evidenced cyclobutenic alcohols *146*.

Alcohols *141* and consequently ketone *140* would be imputable to the isomerization *143* → *144*, but there are no valid explanations for this one sided isomerization.

However this mechanism explains the formation of ketones *138* and *139*.

In summary, and this is important for what we shall see later, cyclohexyne accounts only for ketones *138*, *139* and possibly alcohols *146*.

On the contrary, if we start from the hypothesis of 1,2-cyclohexadiene being the intermediate, the reaction should be written as in Scheme 56 where intermediaries have been written again for more clearness.

This mechanism is consistent with experimental results. Moreover in reactions where *139* and *140* are formed, the influence of the nature of the solvent on ratio *139*/*140* is well explained[98] with the help of what is known about the influence of the solvents on the structure of β diketones or β keto ester enolates[99]. The more cation solvating is the solvent, the more important the proportions of W form. Conversely the less cation solvating is the solvent, the more important the proportion of U or possibly sickle forms. Thus increasing the solvent cation solvating power leads to increased formation of *140*[98]. If we compare 55 and 56 it is clear that ketones *139* and *140* may be explain by mean of cyclohexyne as well as 1,2-cyclohexadiene intermediaries. So, concerning these two types of compounds we have to expect that we will have occasionally some difficulties knowing their true origin.

Moreover, if we examine the overall results obtained, we immediately see that there is a problem to be solved. Indeed, complex base $NaNH_2$-*t*-BuONa in THF and in the presence of an amine, reacts with 1-chlorocyclohexene to give cyclohexyne intermediate. However, in the same solvent, complex base $NaNH_2$-enolates seem to generate chiefly 1,2-cyclohexadiene from the same starting halogeno derivative. Note that one possible hypothesis was that the complex bases used here were able to generate a cyclohexyne and to isomerize it very rapidly to 1,2-cyclohexadiene.

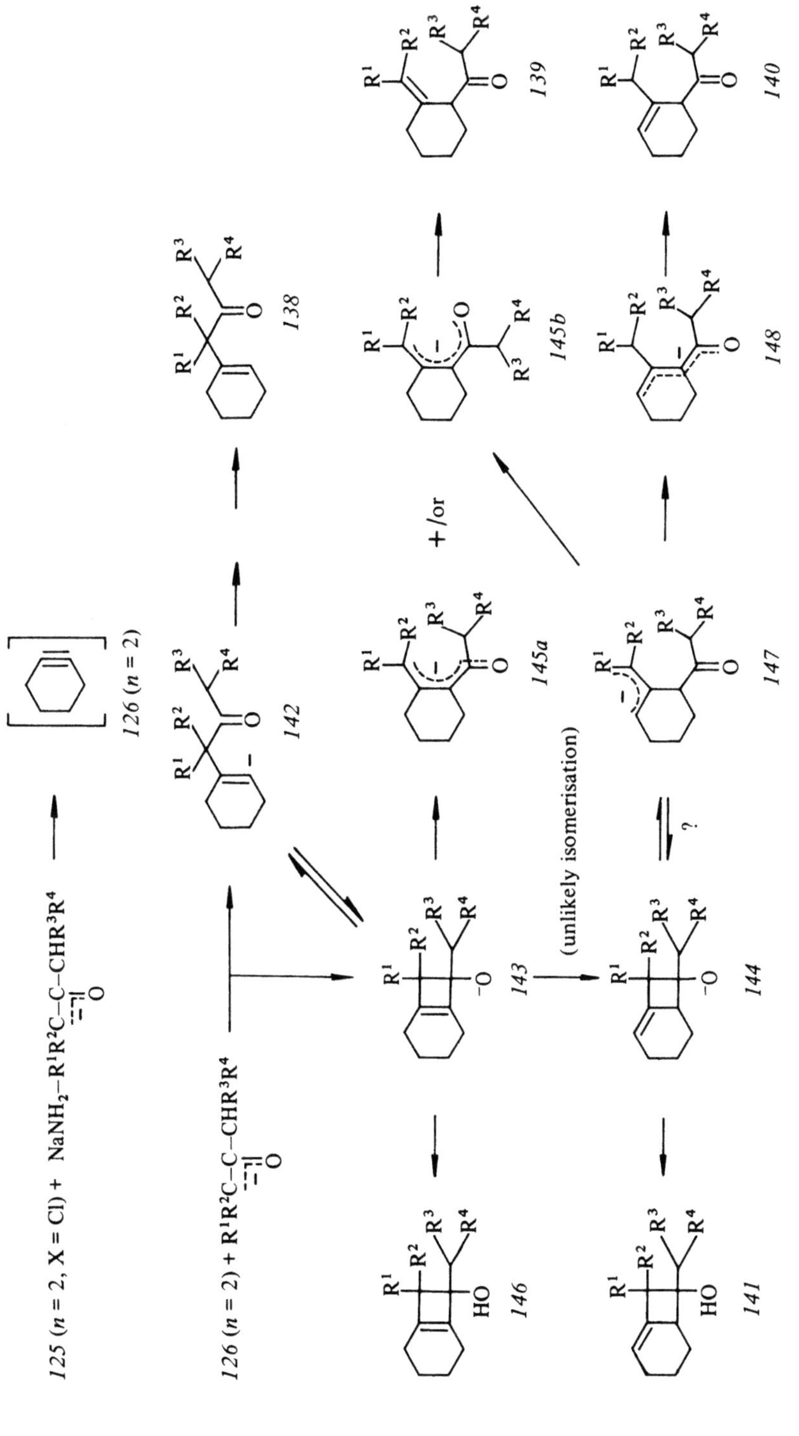
125 (n = 2, X = Cl) + NaNH2–R1R2C–C–CHR3R4
126 (n = 2)
126 (n = 2) + R1R2C–C–CHR3R4
142
138
143
144
145a
145b
146
141
147
148
139
140
+/or
(unlikely isomerisation)
?

Scheme 55

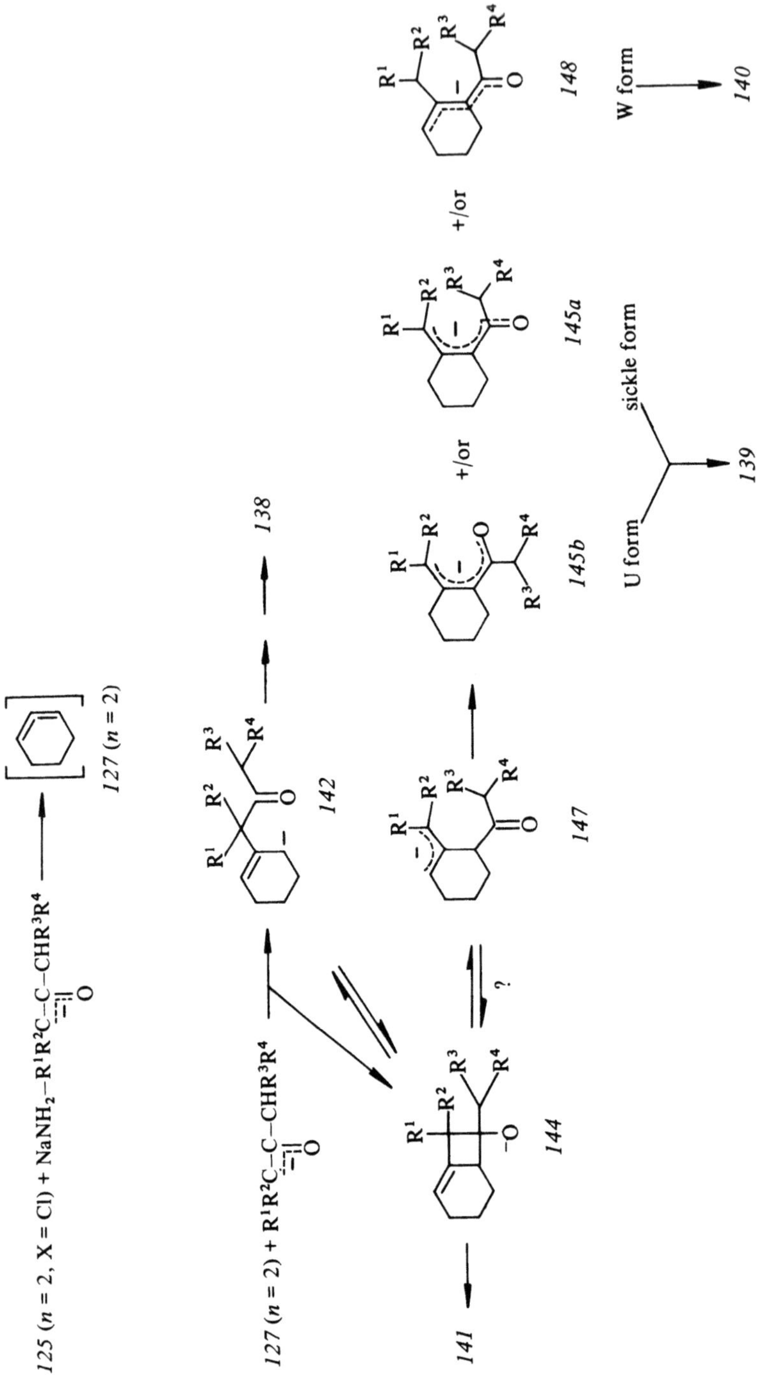
125 (n = 2, X = Cl) + NaNH2–R1R2C–C–CHR3R4
127 (n = 2)
127 (n = 2) + R1R2C–C–CHR3R4
142
138
144
141
147
145b
145a
148
+/or
U form
sickle form
W form
139
140

Scheme 56

To clarify this situation, we studied several substituted chlorocyclohexenes. We shall see that these reactions are not very simple but that it is possible to determine some general useful rules.

First we studied[100] 1,1-dimethyl 2-chloro cyclohexene *149* which can only lead to a cyclohexyne. General formula of compounds which can be formed are given in Scheme 57.

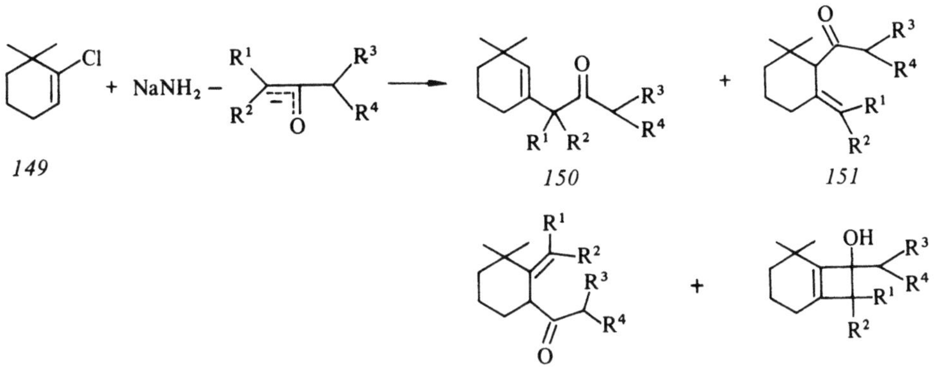

Scheme 57

Of course, here too, the path followed by these condensations depends upon experimental conditions and structure of the starting ketone enolate. Results obtained brought to light the following facts:

Normal ketones *150* (when formed) are always minor components.

When the sodium salts of *153* are not stable enough to give, after hydrolysis, the corresponding alcohols, products formed are essentially or even uniquely transposed ketones *151* and *152*.

These observations and the structure of the compounds formed are in accordance with the intervention of 1,1-dimethyl 2-cyclohexyne without isomerization to the corresponding 1,2-cyclohexadiene.

We can conclude that nucleophilic complex base $NaNH_2$-enolates are able to generate cyclohexynes. Ketone enolates condense on these to give chiefly alcohols *153* and/or ketones *151* and *152*.

Note that this result definitively excludes possible isomerization *143* → *144* of Scheme 55.

Reactivity study of *154* which can only eliminate to 1,2-cyclohexadiene leads to new interesting data[100] (see Scheme 58).

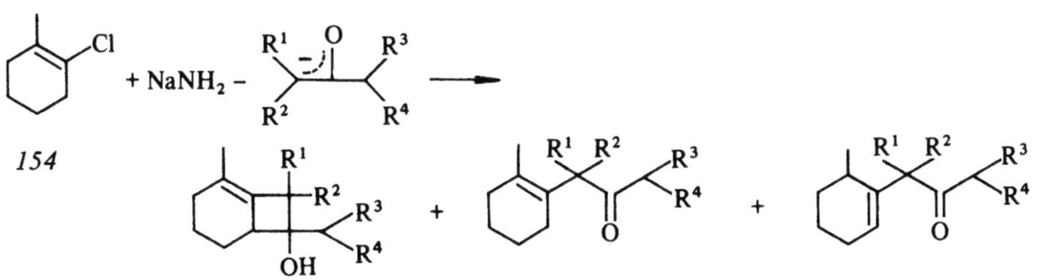

Scheme 58 *155* *156* *157*

This chloro derivative is much less reactive than the others.

All products formed are imputable to nucleophilic attack on the central carbon of the 1,2-diene.
The main products are alcohols *155*.

This means that nucleophilic complex base $NaNH_2$-enolates are also able to generate 1,2-cyclohexadienes and that alcohols *155* arise from condensation of enolates on these intermediates.

Comparison of results obtained with *125* (n = 2, X = Cl) *149* and *154* shows that sodium salts of *153* are less stable than these of *141* or *155*. This observation is general.

We attribute the low reactivity of *154* to a poor tendency to form 1-methyl 1,2-cyclohexadiene. This hypothesis is confirmed by what we show just below.

After that we examined[100)] the reactivity of *158* (Scheme 59) which can eliminate in following two ways:

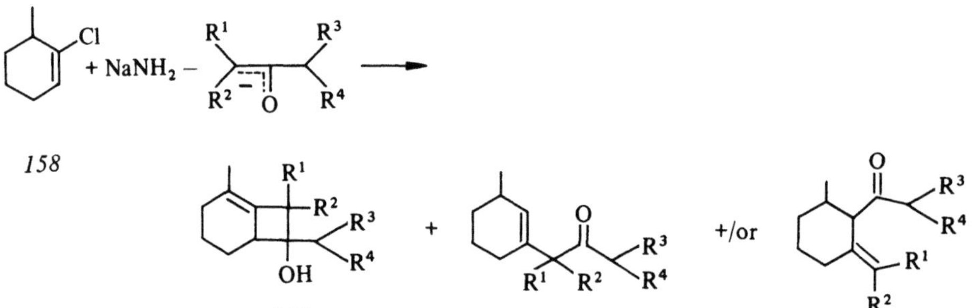

158 *155* *159* (with conjugated isomer if R^1 or R^2 = H) *160*

Scheme 59

159 and *160* are the main products, showing the intervention of a cyclohexyne. Alcohol *155* arising from a cyclohexadiene we can conclude that elimination to 1-methyl 2-cyclohexyne is much more favoured than elimination to 1-methyl 1,2-cyclohexadiene.

Taking into account that *158* is much more reactive than *154*, we confirm what we said before concerning the poor tendency to the formation of methyl substituted cyclohexadiene.

Finally, we studied *161* which can eliminate to a cyclohexyne and an unsubstituted 1,2-cyclohexadiene[100)].

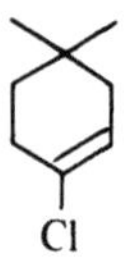

161

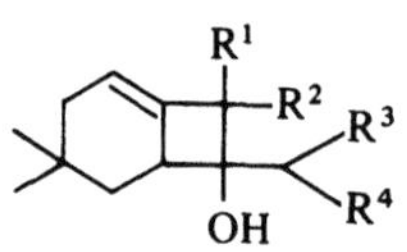

162

The main products formed are methylene cyclobutanols *162*. In other words, *the chief elimination path leads to a cyclohexadiene.* Moreover, comparison of reactivity of two ketone enolates showed that *the amount of minor cyclohexyne de-*

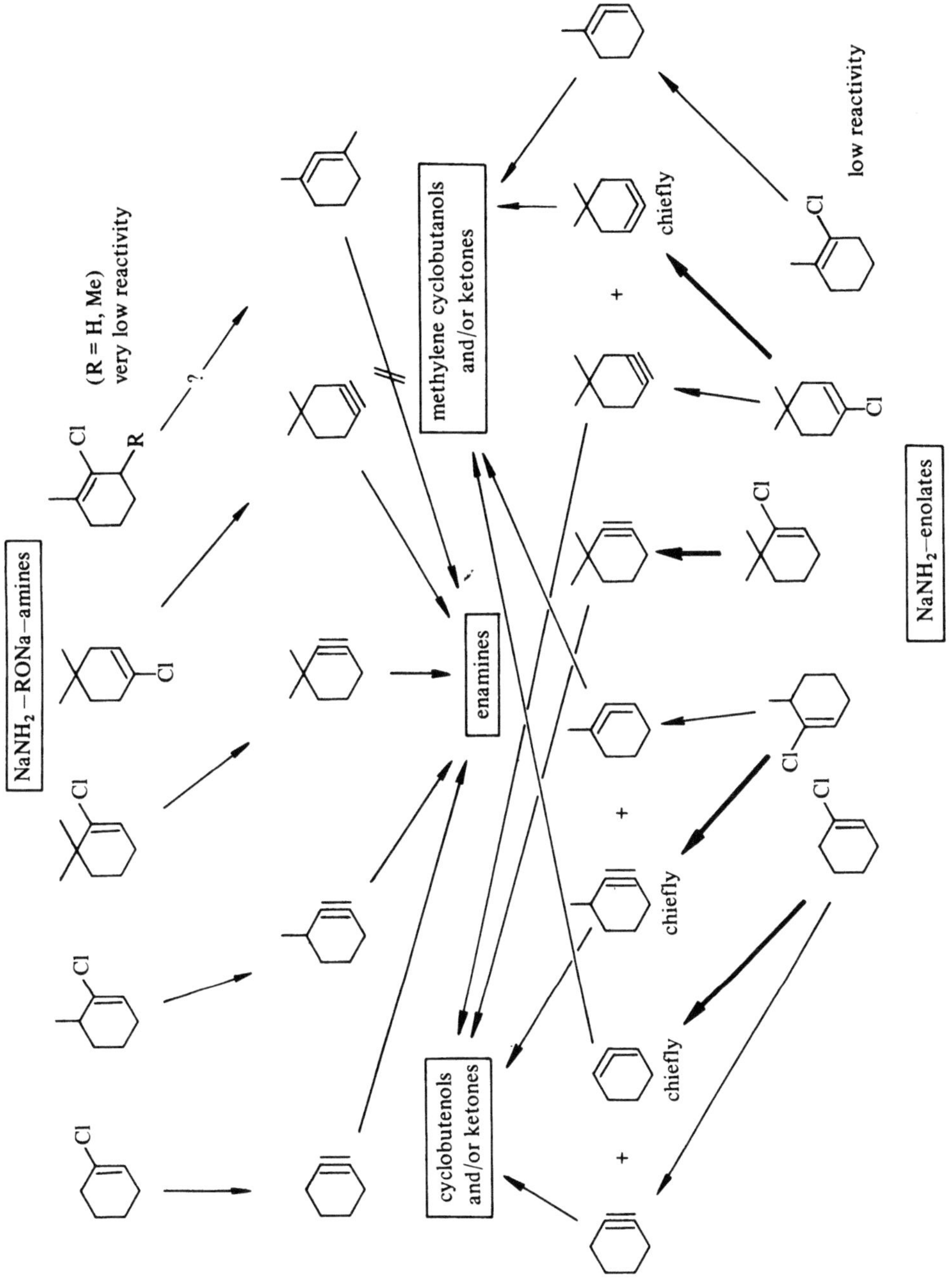
NaNH$_2$–RONa–amines
(R = H, Me)
very low reactivity
?
methylene cyclobutanols
and/or ketones
enamines
cyclobutenols
and/or ketones
chiefly
chiefly
chiefly
low reactivity
NaNH$_2$–enolates
Cl
R
+

Scheme 60

pends upon the nature of the enolate, that is to say, the nature of the nucleophilic complex base.

Now we tentatively can summarize reactivity of chloro vinylic cyclohexenes by Scheme 60 which may be used as a guide in future experiments.

Concerning the reactive intermediates encountered with nucleophilic complex bases, we can conclude to the following formation propensity order:

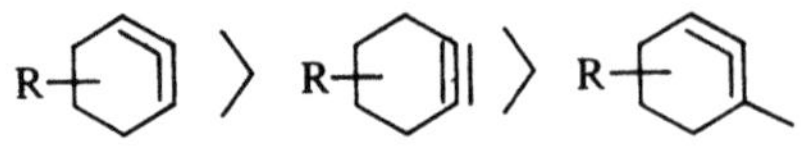

As a final conclusion of this mechanistic discussion we can say that elimination additions in cyclenic series are rather complicated. Although some helpful general rules were now available, we have to be careful in the interpretation of these reactions and we have to remember that one of the most important parameters is the nature of the *couple* base-nucleophile.

Continuing in the condensations of nucleophilic complex base $NaNH_2$-ketone enolates we have to note that, as in arynic reactions, there is a particular behaviour of aromatic ketone enolates. Thus, besides the usual reactions, we found[101] that internal cyclisation may happen as pictured in Scheme 61.

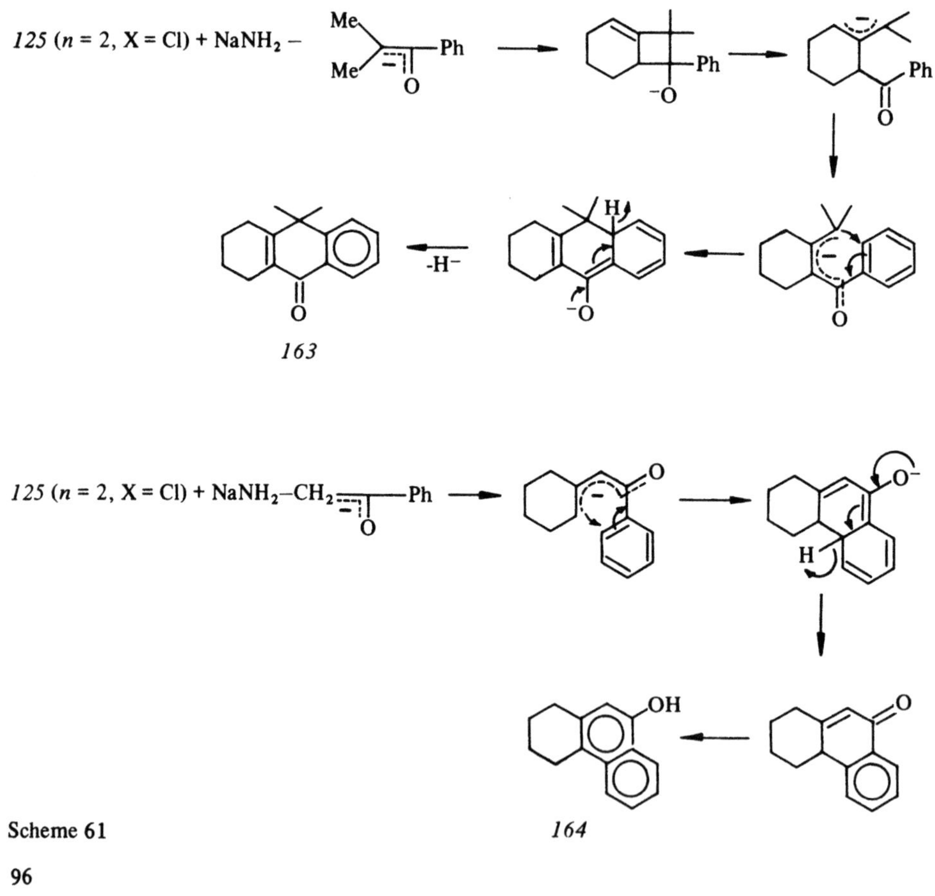

Scheme 61

Of course, as usual, solvent and temperature play an important rôle on the formation of *163* and *164*.

Concerning the synthetic view-point of ketone enolate condensations on halogeno vinylic cyclohexenes, we shall give some examples of products which can be prepared.

In Scheme 62 we have reported of few ketonic structures obtained during our work.

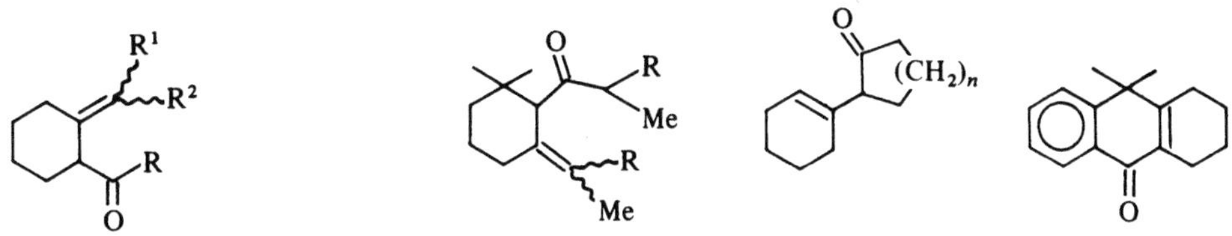

$R^1 = R^2$ = Me R = *i*-Pr (45%)[98] R = H (40%)[100] *n* = 2 (40%)[97] (40%)[101]
R^1 = H R^2 = Me R^3 = Et (45%)[98] R = Me (75%)[100] *n* = 3 (45%)[97]

Scheme 62

Many other ketones may be obtained with more or less good yields. With the help of these reactions we were also able to synthesize numerous alcohols of which we shall give a few examples in Scheme 63.

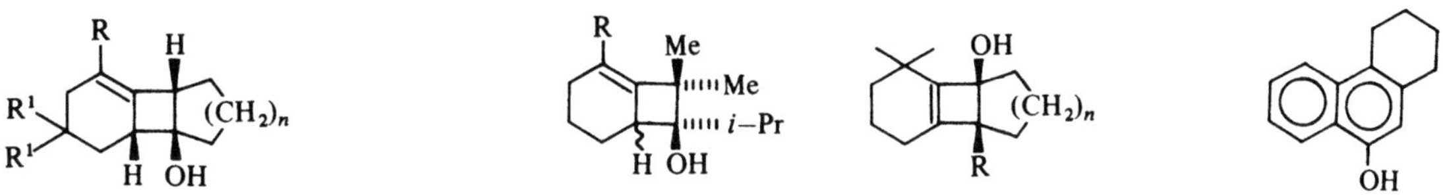

R = R^1 = H *n* = 2, 3, 4 (50–55%)[97, 103] R = H, Me (20–25%)[100] R = H *n* = 1,2(20–35%)[100] 55%[101]
R = Me R^1 = H *n* = 2 (20%)[100] R = Me *n* = 2 (25%)[100]
R = H R^1 = Me *n* = 2 (35%)[100]

Scheme 63

It is noteworthy that at the difference of what we met for arynic reactions, alcohols with a strained ring are isolated with cyclooctanone and with aliphatic ketones. Since the obtaining of these alcohols by us for the first time, a synthesis by photolysis of cyclenones has been published[102]. However, this last method seems, for the present time, much less general and above all much less stereoselective.

Finally we showed[104] that methylene cyclobutenols lead unexpectedly to polycyclic furans *165* by means of reactions pictured in Scheme 64.

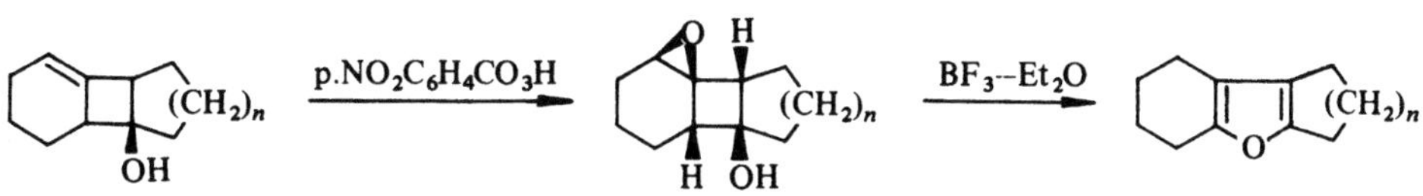

n = 2, 3, 4 70–75% *165* 70–75%

Scheme 64

1-chloro cycloheptene. We saw (Chapter IV.2.2.2) that this compound shows much more propensity to generate 1,2-cycladiene than cyclyne intermediary. Thus we have to expect an increase of methylene cyclobutanols yields relative to those encountered with halogeno cyclohexenes, unless of course, the corresponding alcoholates were not stable.

In fact, reactivity of 1-chloro cycloheptene looks like reactivity of 1-chloro cyclohexene[95, 103, 105] and reactions with nucleophilic complex base $NaNH_2$-enolates lead to compounds similar to the ones isolated in cyclohexenic series. This is exemplified in Scheme 65 with alicyclic ketone enolates.

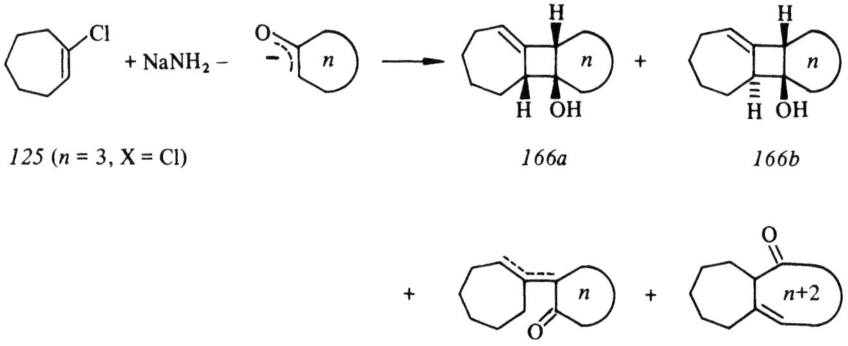

125 ($n = 3$, X = Cl) *166a* *166b* *167* *168*

Scheme 65

We have to underline some particular points. As expected, main products of these condensations generally are of the *166* type. However, here two isomers *166a* and *166b* are formed from alicyclic ketones. Although there were chief compounds when experimental conditions are not too drastic, *trans* derivatives *116b* are less stable than *166a*. Indeed we showed[105] that under basic conditions *166b* isomerizes to *166a* which finally leads to ketones. Taking into accounts the results of Wiberg and his coworkers[106], we tentatively proposed the mechanism pictured in Scheme 66 to explain these reactions.

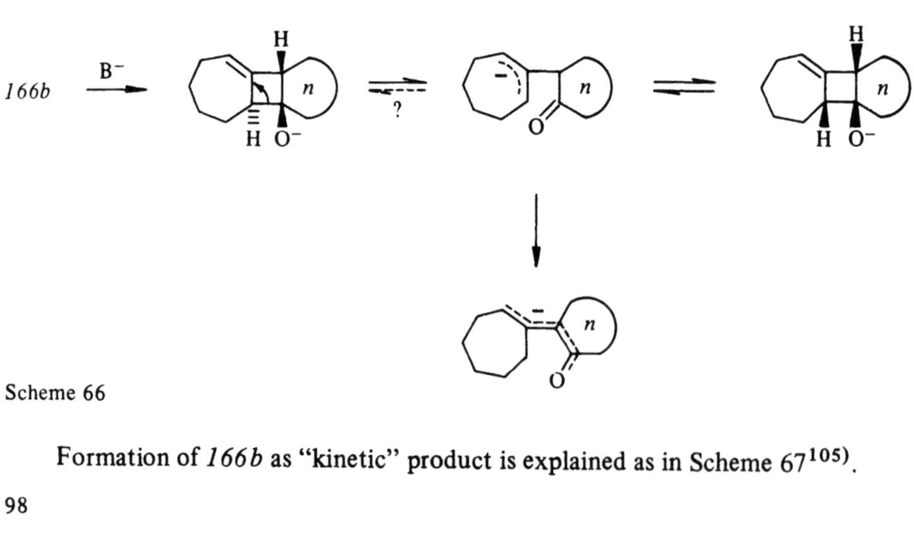

Scheme 66

Formation of *166b* as "kinetic" product is explained as in Scheme 67[105].

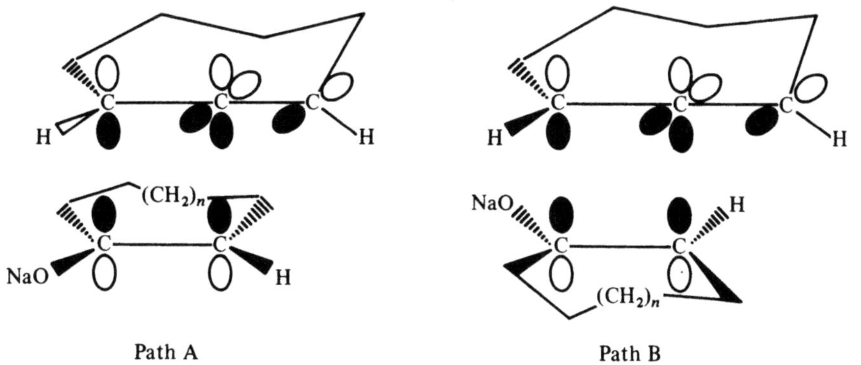

Scheme 67

Reactants can approach one each other following two paths (A and B respectively). As it may be seen, path A is strongly disfavored by steric interactions. On the contrary, path B, which leads to the less thermodinamically stable isomer *166b* does not suffer those inconveniences and thus must be favored. Of course, all we said may be applied to condensations of aliphatic ketone enolates.

Finally, these reactions are useful in synthesis of the new alcohols *166* (as well as their aliphatic counterparts). Overall yields are between 30 and 60%[103, 105]. Moreover, methylene cyclobutanols are so stable in these series that *166* may be obtained for any *n* values. However, when *n* becomes too large, the ketone enolates are mixtures of *cis* and *trans* forms. Thus there are some difficulties in the separation of the mixtures of alcohols obtained[103].

All we have explained in Chapter IV.2. shows the usefulness of complex bases in performing aromatic as well as cyclenic elimination-addition condensations.

Those studies show that mechanisms are sometimes complicated but that it is possible, even in cyclenic series, to derive some rules allowing previsions for future experiments.

Concerning the synthetic applications, it is clear that these reactions have allowed the obtaining of new structures and are useful for the synthesis of many polyfunctional molecules not easily obtainable by more classical ways. Moreover, starting materials are nearly always very simple and often commercially available. One limitation should be perhaps in some cases – the separation of formed products. However now, low- and high-pressure preparative liquid chromatographies become more and more usual in laboratories. Those techniques allow work on a large scale and are no more difficult to handle than other usual methods. Thus these techniques improve the interest of our reactions which lead to elaborate structures from trivial starting molecules.

V. Conclusion

Now we must conclude on complex bases. We hope to have demonstrated that activation of one base by another is a real phenomenon and that, more particularly, sodamide-containing complex bases are useful tools in organic synthesis.

For the present time it is not possible to give precise physical pictures of these phenomena. We cannot give beautiful spectra or numerical data describing the molecular structure of complex bases. Thus, they may appear as muzzy reagents and this may constitute a handicap in convincing of their usefulness. However, looking at all the activating agents studied, general laws appear. We can reasonably predict that for a determined solvent, one activating agent will be better than another in performing a desired reaction. Moreover, those rules will be improved in the near future.

We also have to place our activating agents and their resulting complex bases in regard to more sophisticated base activating agents like crowns[3] and cryptands[4].

Crowns and cryptands are able to increase considerably the basicity of a given base, owing to their extraordinary cationic complexation power. Thus A^-M^+ in the presence of crowns or cryptands will see its M^+ cation strongly complexed leading to A^- as a "nacked ion". In fact, the anion is not completely nacked as it may be seen for example from the use of crowns as transfer agents in phase-transfer catalysis reactions[107].

In complex bases, the anion is much less free. However basicity of certain of our reagents is very high. Moreover, the fact that complex bases are aggregated can be taken as an advantage. Indeed, properties of these reagents depend upon the nature of the activating agent and of the solvent. It is very easy by simple modification of these parameters to "modulate" the nature of the bases and thus to obtain a large "palette" of complex bases with gradual variation properties. Finally, ionic surfaces of complex bases may be also useful in performing certain kinds of reactions.

Now concerning more practical considerations, complex bases present the enormeous advantage of being very easily and simply prepared from commercial cheap starting materials. This can constitute a big advantage even if the results are less spectacular than the ones obtained with crowns or cryptands. As a corollary complex bases can be used on very large scale without inconvenience. Finally, complex bases are not physiologically dangerous and do not necessitate special care. This point has to be noted owing to the uncertainty concerning the physiological activities of crowns and similar reagents[3].

From a general theoretical view-point we think that the concept of synergic effect must be taken into account every time two bases or nucleophiles are together. We have shown that this is important when one of the reagents is soluble and the other is not. But it is clear that this concept applies even when both reagents are soluble or, on the contrary, only very slightly soluble.

Moreover, if this is important in synthesis, we think that this is still much more important in kinetic interpretations of reactions where two bases or nucleophiles intervene successively or competitively.

The phenomenon can also play a role in reactions where a base and a solvent able to produce a base, are brought together. For example if we consider a base in DMSO and if this base is able to generate a small quantity of dimsylsodium, which is the active species: the base, dimsylsodium or the complex base: base-dimsylsodium?

Last, but not least, we shall see in Part II that the concept of activation can be extended to reducing agents.

VI. References

1) Parker, A. J.: Chem. Rev. *69*, 1 (1969)

2) Coetzee, J. F., Ritchie, C. D.: Solute-solvent interactions. New York: Marcel-Dekker 1969

3) See for example Gokel, G. W., Durst, H. D.: Synthesis *1976*, 168 and references cited therein

4) Lehn, J.-M.: Structure and bonding *16*, 1 (1973)

5) Wittig, G., Pieper, G., Fuhrmann, G.: Ber. Deut. Chem. Ges. *73*, 1193 (1940)

6) Roberts, J. D., Simmons, H. E., Carlsmith, L. A., Vaughan, C. W.: J. Am. Chem. Soc. *75*, 3290 (1953); Roberts, J. D., Semenow, D. A., Simmons, H. E., Carlsmith, L. A.: J. Am. Chem. Soc. *78*, 601 (1956)

7a) Hoffmann, R. W.: Dehydrobenzene and cycloalkynes. Weinheim: Verlag Chemie 1967

b) Viehe, H. G.: Chemistry of acetylenes. New York: Marcel Dekker 1969

8) Wittig, G.: Angew. Chem. Int. Edit. Engl. *4*, 731 (1965)

9) Caubère, P., Loubinoux, B.: Bull. Soc. Chim. Fr. *1968*, 3008

10) Caubère, P., Loubinoux, B.: Bull. Soc. Chim. Fr. *1968*, 3857

11) Caubère, P., Loubinoux, B.: Bull. Soc. Chim. Fr. *1969*, 2483, see also Caubère, P.: Acc. Chem. Res. *7*, 301 (1974)

12) Cuvigny, T., Normant, H.: Bull. Soc. Chim. Fr. *1965*, 1972

13) Caubère, P., Coudert, G.: Bull. Soc. Chim. Fr. *1971*, 2234

14) Coudert, G.: Thèse de Doctorat es Sciences, Nancy 1974

15) Normant, H., Cuvigny, T., Normant, J., Angelo, B.: Bull. Soc. Chim. Fr. *1965*, 3441

16) Ndebeka, G.: Thèse de Doctorat de Spécialité, Nancy, 1977

17) Biehl, E. R., Hsu, K. C., Nich, E.: J. Org. Chem. *35*, 2454 (1970)

18) Caillet, A., Bauer, D.: Journées d'Etudes Div. Chim. Anal. Org. Soc. Chim. Fr. 1974; Caillet, A.: Thèse de Doctorat es Sciences, Paris 1974

19) Unpublished works

20) Sicher, J.: Angew. Chem. Int. Ed. Engl. *11*, 200 (1972) and references cited therein

21) Bartsch, R. A.: Acc. Chem. Res. *8*, 239 (1975)

22) Morton, A. A., Claff Jr., C. E., Collins, F. W.: J. Org. Chem. *20*, 428 (1955); Benkeser, R. A., Cummino, T. F., Wen-Hong Tong: J. Am. Chem. Soc. *90*, 4366 (1968)

23) Lochmann, L., Paspisil, J., Vodnansky, J., Trekuval, J., Lim, D.: Coll. Czech. Chem. Commun. *30*, 2187 (1965); Peascoe, W., Applequist, D. E.: J. Org. Chem. *38*, 1510 (1973); Tadashi Narita, Teiji Tsuruta: J. Organomet. Chem. *30*, 289 (1971)

24) Koster, R., Arora, S., Binger, P.: Angew. Chem. Int. Ed. Engl. *8*, 205 (1969)

25) Salaun, J. R., Conia, J. M.: J. Chem. Soc. Chem. Commun. *1971*, 1579; Salaun, J. R., Champion, J., Conia, J. M.: Organic Synthesis *53*, 1825 (1973)

26) Le Bel, N. A.: Adv. Alicyclic Chem. *3*, 196 (1971); Smith, H.: Chemistry in nonaqueous ionizing solvents. New York: Interscience 1963

27) Maitte, P.: Bull. Soc. Chim. Fr. *1959*, 499 and references cited therein

28) Caubère, P., Coudert, G.: J. Chem. Soc. Chem. Commun. *1972*, 1289

29) Guillaumet, G., Lemmel, V., Coudert, G., Caubère, P.: Tetrahedron *30*, 1289 (1974)

30) Beltrame, P., Biale, G., Lloyd, D. J., Parker, A. J., Ruane, M., Winstein, S.: J. Am. Chem. Soc. *94*, 2240 (1972) and references cited therein

31) Caubère, P., Moreau, J.: Tetrahedron *25*, 2469 (1969)

32) Caubère, P., Moreau, J.: Tetrahedron *26*, 2637 (1970)

33) Davis, D. D., Ansari, G. G.: J. Org. Chem. *35*, 4285 (1970) and references cited therein

34) Caubère, P., Coudert, G.: Tetrahedron *28*, 5635 (1972)

35) Caubère, P., Coudert, G.: Bull. Soc. Chim. Fr. *1973*, 3067

36) Ram K. Sharma: Int. Sci. Chem. Rep. *1*, 101 (1967)

37) Caubère, P., Derozier, N.: Bull. Soc. Chim. Fr. *1969*, 1737

38) Caubère, P., Loubinoux, B.: C. R. Hebd. Séances Acad. Sci. Ser. C., *264*, 1887 (1967); Caubère, P.: Bull. Soc. Chim. Fr. *1967*, 3446

39) Caubère, P.: Bull. Chim. Fr. *1967*, 3451

40) Huisgen, R.: Organometallic Chemistry, Reinhold *1960*, 36

[41] Heaney, H.: Chem. Rev. *62,* 81 (1962)
[42] Scott, F. L., Oesterling, R. E.: J. Am. Chem. Soc. *82,* 5247 (1960)
[43] Caubère, P., Hochu, M. F.: Bull. Soc. Chim. Fr. *1969,* 2854
[44] Effenberger, F., Auer, E., Fischer, P.: Chem. Ber. *103,* 1440 (1970)
[45] Caubère, P., Lalloz, L.: Bull. Soc. Chim. Fr. *1974,* 1983
[46] Caubère, P., Lalloz, L.: Bull. Soc. Chim. Fr. *1974,* 1989
[47] Caubère, P., Lalloz, L.: Bull. Soc. Chim. Fr. *1974,* 1996
[48] Hine, J., Langford, P. B.: J. Org. Chem. *27,* 4149 (1962)
[49] Wotiz, J. M., Huba, F.: J. Org. Chem. *24,* 595 (1959)
[50] Wittig, G., Merkle, W.: Chem. Ber. *75,* 1491 (1942); Biehl, E. R., Nieh, E., Hsu, K. C.: J. Org. Chem. *34,* 3995 (1969)
[51] Moyer Jr., C. E., Bunnett, J. F.: J. Am. Chem. Soc. *85,* 1891 (1963); J. F. Bunnett, C. E., Moyer Jr., C. E.: J. Am. Chem. Soc. *93,* 1183 (1971); Bunnett, J. F., Scorano, B.: J. Am. Chem. Soc. *93,* 1190 (1971); Bunnett, J. F., McLennan, B. J.: J. Am. Chem. Soc. *93,* 1198 (1971); Bunnett, J. F.: Acc. Chem. Res. *5,* 139 (1972)
[52] Biehl, E. R., Patrizi, R., Reeves, P. C.: J. Org. Chem. *36,* 3252 (1971)
[53] Lalloz, L., Caubère, P.: Synthesis *1975,* 657
[54] Moreau-Hochu, M. F., Caubère, P.: Tetrahedron *33,* 955 (1977)
[55] Leake, W. W., Levine, R.: J. Am. Chem. Soc. *81,* 1169 (1959)
[56] Gore, T. S., Yadav, V. K.: Indian J. Chem. *1,* 321 (1963) (Chem. Abstr. *59,* 12690a) (1963)
[57] See for example Rossi, R. A., Bunnett, J. F.: J. Org. Chem. *38,* 3020 (1973); Bunnett, J. F., Sundberg, E. J.: Chem. Pharm. Bull. *23,* 2620 (1975) and references cited in those articles
[58] Caubère, P., Guillaumet, G.: Bull. Soc. Chim. Fr. *1972,* 4643
[59] See for example Normant, H.: Bull. Soc. Chim. Fr. *1965,* 791 and references cited therein
[60] Caubère, P., Guillaumet, G.: Bull. Soc. Chim. Fr. *1972,* 4649
[61] Bumgardner, C. L., McDaniel, K. G.: J. Am. Chem. Soc. *91,* 6821 (1969) and references cited therein
[62] Burley, J. N., Young, R. N.: J. Chem. Soc. C, *1971,* 3780
[63] Caubère, P., Derozier, N., Loubinoux, B.: Bull. Soc. Chim. Fr. *1971,* 302
[64] Caubère, P., Guillaumet, G., Mourad, M. S.: Tetrahedron Lett. *1971,* 4673; Caubère, P., Mourad, M. S., Guillaumet, G.: Tetrahedron *29,* 1851 (1973)
[65] Caubère, P., Mourad, M. S.: J. Chem. Soc. Chem. Commun. 1972, 1238; Caubère, P., Mourad, M. S.: Tetrahedron *30,* 3439 (1974)
[66] Viriot-Villaume, M. L., Carre, C., Caubere, P.: Tetrahedron Lett. *1974,* 3301
[67] Cava, M. P., Muth, K.: J. Am. Chem. Soc. *82,* 652 (1960)
[68] Caubère, P., Guillaumet, G., Mourad, M. S.: Tetrahedron *29,* 1857 (1973)
[69] Caubère, P., Guillaumet, G., Mourad, M. S.: Tetrahedron *28,* 95 (1972)
[70] Caubère, P., Mourad, M. S., Guillaumet, G.: Tetrahedron *29,* 1843 (1973)
[71] Allard, M., Levisalles, J.: Bull. Soc. Chim. Fr. *1972,* 1926
[72] Mourad, M. S.: Thèse de Doctorat es Sciences, Nancy 1974
[73] Caubère, P., Guillaumet, G., Mourad, M. S.: Bull. Soc. Chim. Fr. *1973,* 3493
[74] Carre, C., Viriot-Villaume, M. L., Caubère, P.: Synthesis *1977,* 48
[75] Carre, C.: Thèse de Doctorat de Spécialité, Nancy 1975
[76] Arnold, B. J., Mellows, S. M., Sammes, P. G., Wallace, T. W.: J. Chem. Soc. Perkin Trans. *1,* 401 (1974)
[77] Courtois, A., Protas, J., Mourad, M. S., Caubere, P.: Acta Crystall. B, *31,* 1907 (1975)
[78] Caubère, P., Mourad, M. S.: Tetrahedron Lett. *1973,* 2221
[79] Brunet, J. J., Essiz, M., Caubère, P.: Tetrahedron Lett. *1974,* 871
[80] Sammes, P. G., Wallace, T. W.: J. Chem. Soc. Chem. Commun. *1973,* 524
[81] Sammes, P. G., Wallace, T. W.: J. Chem. Soc. Perkin Trans. *1,* 1377 (1975)
[82] Essiz, M., Coudert, G., Guillaumet, G., Caubère, P.: Tetrahedron Lett. *1976,* 3185
[83] Caubère, P., Lalloz, L.: J. Org. Chem. *40,* 2853 (1975)
[84] Caubère, P., Lalloz, L.: J. Org. Chem. *40,* 2859 (1975)
[85] Caubère, P., Mourad, M. S.: Bull. Soc. Chim. Fr. *1974,* 1415

86) Scardiglia, F., Roberts, J. D.: Tetrahedron *1*, 343 (1957); Montgomery, L. K., Roberts, J. D.: J. Am. Chem. Soc. *82*, 4750 (1960); Wittig, G., Krebs, A.: Chem. Ber. *94*, 3260 (1961); Montgomery, L. K., Scardiglia, F., Roberts, J. D.: J. Am. Chem. Soc. *87*, 1917 (1965); Wittig, G., Fritze, P.: Angew. Chem. Int. Ed. Engl. *5*, 846 (1966); Montgomery, L. K., Applegate, L. E.: J. Am. Chem. Soc. *89*, 2952 (1967)
87) Bottini, A. T., Clause, A. O., Crelier, A. M., Applegate, L. E.: J. Am. Chem. Soc. *89*, 3453 (1967); Montgomery, L. K., Applegate, L. E.: J. Am. Chem. Soc. *89*, 5305 (1967)
88) Mohanakrishnan, P., Trayal, S. R., Viadyanathaswamy, R., Devaprabhakara, D.: Tetrahedron Lett. *1972*, 2871; Bottini, A. T., Corson, F. P., Fitzgerald, R., Frost II, K. A.: Tetrahedron *28*, 4883 (1972) and references cited therein; Bottini, A. T., Frost II, K. A., Anderson, B. R., Dev, V.: Tetrahedron *29*, 1975 (1973) and references cited therein; Freeman, P. K., Hardy, T. A.: Tetrahedron Lett. *1973*, 3317; Bottini, A. T., Anderson, B.: Tetrahedron Lett. *1973*, 3321; Bottini, A. T., Hilton, L. L., Plott, J.: Tetrahedron *31*, 1997 (1975); Bottini, A. T., Hilton, L. L.: Tetrahedron *31*, 2003 (1975); Bottini, A. T., Anderson, B. R., Dev, V., Frost Jr., K. A.: Tetrahedron *32*, 1613 (1976)
89) Ball, W. J., Landor, S. R.: J. Chem. Soc. *1962*, 2298
90) Caubère, P., Brunet, J. J.: Tetrahedron Lett. *1969*, 3323
91) Caubère, P., Brunet, J. J.: Bull. Soc. Chim. Fr. *1970*, 2418
92) Corson, F. P.: Ph. D. University of California, 1967
93) Viehe, H. G.: J. Chem. Soc. Annual Reports B, *64*, 152 (1967); Viehe, H. G., Delavarenne, Y. D.: Chem. Ber. *103*, 1209 (1970)
94) Caubère, P., Brunet, J. J.: Tetrahedron *27*, 3515 (1971)
95) Brunet, J. J., Fixari, B., Caubère, P.: C. R. Hebd. Séances, Acad. Sci. Ser. C, *276*, 1045 (1973)
96) Brunet, J. J., Fixari, B., Caubère, P.: Tetrahedron *30*, 2931 (1974)
97) Caubère, P., Brunet, J. J.: Tetrahedron *28*, 4835 (1972)
98) Caubère, P., Brunet, J. J.: Tetrahedron *28*, 4847 (1972)
99) Zaugg, H. E., Schaeffer, A. D.: J. Am. Chem. Soc. *87*, 1857 (1965); Rhoads, S. J., Holder, R. W.: Tetrahedron *25*, 5443 (1969)
100) Fixari, B., Brunet, J. J., Caubère, P.: Tetrahedron *32*, 927 (1976)
101) Caubère, P., Brunet, J. J.: Tetrahedron *28*, 4859 (1972)
102) Cookson, R. C., Rogers, N. R.: J. Chem. Soc. Chem. Commun. *1972*, 809; Dalton, J. C., Chan, H. F.: J. Am. Chem. Soc. *95*, 4085 (1973); Cookson, R. C., Rogers, N. R.: J. Chem. Soc. Perkin Trans. *1*, 1037 (1974); Engel, P. S., Schexnayder, M. A.: J. Am. Chem. Soc. *97*, 145 (1975)
103) Brunet, J. J., Fixari, B., Caubère, P.: Tetrahedron *30*, 1237 (1974)
104) Loubinoux, B., Viriot-Villaume, M. L., Chanot, J. J., Caubère, P.: Tetrahedron Lett. *1975*, 843
105) Brunet, J. J., Fixari, B., Caubère, P.: Tetrahedron *30*, 1245 (1974)
106) Wiberg, K. B., Hiatt, J. E., Hseih, K.: J. Am. Chem. Soc. *92*, 544 (1970)
107) Sasaki, T., Eguchi, S., Ohno, M., Nakata, F.: J. Org. Chem. *41*, 2408 (1976) and references cited therein

Received February 3, 1977

Complex Bases and Complex Reducing Agents
New Tools in Organic Synthesis

II. Activation of Sodium Hydride
Sodium Hydride Containing Complex Reducing Agents

Paul Caubère

Université de Nancy I, Laboratoire de Chimie Organique I (Associated to CNRS),
Case Officielle 140, 54037 Nancy Cedex, France

Table of Contents

1. Introduction

Complex reducing agents are more recent reagents than complex bases. Their story begins some years ago with studies on sodium hydride reducing properties. Generally this hydride is used in organic chemistry as a base[1]. However hydride ion H^- is a soft base[2] and we thought that it would be able to reduce organic compounds. We formulated this hypothesis after having observed[3] the coupling reaction pictured in Scheme 1.

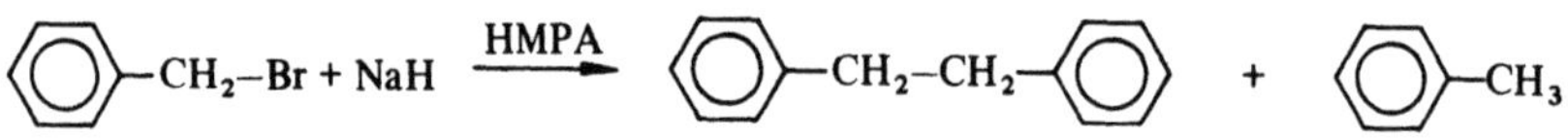

1 *2* *3*

Scheme 1

As a matter of fact, if NaH had played the rôle of a classical base, the product formed would have been *trans* stilbene[4].

We demonstrated further that in the coupling reactions, H^- gives electrons to benzyl bromide, leading to benzyl anion[5]. This step of the reaction is generally symbolysed by a simultaneous electrons transfer. Of course it must be understood as pictured in Scheme 2.

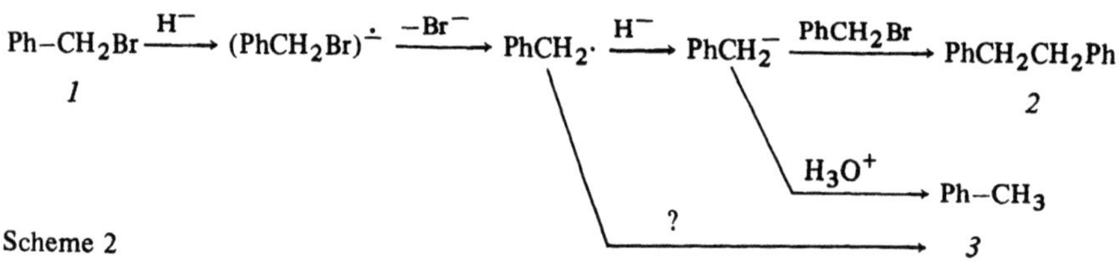

Scheme 2

With benzyl halides other than *1*[5, 6] *trans* stilbenes may be formed by reaction with NaH when the halogen is not soft enough and the aromatic ring bears an electron withdrawing group (Scheme 3).

X = Br, Z = o.F, p.CN, o.NO_2
X = Cl, Z = p.Cl, o.Cl, p.NO_2

Scheme 3

In other words, H^- in HMPA is able to attack halogen as well as hydrogen following the softness of the first and the "acidity" of the second.

After having obtained those results, two questions were formulated:

1. What is the reactivity of NaH–RONa in HMPA?
2. Is it possible to activate NaH by alcoholates and are the reactions performed in HMPA, with NaH alone, feasible in THF with the help of NaH–RONa? We shall briefly answer these questions in the next chapter.

II. Activation of NaH by Alcoholates NaH–RONa Reagents

II.1. Reactions of NaH–RONa with Benzyl Halides

First we studied reactivity of unsubstituted benzyl halides[6, 7]. Benzyl iodide is too sensitive to nucleophilic attack to give valuable informations. On the contrary, benzyl chloride and bromide gave us interesting results as may be seen in Scheme 4.

$$PhCH_2X \longrightarrow PhCH_2X + PhCH_2{-}CH_2{-}Ph + PhCH_3 + Ph{-}CH{=}CH{-}Ph + PhCH_2OR$$

X	Reagent	$PhCH_2X$	$PhCH_2{-}CH_2{-}Ph$	$PhCH_3$	$Ph{-}CH{=}CH{-}Ph$	$PhCH_2OR$
X = Cl	NaH–HMPA	10%	50%	18%	0%	–
	NaH-t-AmONa–HMPA	0%	0%	0%	80%	10%
	NaH–THF	90%	traces	0%	0%	–
	NaH-t-AmONa–THF	0%	90%	0%	0%	traces
X = Br	NaH–HMPA	0%	70%	18%	0%	–
	NaH-t-AmONa–HMPA	0%	0%	0%	15%	75%
	NaH–THF	90%	traces	0%	0%	–
	NaH-t-AmONa–THF	0%	72%	0%	0%	15%

Scheme 4

Taking into account that *t*-AmONa in HMPA leads chiefly to benzyl tertioamyl ether and that this alcoholate shows very low reactivity in THF, we can conclude that a real activation of NaH by *t*-AmONa takes place. From these results, the following points have to be underlined:

Nucleophilic power of the alcoholate is partially masked in HMPA. This is clearly shown by the reactions performed with benzyl chloride. However, nucleophilic properties are still sufficient to substitute benzyl bromide.

t-AmONa enhances the proton abstracting power of NaH in HMPA and its "halogen abstracting power" in THF. Note that NaH alone is unreactive in this latter solvent.

Finally, a strong similarity of behaviour is observed between NaH in HMPA and NaH-*t*-AmONa in THF. However, NaH-*t*-AmONa is much more reactive towards halogens than NaH. This last observation is well illustrated by study of the reactivity of bromo diphenylmethane (Scheme 5).

$$^{7)}\ \underset{85\%}{Ph_2CH{-}CHPh_2} \xleftarrow[\text{THF}]{\text{NaH-}t\text{-AmONa}} Ph_2CHBr \xrightarrow[\text{HMPA}]{\text{NaH}} \underset{90\%}{Ph_2C{=}CPh_2}\ ^{5)}$$

Scheme 5

All the conclusions given above were confirmed with substituted benzyl halides[6, 7]. It is very important to note that we find again here what we found for the activation of $NaNH_2$ (see Part I on complex bases).

II.2. Reactions of NaH–RONa with gem Dihalogeno Cyclopropanes

During our first work on the potential reducing properties of NaH, we had shown that NaH is able to reduce nonenolisable ketones[8] as well as some other functional groups[9]. Note that a few examples of such reductions performed in more or less good conditions had been described in the literature[1, 10].

With certain aromatic ketones, we showed that H^- is able to transfer its electrons "one after one" and the intervention of cetyl radicals. This is briefly illustrated in Scheme 6[8].

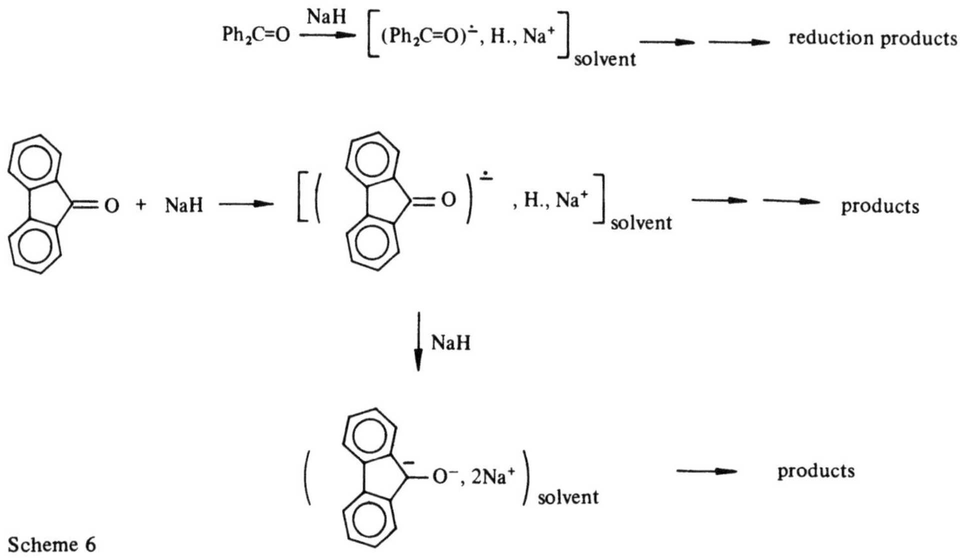

Scheme 6

We had suspected the propensity of H^- to mono-electronic transfers since we had discovered that *addition of a mixture of HMPA–H_2O to a suspension of NaH in HMPA leads to the formation of blue solutions of solvated electrons!*[5]

Thus we thought that like radical reducing agents (for example the well-known Bu_3SnH[11]) NaH was able to reduce gem dihalogeno cyclopropanes. As a matter of fact, those reductions are performable in HMPA and we interpreted the reaction by the intervention of radical species with a solvent cage effect[12]. However, in the conditions where the reductions were performed, some side reactions can take place. These latters cannot be avoided by changing HMPA for THF because in this solvent NaH does not react with gem dihalogeno cyclopropanes.

Taking into account the results obtained with benzyl halides, we tried to improve these reductions with the help of NaH-t-AmONa in THF. As it may be seen in Scheme 7, our hopes were satisfied[12].

It is clear that in THF, *t*-AmONa considerably enhances the reducing power of NaH and the yields obtained compare favorably with those in HMPA. Moreover, propensity to reduce rather than to eliminate is stronger for NaH-t-AmONa in THF than for NaH in HMPA. This conclusion is clearly illustrated by the reactions pictured in Scheme 8[12].

$n = 2$ R = H	NaH–HMPA	55%	
	NaH-t-AmONa–THF	66%	
$n = 2$ R = Me	NaH–HMPA	60%	
	NaH-t-AmONa–THF	70%	
$n = 3$ R = H	NaH–HMPA	50%	
	NaH-t-AmONa-THF	60%	
$n = 4$ R = H	NaH–HMPA	0–40%[a])	20–70%[1])
	NaH-t-AmONa–THF	60%	30%

[1]) Reaction with little reproducibility.

Scheme 7

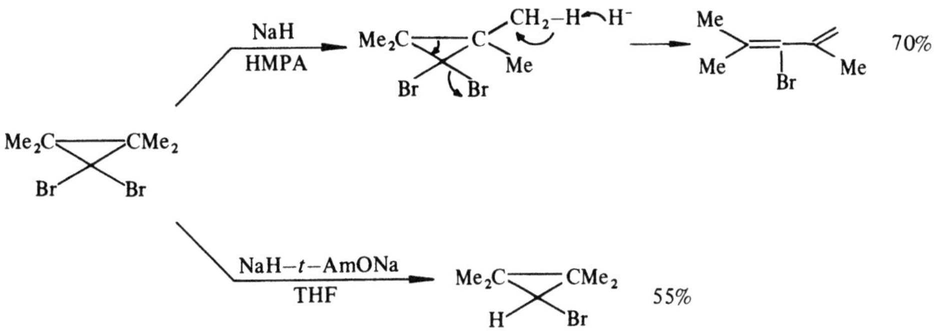

Scheme 8

II.3. Limitations of NaH and NaH–RONa as Reducing Agents

As we said before, in our opinion H^- is able to react by way of monoelectronic transfers and very often apparent simultaneous transfers of two electrons must rather be very fast consecutive mono-electronic transfers.

Thus H^- should be a good reducing species in solution and NaH a good, cheap and easily handled reagent. However, this salt suffers a large inconvenience: it is very insoluble in most organic solvents. HMPA would be able to help in outlining this difficulty but in this solvent the affinity of H^- for the proton is too high and limits its use. On the other hand, NaH in THF is very insoluble, but in the presence of sodium alcoholate, aggregates must be formed which enhance the "dissolution", as for $NaNH_2$. In those aggregates, electrons must be more movable and more able to perform reductions. In other words, aggregates NaH–RONa constitute "tanks" of movable electrons.

However, activated sodium hydride also suffers some inconveniences. Thus even in THF basic properties are not low enough to allow reduction of substrates possessing too much acidic hydrogens. For example, enolisable ketones are not reduced but enolized. Finally the "quality" of commercial NaH is very important for the reproducibility of those reactions.

Our investigations to outline those large inconveniences led us to "complex reducing agents". In fact, some preliminary experiments performed with other goals, showed us that NaH does not react easily with cupric or cuprous salts. We thought that if NaH–RONa really were good movable electron "tanks", it should be possible to perform reductions of metalic salts. Our first experiments showed that this was true. Thus when a cupric or cuprous salt is added to NaH-*t*-AmONa in THF at 50–60 °C, a brown-black coloration appears and the reagent thus obtained has reducing properties. We called the mixtures NaH–RONa–MX_n complex reducing agents.

III. Reactions of NaH-*t*-AmONa–MX_n (M = Ni, Co, Cu) with Halogeno Compounds

III.1. Halogeno Aromatic Derivatives

We chose to use reductions of these compounds as test reactions and we tried first to reduce halogeno naphthalenes. Quickly we observed that constituent ratios of complex reducing agents cannot be randomly taken. Thus, without data on the nature of the reducing agents, we determined empirically the best ratios giving the best yields in our conditions. Scheme 9 summarizes our main results[13].

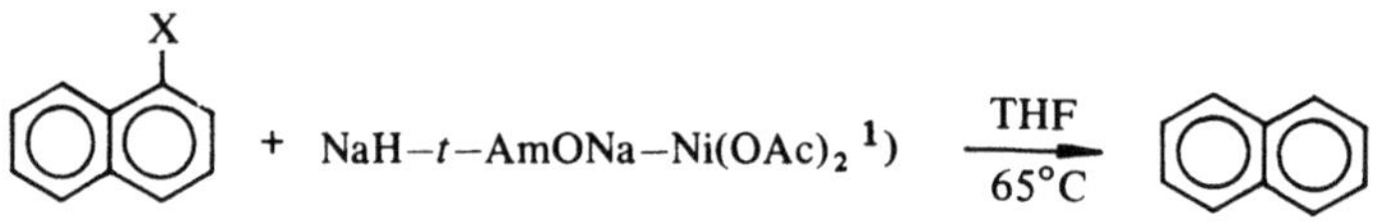

X = F	M = Ni	6 h	94–96%
	Co	very slow	–
	Cu	no reaction	–
X = Cl	M = Ni	1 h	95–97%
	Co	18 h	95%
	Cu	very slow	–
X = Br	M = Ni	1 h	85–90%
	Co	4–5 h	90–95%
	Cu	17 h	60–65% (+ binaphthyl)

1) NaH/*t*-AmONa/$M(OAc)_2$/RX = 4/2/1/1 M = Ni, Co
6/2/1/1 M = Cu

Scheme 9

Each pair of constituents (*i.e.* NaH-*t*-AmONa, NaH–$M(OAc)_2$ and *t*-AmONa–$M(OAc)_2$) are poorly reactive on the mentioned halogeno naphthalenes.

Examination of the results shows that reducing properties of complex reducing agents depend upon the nature of the metal.

Nickel is the most efficient reagent. On the other hand, more selective reductions should be expected with Co and Cu. With this latter which is the classical metal of Ullman reactions[14], formation of binaphthyl is observed. This point is of interest because it shows that it would be possible to obtain new coupling reagents from complex reducing agents.

Observation that our reagents were able to reduce even fluoronaphthalene, led us to try to reduce substituted halogenobenzenes with difficulty (or not at all) reduced by NaH[15]. Thus we attempted to reduce several of these compounds with our best complex reducing agent[13]. As it may be seen on Scheme 10 results were more than satisfying.

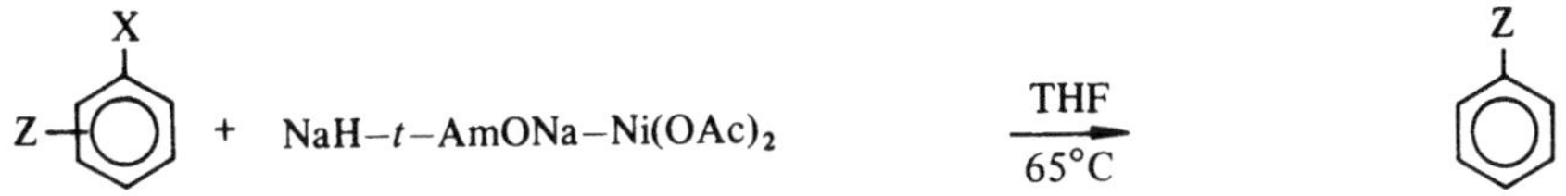

X = Cl Z = 2-Me, 3-Me, 4-Me
X = Br Z = 4-Me, 2-MeO, 4-MeO, 2-COOH, 3-COOH } 1–2 h 85–95%

X = F Z = 4-MeO
X = Br Z = 4-COOH } 12–18 h 85–95%

Scheme 10

Now if we compare the results given in Schemes 9 and 10 with those obtained by Nelson and Gribble[15] with NaH in THF, it is easily seen that complex reducing agents are much more powerful. For example, reaction of NaH with 2-bromobenzoic acid in THF needs eight days to give only a mixture of starting material and benzoic acid.

Those results being obtained, we wanted to examine reductions of gem dihalogeno cyclopropane with the goal of comparing NaH in HMPA, NaH-*t*-AmONa in THF and complex reducing agents. Moreover we hoped to get some more informations about our new reagents.

Then we first studied 7,7-dibromo and 1-methyl 7,7-dibromo norcaranes[16] from which we obtained the general feature given in Scheme 11.

Some comments have to be made:

With the conditions given here, NaH-*t*-AmONa reduces slowly and yields are strongly dependent upon the NaH "quality". This latter is of much less consequence with complex reducing agents.

"NaH–MX_n" does not react. Likewise *t*-AmONa–MX_n does not reduce. We must notice a curious exception for *t*-AmONa–$Ni(OAc)_2$ when the ratios alcoholate/nickel salt/bromo compound are worth 4/1/1. Yields obtained are lower than with complex reducing agents and reductions are essentially observed for the reagent ratios given. Moreover, from all our studies it appears that these reductions are far from general. Owing to the results of Whitesides and colleagues[17] we have to suppose that here the reducing agent is the nickel *tert*-amylate, and that in some cases these salts are able to give their electrons.

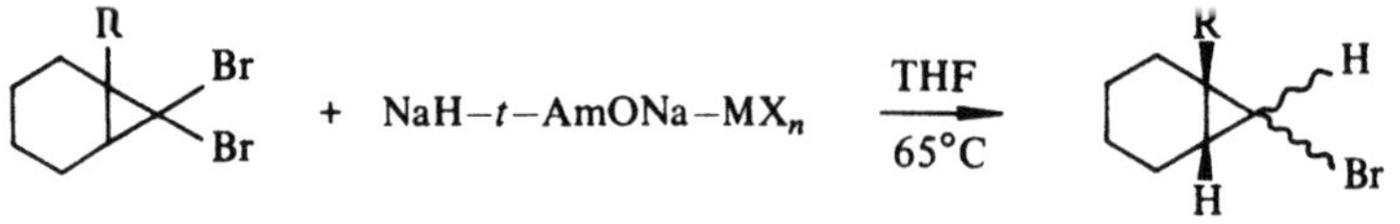

R	MX_n	Bromide/NaH/t-AmONa/MX_n				t(h)[1])	*cis/trans*	Yield %
H	$Ni(OAc)_2$	1	/1	/1	/1	4–5	74/26	80–85
H	$Co(OAc)_2$, $Cu(OAc)_2$, $CoCl_2$, $CuCl_2$, Cu_2Cl_2, Cu_2Br_2	1	/4	/2	/1	1–2	∿70/30	between 55 and 70
H	Cu_2I_2	1	/4	/2	/1	16	60/40	30
CH_3	$Ni(OAc)_2$	1	/1	/1	/1	3–4	66/34	80
CH_3	$Cu(OAc)_2$	1	/4	/2	/1	2 1/2	66/34	66
CH_3	Cu_2Cl_2	1	/4	/2	/1	8	70/30	60–66

[1]) Time needed for the total disparition of starting material.

Scheme 11

Coming back to complex reducing agents, we see as we observed with aromatic halogeno compounds, that the nature of the metal used is very important not only on the reactivity of the reagents but also on the needed ratios for the constituents.

Sometimes, formations of hydrocarbons proceeding from full reductions are observed. Yields always are low with Co and Cu reagents. With Ni reagents they can reach substantial values when the NaH ratio is increased. This is of interest for further investigations on selective reductions.

A few experiments performed with NaD show no deuterium incorporation. This observation and the values of the *cis/trans* ratios compared with what is given in the literature[11, 18]) lead one to conclude that reductions are radical reactions with hydrogen provided by the solvent.

Those reductions were also performed on some more dibromo cyclopropanes

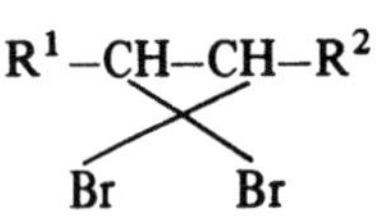

(R^1 = H $R^2 = C_6H_{13}$; [R^1, R^2] = $(CH_2)_5$, $(CH_2)_6$

with very good yields[16]).

Thus we can say that complex reducing agents are much more efficient than NaH or even NaH-*t*-AmONa. This is confirmed by the reaction pictured in Scheme 12[16]).

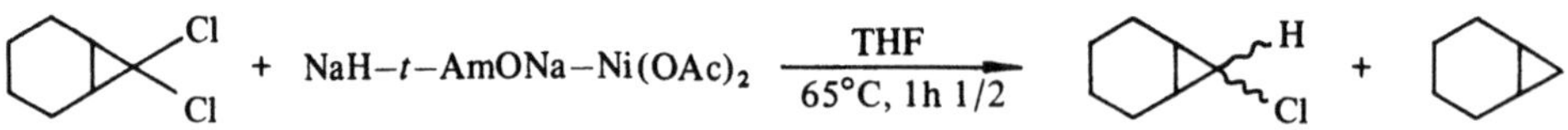

Scheme 12 — 60–65% — 5–10%

It is well known[19]) that dichloronorcarane is scarcely reduced by classical reducing agents. Thus the result given in Scheme 12 underlines the usefulness of complex reducing agents. Note that use of $Co(OAc)_2$ instead of $Ni(OAc)_2$ eliminates nearly completely norcarane formation but the yields of chloronorcarane are only 40 to 45%.

Finally it must be noted that we also showed[16] that metallic salts may be used more or less catalytically. Thus when the molecular ratios dibromonorcarane/metallic salts are worth 8 for $Ni(OAc)_2$ and 4 for $Co(OAc)_2$ or $Cu(OAc)_2$ mono reductions are performed with 60–70% and 45–50% yields respectively.

In conclusion, Ni, Co and Cu sodium hydride-containing complex reducing agents are good new reagents allowing reductions of halogeno compounds. Further investigations should show us that those results, extended to other metals, should be useful in selective reductions of polyhalogeno derivatives.

III.2. Reactions of NaH-*t*-AmONa–$Ni(OAc)_2$ with Ketones

Results shown earlier show that complex reducing agents may be useful tools in performing reductions. Nevertheless a serious limitation would be anticipated with base-sensitive substrates if the classical basic properties of sodium hydride and sodium alcoholate were preserved in those reagents.

However, we showed earlier (Part I) that in nucleophilic complex bases the nucleophilic properties are partially masked. Thus by analogy it was not abnormal to expect that complex reducing agents were constituted by new metallic species where NaH and *t*-AmONa would have lost most part of their basic properties. In fact our expectation was satisfied.

Thus as it may be seen from the results described in Scheme 13, the basic character of NaH and *t*-AmONa is strongly masked in complex reducing agents[20].

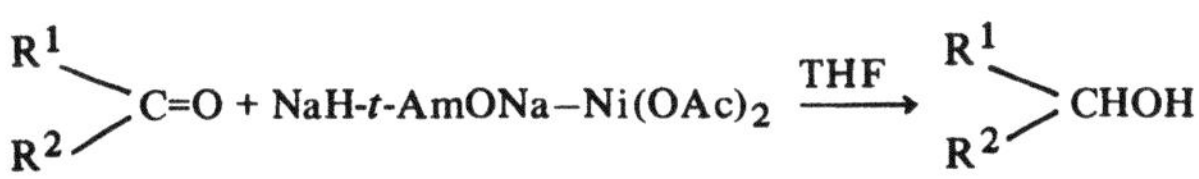

$R^1 = R^2$ = *n*-Bu, *t*-Bu; R^1 = Me R^2 = Ph; $[R^1, R^2] = (CH_2)_5$	95–100%
$[R^1, R^2]$ = *t*-Bu–$CH(CH_2-CH_2)_2$	55%

Scheme 13

Of course we verified (on phenyl methylketone) that a complex reducing agent does not reduce ketone enolates.

It is noteworthy that the reagent is efficient even with the strongly hindered ditertbutyl ketone. However, we did one intriguing observation:

Conversion to alcohol passes through a maximum, then decreases; *i.e.* a reoxydation occurs! We explained this observation assuming (Scheme 14) that during the reduction, formation of sodium as well as nickel alcoholates occurs (oxydation state of nickel being unknown, we shall write it "Ni").

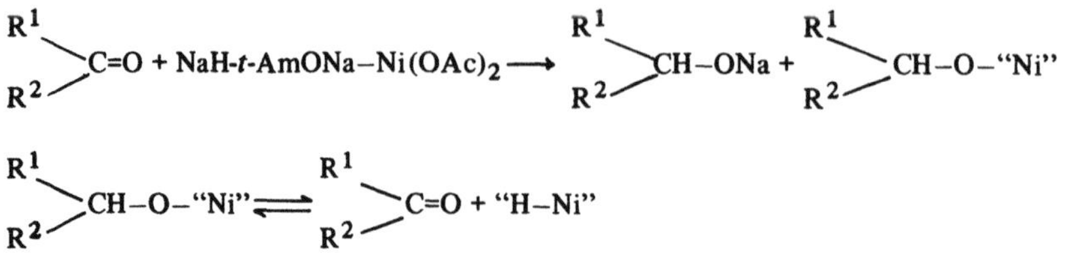

Scheme 14

Of course, nickel alcoholate can give a classical β elimination[21] leading to starting ketone and nickel hydride species. These latter reduce again or decay irreversibly thus explaining the reappearance of the ketone.

This hypothesis is strongly supported by the following experiments[22]: Commercial mixture of 4-tertbutylcyclohexanols (*cis/trans* = 32/68) or their sodium salts submitted to the action of NaH-*t*-AmONa–Ni(OAc)$_2$ leads to formation of 4-tertbutyl cyclohexanone. In this "oxydation" the *cis* isomer disappears more quickly than the *trans* one, in accord with the lower stability of its alcoholate. This kind of "oxydation" had already been observed by Eliel and his co-workers[23] during the reduction of ketones on nickel catalysts.

Finally we thought that low reduction yields obtained with 4-*tert* butyl cyclohexanone (see Scheme 13) was a consequence of this reoxydation. If such was the case, this side reaction should be avoided by addition of a metallic salt able to complex the oxygen of the alcoholate but unable to give β hydride elimination.

Indeed, addition of alkali halides and particularly LiCl or LiBr allows the reduction of 4-*tert* butyl cyclohexanone with nearly quantitative yields[20]. Note that with other ketones, addition of lithium salt improves the reproducibility of their reduction by strongly postponing the reoxidation.

III.3. Reaction of NaH–RONa–Ni(OAc)$_2$ with Unsaturated Hydrocarbons

Successes obtained with ketones led us to examine possible alkene reductions. From numerous experiments[20, 22] we found that the solvent is important in these reactions and that anisole is a good one. Of course reduction rates strongly depend upon the nature of the ethylenic compound. Thus the following reactivity order has been established:

n-Bu–CH=CH$_2$ > *trans n*-Pr–CH=CH–CH$_3$, n-Hex–CH=CH$_2$ > *trans n*-Pent–CH=CH–CH

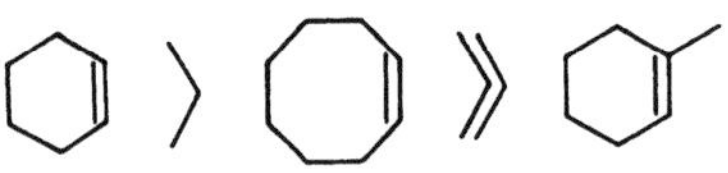

Yields are between 90 and 100% except for 1-methyl cyclohexene. Thus it appears that the complex reducing agent is very sensitive to steric hindrance and will be able to perform selective reductions.

Finally, when methylene cyclohexan is submitted to the action of NaH-*t*-AmONa–$Ni(OAc)_2$ reduction yield is only 75% and 1-methyl cyclohexene appears. This isomerization as well as the reductions performed strongly suggest[21, 24] the presence of nickel hydride species in the complex reducing agent.

This suggestion is supported by the results with alkynes[20]. Thus, as it may be seen in Scheme 15 1-alkynes may be reduced selectively to alkenes or alkanes following the nature of the activating agent of complex reducing agent, solvent, and temperature used.

$$R{-}C{\equiv}C{-}H + NaH{-}R'ONa{-}Ni(OAc)_2 \longrightarrow R{-}CH{=}CH_2 + R{-}CH_2{-}CH_3$$

				$R{-}CH{=}CH_2$	$R{-}CH_2{-}CH_3$
R = *n*-Bu	R′ = *t*-Am	20–25°	THF	75%	15%
	R′ = *t*-Bu	65°	anisole	–	100%
R = *n*-Hex	R′ = *t*-Am	20–25°	THF	100%	–
	R′ = *t*-Am	65°	anisole	–	80%

Scheme 15

Under certain conditions, some 2-alkenes appear during these reductions. This is another argument in favour of nickel-hydride intervention. This hypothesis is also strongly supported by the results obtained with disubstituted alkynes (Scheme 16)[20].

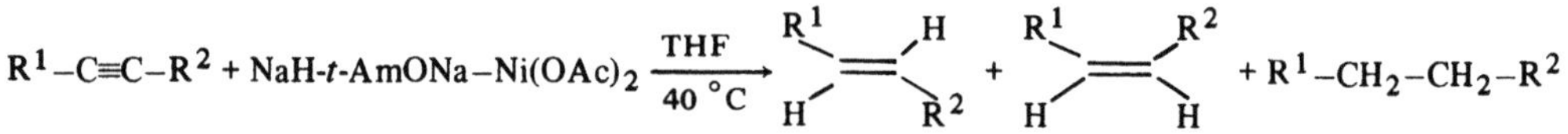

		trans	cis	alkane
R^1 = Me R^2 = *n*-Pr	(42 h)	traces	95–99%	traces
$R^1 = R^2$ = Ph	15–20 h	10–15%	63–67%	13–17%
	120 h	45%	0%	55%

Scheme 16

It is clear that these reductions are *syn* additions, although with diphenyl acetylene isomerization to the more stable *trans* stilbene cannot be entirely suppressed. This stereoselectivity was also expected from nickel hydride species[24].

Finally all our results concerning reactions of complex reducing agents with unsaturated hydrocarbons can be interpreted by the mechanism pictured in Scheme 17.

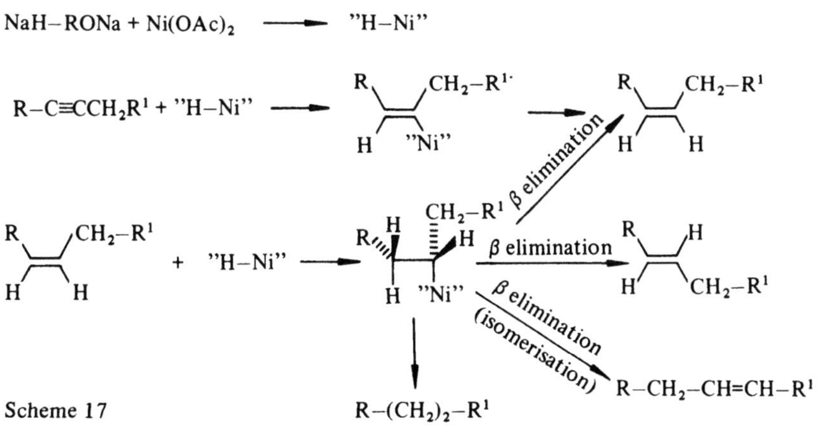

Scheme 17

This kind of mechanism is currently given to explain reactions observed between metallic hydrides and unsaturated hydrocarbon (see for example[24]). Of course the exact nature of complex reducing agents remains to be thoroughly investigated.

III.4. Use of Complex Reducing Agents in Carbonylation Reactions

As we said earlier, we do not know the structure and of course the mechanism of the formation of complex reducing agents. However, we formulated the hypothesis that at some step of their synthesis, the role of NaH–RONa is to give electrons to metallic salts leading to low-valent metal species which evolve towards metallic hydrides.

If this hypothesis were true, the same reduction performed in the presence of appropriate ligands should lead to new metallic complexes which should have nothing to do with complex reducing agents. This was successfully verified[25]. Thus when $Co(OAc)_2$ is added to a suspension of NaH-*t*-AmONa in *THF under carbon monoxide at atmospheric pressure and between 40–60 °C,* a dark blue color develops. The heterogeneous medium thus obtained shows strong infrared absorption (2130, 2060, 2010 and 1890 cm^{-1}). Those spectra show the presence of carbonyl metallic species[26]. Moreover we were able to carbonylate halogeno aromatic compounds (Scheme 18)[25].

$$R^1\text{–}C_6H_4\text{–}X + \text{NaH-}t\text{-AmONa–}Co(OAc)_2\text{–}CO \xrightarrow[\text{2) } H_3O^+ \text{ or EtOH then } H_3O^+]{\text{1) 60–65 °C, THF}} R^1\text{–}C_6H_4COOR^2 + R^1C_6H_4COO$$

(overall yields: 100%)

$R^1 = H$ X = I, Br
$R^1 = p\text{-}CH_3$ X = Br

Scheme 18

Finally, preliminary experiments show that Ni and Fe salts are also able to produce carbonyl species in the same conditions.

These results are particularly interesting. Indeed, it appears from the literature that there are only a few methods allowing the preparation of carbonyl metals in aprotic solvent from metallic salts, under CO at atmospheric pressure and at low temperature[27].

Moreover, our conditions compare favorably with those we can find in the literature about the carbonylation of halogeno aromatic compounds[27c, 28].

These results are only preliminary, but it is already clear that our reagents have noticeable reactivity.

III. 5. A Few More Complex Reducing Agent and Conclusions

During the past few years metallic reagents obtained by reaction between metallic salts and various hydrides become more and more numerous[29]. It has even been recently published that $NaH\text{–}FeCl_3$ and $NaH\text{–}FeCl_2$ have very special reducing properties[30]. However, these authors are forced to use a very large excess of metal salt and sodium hydride relative to the substrate to be reduced. We have to underline that there is always a risk in using too large an excess of NaH. Indeed, this reagent can contain a small quantity of metallic sodium which can modify the path of the reactions. Note that sodium metal can be easily evidenced with the help of the coloration it gives with HMPA[31].

All we described in preceding chapters shows that NaH–RONa reacts with metallic salts giving also a new kind of reducing agents. What is their interest?

First they are very simple to prepare and they are not expensive. They constitute a new use of sodium hydride in organic chemistry and enhance the value of this very current compound. Moreover very large excesses of NaH or metallic salts are not needed.

Secondly properties of complex-reducing agent may be varied infinitely, allowing a very broad range of reducing agents. Indeed, we saw that properties of complex bases vary strongly with the nature of the activating agent, the solvent and the relative ratios of the constituents; with complex reducing agents the same parameters intervene. Moreover, we can still modulate their properties by changing the nature of the metallic salts. Concerning this parameter, our recent results[32] show that the following metal derivatives are able to give complex reducing agents: $(Ph_3P)_2NiCl_2$, $FeCl_3$, $CrCl_2$, $ZrCl_4$, $ZnCl_2$, $MoCl_5$, WCl_6, $Mn(OAc)_2$, $Pd(OAc)_2$, VCl_3.

All those properties are very important in view of selective reductions. We shall briefly illustrate in Scheme 19 what we have just said, with the help of a few of our most recent results[32].

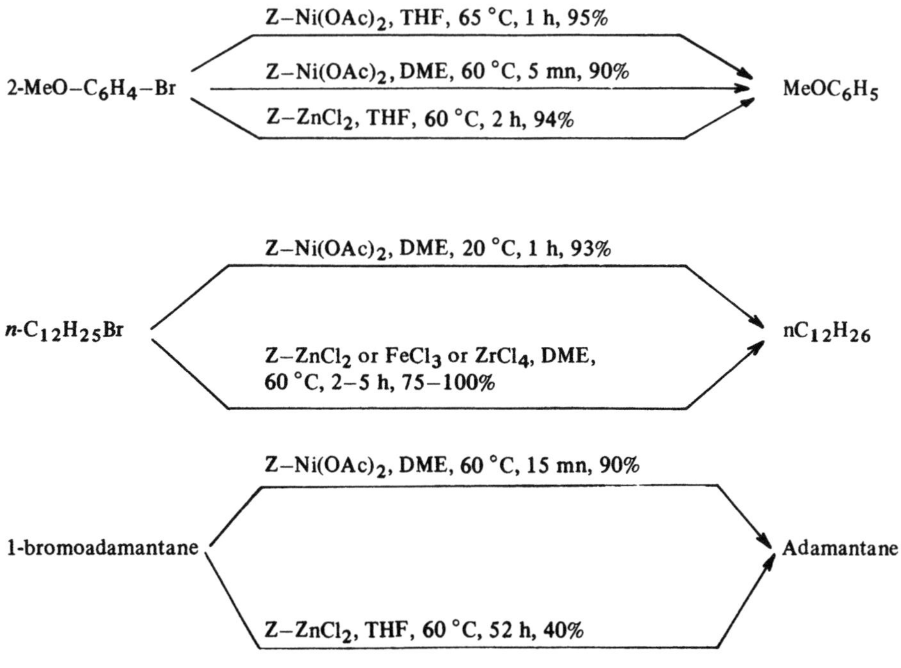

Z = NaH-*t*-AmONa

Scheme 19

Note that following the salt used there is no or only little elimination with bromododecane and that complex reducing agents are able to reduce aromatic aliphatic primary and tertiary halogeno derivatives.

Thirdly by addition of appropriate ligands during the reaction between NaH–RONa and MX_n it is possible to obtain new complex reagents able to perform many reactions other than reductions.

In conclusion, all the work until now on complex reducing agents is only the opening of a new large field of investigations. Of course many points remain obscure and have to be studied. However, from our actual experiments we can say that we will publish in near future some new results, not only in the reduction area but also in the field of the numerous organic reactions by means of transition metal complexes.

IV. General Conclusion

From several view-points the problem of the "synergy" of two reagents upon formation of mixed aggregates is fascinating. This concept applies to heterogeneous as well as homogeneous systems.

As we said before, the chief inconvenience is that the phenomena cannot be easily measured and we can only give imperfect pictures of them. Moreover, for the future, if we can imagin physicochemical experiments leading to better knowledge of homogeneous medium, it is easy to understand that this will be much more difficult for heterogeneous reagents like complex bases or complex reducing agents. Thus, for the present, we have to push on with the help of very imperfect pattern that we progressively and empirically modify in the function of the results of synthetic experiments.

A consequence is that this kind of chemistry may appear as being too close to "Cartesian spirits" and looking a bit like "alchemy". However, we must do with what we have. We cannot ignore experimental observations when we practice an "experimental science"!

In fact, in my opinion, the concept of activation should be the starting point for many new investigations. It should lead to new reagents for synthesis as well as new reaction interpretations.

Acknowledgments. I express my sincere appreciation to my co-workers, who are named in the references. I am grateful to them for their experimental and intellectual contributions to our work. I thank the Centre National de la Recherche Scientifique and the Délégation Générale à la Recherche Scientifique et Technique (Comité ASCO) for financial support.

Note Added in Proof:

Since this article has been written, some news results have been obtained in complex bases as well as in complex reducing agents fields. We shall briefly report here the new applications.

I. Complex Bases as Triggering Agents for Anionic Polymerizations[33)]

Taking account of the special properties of sodamide containing complex bases, it might be thought that they would be of interest as initiating agents of anionic polymerization reactions. The usefulness of numerous complex bases, of which activating agents had been evidenced by the benzylation and carbonatation reaction tests (see Part one, II.4.), was initially investigated in styrene polimerization.

We found that convenient activating agents belong to very different classes of compounds. We give in Scheme 1 a few examples of the activating agents capable of giving "initiating complex bases" with $NaNH_2$.

Polyalcoxyglycols or glycol monoethers:	$EtO(CH_2)_2OH$, $R(OCH_2CH_2)_2OH$ (R = Et, Bu), 1,2-$(MeO)(HO)C_6H_4$
Primary alcohols:	Pr–OH, $Me_3C–CH_2OH$, $Me–(CH_2)_3–C(Me)_2–CH_2OH$, $Me–(CH_2)_{10}–CH_2OH$
Secondary alcohols:	i·PrOH, Et_2CHOH, $(iPr)_2CHOH$
Tertiary alcohols:	*t*·BuOH, $Et_2(Me)COH$, $Me(CH_2)_7–C(Me)_2OH$
Glycols:	$Me_2C(OH)(CH_2)_2C(OH)Me_2$

Scheme 1. Examples of alcohols of which corresponding sodium salts were used as $NaNH_2$ activating agents in polymerization reactions.

Characteristic features of these polymerizations are that yields are medium to very good and that they are performable in tetrahydrofuran (THF) as well as in toluene. Moreover, it is noteworthy that, all other things being equal, polymer properties seem to depend on the nature of the activating agents.

Of course, if the only polymerizations performable were those of styrene, complex bases would be of little interest in the anionic polymerization field.

In fact, this is not the case and we were able to polymerize a very wide range of monomers in various conditions. A few examples of them are given in Scheme 2.

Monomer	$NaNH_2$ Activating Agent	Solvent
Vinyl pyridine	$Bu–(OCH_2CH_2)_2ONa$	THF, $Ph–CH_3$
Divinyl benzene	$Bu–(OCH_2CH_2)ONa$	THF, $Ph–CH_3$
1-Octene	$Bu–(OCH_2CH_2)_2ONa$, 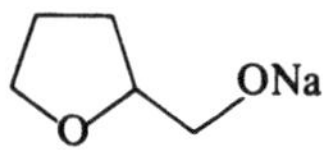	THF, $Ph–CH_3$
Butadiene	$Et–(OCH_2CH_2)_2ONa$	THF
Isoprene	*t*·BuONa, $Et–(OCH_2CH_2)_2ONa$	THF, $Ph–CH_3$
Dimethyl butadiene	$Bu–(OCH_2CH_2)_2ONa$, 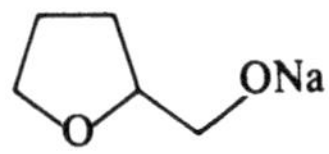	THF, $Ph–CH_3$
Ethylene oxyde	$Bu–(OCH_2CH_2)_2ONa$	THF
Popylene oxyde	$Bu–(OCH_2CH_2)_2ONa$	THF
Propylene episulfur	$Et–(OCH_2CH_2)_2ONa$, $Bu–(OCH_2CH_2)_2ONa$	THF

Scheme 2. Examples of monomers polymerized and of the initiating complex bases used

These examples are representative of only a small part of all our experiments. However we have shown that Complex Bases are also very interesting cheap reagents in the *initiation* of the very important anionic polymerization reactions.

II. Complex Reducing Agents

II.1. "Sodium Hydride Iron" Containing Complex Reducing Agents[34]

We mentioned (Part two III.5.) that $FeCl_3$ was one of the metallic salts capable of giving a complex reducing agent: NaH–RONa–$FeCl_3$. On the other hand, we cited the interesting works of Fujisawa, Sugimoto and Ohta (Part two Ref.[30]) concerning the reducing power of NaH–$FeCl_3$ and NaH–$FeCl_2$. However, the reactions performed by these authors required an excess of NaH as well as iron salt. Moreover, most of their results were obtained with $FeCl_2$, a reagent less available pure from commercial sources than $FeCl_3$ and less easily storable without alteration. It might be thought that most of these disadvantages would be removed by using the complex reducing agent NaH-*t*-BuONa–$FeCl_3$. This hypothesis was verified and we showed that our reagent is able to reduce ethylenic as well as acetylenic hydrocarbons as exemplified in Scheme 3.

$$R{-}CH{=}CH{-}R^1 + NaH\text{-}t.BuONa{-}FeCl_3 \xrightarrow{THF} R{-}CH_2{-}CH_2{-}R^1 \quad (50{-}95\%)$$

R = H R^1 = C_4H_9, C_6H_{13}, Ph
R = CH_3 R^1 = C_3H_7 (*cis* and *trans*)
[R, R^1] = $(CH_2)_4$, $(CH_2)_6$

$$R{-}C{\equiv}C{-}R^1 + NaH\text{-}t.BuONa{-}FeCl_3 \xrightarrow{THF} R{-}CH_2{-}CH_2{-}R^1 \quad (95{-}98\%)$$

R = H R^1 = C_6H_{13}, Ph
R = CH_3 R^1 = C_3H_7

Scheme 3

With ethylenic and acetylenic derivatives, the ratios NaH/*t*.BuONa/$FeCl_3$/Substrate were, respectively, 5,33/2,66/1,33/1 and 10,66/5,33/2,66/1. These ratios, as well as our reaction conditions, are more advantageous than that described in the Japanese works.

Finally we also showed that, contrary to NaH–$FeCl_2$ or NaH–$FeCl_3$, the iron containing complex reducing agent does not (or only very slightly) reduce carbonyl substrate. Thus, selective reductions are performable as exemplified in Scheme 4.

$$C_6H_{13}{-}CH{=}CH_2 + R{-}CO{-}R^1 + NaH\text{-}t.BuONa{-}FeCl_3 \xrightarrow{THF} \underset{(95-98\%)}{C_8H_{18}} + \underset{(90-95\%)}{R{-}CO{-}R^1} + \underset{(0-5\%)}{R{-}CHOH{-}R^1}$$

Scheme 4

Selective reductions seem also performable on dienic substrates. Thus, 4-vinyl cyclohexene was reduced to 4-ethyl cyclohexene in 98% yield.

These results confirm the advantage of using activated NaH–RONa instead of NaH alone.

II.2. Selective Reduction of Halide Derivatives[35]

We showed in Part two (III.5.) that complex reducing agents are able to reduce aliphatic halide derivatives. Our recent experiments confirm that primary, secondary and tertiary halide derivatives are reduced and that the reducing ability of the complex reducing agents towards these substrates depends on the nature of the metallic salts used.

Thus, selective reductions were possible. This interesting property is exemplified in Scheme 5.

$$R^1{-}X^1 + R^2{-}X^2 \xrightarrow{\text{NaH-}t\text{-AmONa}-MX_n} R^1{-}H + R^1{-}X^1 + R^2{-}H + R^2{-}X^2$$

R^1	X^1	R^2	X^2	MX_n	Solv.	T °C	t(mn)	R^1H	$R^1{-}X^1$	$R^2{-}H$	$R^2{-}X^2$
C_8H_{17}	Br	Pr(Bu)CH	Br	$ZnCl_2$	THF	40	360	100%	0%	0%	100%
C_8H_{17}	Br	C_8H_{17}	Cl	$Ni(OAc)_2$	DME	60	3	100%	0%	0%	100%
C_8H_{17}	Br	$C_6H_5CH_2$	Cl	$Ni(OAc)_2$	DME	20	30	0%	90%	85%[1])	0%

[1]) Formations of 15% of bibenzyl are registered.

Scheme 5

II.3. Coupling and Oligomerisation Reactions[35])

Testing the reagent NaH-*t*.AmONa–$(Ph_3P)_2NiCl_2$ in the 1-bromo naphthalene reduction, we observed a noticeable formation of binaphthyl. It was assumed that Ph_3P was accountable for this side reaction and that the preparation of complex reducing agent NaH-*t*.AmONa–$Ni(OAc)_2$ in the presence of Ph_3P would lead to new reagent capable of coupling organic halides. Preliminary experiments showed that this was the case, but some side reactions appeared with the phosphine ligands. Changing them to bipyridyl eliminates this inconvenience and we were able to perform the reactions summarized in Scheme 6.

$$Ar{-}X \xrightarrow[\text{THF}]{\text{NaH-}t\text{.AmONa}-Ni(OAc)_2-\text{Bipyr}} \underset{(70-85\%)}{Ar{-}Ar} + \underset{(15-25\%)}{Ar{-}H}$$

ArX = 1-BrC_8H_7; 2-$Cl(CH_3)C_6H_4$; 4-$Cl(CH_3)C_6H_4$

Scheme 6

For the present, we have not optimised the reaction conditions. However it is clear that addition of a potential ligand during the preparation of a complex reducing agent leads to a poor reducing but good coupling reagent.

Our present hypothesis is that without additional ligand, NaH-*t*.AmONa reacts with metallic salts to give some hydride reducing species. On the contrary, in the presence of additional ligand, the reaction takes another way and leads to low valent metallic species with poor reducing properties.

Note that this hypothesis is strongly supported by the oligomerisation reactions reported in Scheme 7.

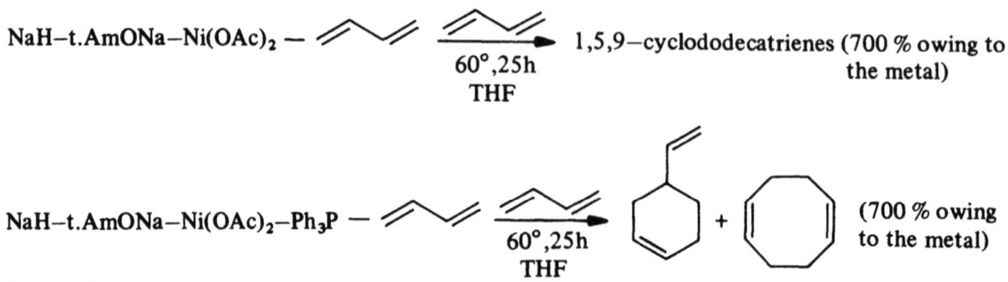

Scheme 7

In those experiments, the complex reducing agent was prepared in THF in the presence of the additional ligands and the butadiene then added to the reagent thus obtained. Taking into account the data from the literature (see for example[36]) it is clear that some low valent nickel species (with few reducing properties) are formed during the preparation of the catalyst.

II.4. Complex Reducing Agents as a Source of New Heterogeneous Hydrogenation Catalysts[37]

We said (Part two III.3.) that hydride species are certainly involved in complex reducing agents. On the other hand, surface hydride formation is one of the postulated steps in catalytic hydrogenations[38]. These two observations led us to examine complex reducing agents as new sources of hydrogenation catalysts. Note that a wide range of these kind of catalysts was obtained by reaction of a transition metal salt with a reducing system[39].

The results obtained, until now, concern nickel containing NaH of complex reducing agent. Thus, the neutralisation by *t.*AmOH of the remaining NaH-*t*.AmONa–Ni(OAc)$_2$ prepared in THF led to a *non pyrophoric* black suspension (called NiC) showing catalytic properties in hydrogenation reactions under normal pressure.

Hydrogenation of α olefines, as well as internal and cyclic olefines, is possible. However the catalyst is very sensitive to the olefin structure and, for example, trisubstituted ethylenic derivatives are hardly reduced.

From a comparative experiment (performed with 1-octene) we can conclude that NiC is as active as Brown's P_1 or P_2 catalysts[40] and thus more active than current Raney nickel.

Of course, acetylenic hydrocarbons are also easily hydrogenated on NiC. Interestingly, the yield of the ethylenic derivatives formed by semihydrogenation in the reaction medium reached 78–98% before the start of their own hydrogenation. This observation shows that NiC is a selective hydrogenation catalyst.

From our first results it also appears that NiC is able to perform the semihydrogenation of dienes. Thus, for example, 4-vinyl cyclohexene is quantitatively hydrogenated to 4-ethyl cyclohexene.

In conclusion, all these new results show the wide range of applications of complex reducing agents and the very large investigation fields now open.

Acknowledgments. I thank the Centre National de la Recherche Scientifique, the Délégation Générale à la Recherche Scientifique et Technique (Comité ASCO), Pechiney Ugine Kuhlmann, the Société Nationale des Poudres et Explosifs and the Nobel Hoechst Society for their financial support.

V. References

1) See for example Plesek, J., Hermanck, S.: Sodium Hydride, Iliffe Books L.T.D. 1968
2) Pearson, R. G., Songstad, J.: J. Am. Chem. Soc. *89*, 1827 (1967)
3) Caubère, P.: Bull. Soc. Chim. Fr. *1966*, 1293
4) See for example Smith, H.: Organic reactions in liquid ammonia. New York: Interscience Wiley 1963; Hamma, S. B., Wideman, L. G.: Chem. Ind. (London) *1968*, 486
5) Caubère, P., Moreau, J.: Tetrahedron *25*, 2469 (1969)
6) Caubère, P., Moreau, J.: Bull. Soc. Chim. Fr. *1970*, 1986
7) Caubère, P., Moreau, J.: Tetrahedron *26*, 2637 (1970)
8) Caubère, P., Moreau, J.: Bull. Soc. Chim. Fr. *1971*, 3270
9) Caubère, P., Moreau, J.: Bull. Soc. Chim. Fr. *1971*, 3276
10) Swamer, F. W., Hauser, C. R.: J. Am. Chem. Soc. *68*, 2647 (1946); Pierre, J. L., Arnaud, P.: Bull. Soc. Chim. Fr. *1967*, 2107; McConaghy, J. S., Bloomfield, J. J.: J. Org. Chem. *33*, 3425 (1968); Rome, D. W., Johnson, B. L.: Tetrahedron Lett. *1968*, 6053; Bumgardner, C. L., McDaniel, K. G.: J. Am. Chem. Soc. *91*, 6821 (1969); Lewis, G. E.: J. Org. Chem. *30*, 2433 (1965); Gladych, J. M. Z., Hornby, R.: Chem. Ind. (London) *1970*, 652 and references cited therein
11) Kuivila, H. G.: Synthesis *1970*, 499 and references cited therein
12) Moreau, J., Caubère, P.: Tetrahedron *27*, 5741 (1971)
13) Guillaumet, G., Mordenti, L., Caubère, P.: J. Organomet. Chem. *92*, 43 (1975)
14) Fanta, P. E.: Synthesis *1974*, 9
15) Nelson, R. B., Gribble, G. W.: J. Org. Chem. *39*, 1425 (1974)
16) Guillaumet, G., Mordenti, L., Caubère, P.: J. Organomet. Chem. *102*, 353 (1975)
17) Whitesides, G. M., Sadowski, J. S., Lilburn, J.: J. Am. Chem. Soc. *96*, 2829 (1974)
18) Barlet, R., Vo-Quang, Y.: Bull. Soc. Chim. Fr. *1969*, 3729; Groves, J. T., Way Ma, K.: J. Am. Chem. Soc. *96*, 6527 (1974) and references cited in those publications
19) Seyferth, D., Yamazaki, H., Alleston, D. L.: J. Org. Chem. *28*, 703 (1963)
20) Brunet, J. J., Mordenti, L., Loubinoux, B., Caubère, P.: Tetrahedron Lett. *1977*, 1069
21) Fahey, D. R.: J. Organomet. Chem. Rev. A, *7*, 245 (1972)
22) Caubère, P.: Gordon Conference: Organic reactions and processes. New Hampton 1976; Brunet, J. J., Mordenti, L., Caubère, P.: Results to be published
23) Eliel, E., Shroeter, S. H.: J. Am. Chem. Soc. *87*, 5031 (1965)
24) Jolly, P. W., Wilke, G.: The organic chemistry of nickel. Vol. 2. London: Academic Press 1975
25) Loubinoux, B., Fixari, B., Brunet, J. J., Caubère, P.: J. Organomet. Chem. *105*, C 22 (1976)
26) Collins, C. J., Christie, J. B.: J. Am. Chem. Soc. *82*, 1254 (1960); Edgell, W. F., Yang, M. T., Koizumi, N.: J. Am. Chem. Soc. *87*, 2563 (1965); Bor, G., Noack, K.: J. Organomet. Chem. *64*, 367 (1974); Bor, G.: J. Organomet. Chem. *94*, 181 (1975); Shriver, D. F.: Acc. Chem. Res. *3*, 231 (1970)
27a) Wender, I., Pino, P.: Organic synthesis *via* metal carbonyls. New York: Interscience 1968
b) Orchin, M., Rupilius, W.: Cat. Rev. *6*, 85 (1972)
c) Cassar, L., Chiusoli, G. P., Guerrieri, F.: Synthesis *1973*, 509
d) Ellis, J. E.: J. Organomet. Chem. *86*, 1 (1975)
28) Corey, E. J., Hegedus, L. S.: J. Am. Chem. Soc. *91*, 1233 (1969) and references cited therein; Nakayama, M., Mizoroki, T.: Bull. Chem. Soc. Jap. *44*, 508 (1971); King, R. B.: Acc. Chem. Res. *3*, 417 (1970)

29) See for example Rerick, M. N.: in Reduction, Augustine, R. L., Marcel Dekker 1968; McMurry, J. E., Fleming, M. P.: J. Am. Chem. Soc. *96,* 4708 (1974); McMurry, J. E., Fleming, M. P.: J. Org. Chem. *40,* 2555 (1975); Semmelhack, M. F., Stauffer, R. D.: J. Org. Chem. *40,* 3619 (1975); Ashby, E. C., Lin, J. J.: Tetrahedron Lett. *1975,* 4453; Chum, P. W., Wilson, S.: Tetrahedron Lett. *1976,* 15 and references cited in those publications
30) Fujisawa, T., Sugimoto, K., Ohta, H.: J. Org. Chem. *41,* 1667 (1976); Fujisawa, T., Sugimoto, K., Ohta, H.: Japan Kokai 7691, 202 Cl. C 07 C 31/02 (Chem. Abstr. *85,* 142799 w (1976)); Fujisawa, T., Sugimoto, K., Ohta, H.: Chem. Lett. *1976,* 581
31) Normant, H.: Bull. Soc. Chim. Fr. *1968,* 791
32) Vanderesse, R.: D. E. A. Nancy 1976; Loubinoux, B., Vanderesse, R., Caubère, P.: Results to be published
33) Boileau, S., Raynal, S., Lecolier, S., Coudert, G., Ndebeka, G., Caubère, P.: to be published
34) Brunet, J. J., Caubère, P.: Tetrahedron Letters, *1977,* 3947
35) Loubinoux, B., Vanderesse, R., Caubère, P.: Tetrahedron Letters, *1977,* 3951
36) Jolly, P. W., Wilke, G.: The organic chemistry of nickel. Vol. 2. New York-London: Academic Press: 1975
37) Brunet, J. J., Gallois, P., Caubère, P.: Tetrahedron Letters, *1977,* 3955
38) Jolly, P. W., Wilke, G.: The organic chemistry of nickel. Vol. 1. Academic Press: 1974
39) See for example: Takegami, Y., Ueno, T., Fujii, T.: Bull. Chem. Soc. Japan, *38,* 1279 (1965); Wade, R. C., Holah, D. G., Hughes, A. N., Hui, B. C.: Catal. Rev. Sci. Eng. *14* (2), 211 (1976); Urushibara, Y.: Bull. Chem. Soc. Japan, *25,* 280 (1952); Motoyama, I.: Bull. Chem. Soc. Japan *33,* 232 (1960); Watt, G. W., Roper, W. F., Parker, S. G.: J. Am. Chem. Soc. *73,* 5791 (1951) and ref. cited therein; Leprince, J. B., Collignon, N., Normant, H.: Bull. Soc. Chim. France *3–4,* 367 (1976)
40) Brown, H. C., Brown, C. A.: J. Am. Chem. Soc. *85,* 1005 (1963); Brown, C. A.: J. Org. Chem. *35,* 1900 (1970); Brown, C. A., Ahuja, W. K.: J. Org. Chem. *38,* 2226 (1973)

Received February 3, 1977, September 21, 1977

Aromatic and Heteroaromatic Compounds by Electrocyclic Ring-Closure with Elimination

Prof. Dr. J. Christian Jutz

Organisch-Chemisches Institut der Technischen Universität München, Lichtenbergstr. 4, 8046 Garching, Germany

Table of Contents

I. Introduction

In the past the syntheses of polycyclic aromatics mostly required multistep and often laborious procedures[1]. Among these classical methods the Diels-Alder cycloadditions still play an important role in the preparation of aromatics.

However, the situation may now be changing a little. The generally accepted Woodward-Hoffmann rules[2] not only allow the interpretation of some long-known but formerly not intelligible reactions, but they have also stimulated the efforts to use other, new types of pericyclic reactions for the synthesis of aromatics and heteroaromatics. As an example of this new trend we may mention the "photodehydrocyclization" of stilbenes, stilbenelike compounds[3a], and related aryl or hetaryl polyenes[3c]. This important, very versatile reaction implies a nonthermal, photochemically induced electrocyclic hexatriene-cyclohexadiene ring-closure followed by an oxidative elimination of hydrogen, and opens the approach to a wide range of polycyclic aromatics and heteroaromatics, *e.g.*, helicenes[3b], which would otherwise be partly inaccessible.

Another, recently developed method to synthesize a broad spectrum of oligo- and polycyclic aromatics and heteroaromatics in a surprisingly simple manner is likewise based on an electrocyclic – but thermal – hexatriene-cyclohexadiene ring-closure combined with an elimination reaction. This new synthetic method and its scope will be the topic of the following report.

II. General Considerations

A. Ziegler-Hafner's Synthesis as a Prototype for an Electrocyclic Ring-Closure with Elimination

A prototype for the new cyclization reaction is the Ziegler-Hafner synthesis of azulene (*3*)[4] by heating the fulvenoid decapentaene (*1*). Woodward[2a] was the first to argue that a thermal, electrocyclic [π 10 *s*] ring-closure could be involved in this thermolytic process. Of course it seems likely that the nonisolable intermediate (*2*) may be formed by an intramolecular electrophilic attack of atom 10 to the electron-rich position 1 in compound (*1*). Indeed a distinction between the two different reaction modes is mechanistically irrelevant in this case. But only its interpretation as an electrocyclic process allows the generalization to [π 6 *s*] cyclizations too.

The ring-closure of (*1*) to (*2*) depends on an effective overlap of the orbitals at atoms 1 (*4*) and 10 in the transition state, and can therefore only proceed from the proper configuration and conformation (*1b*). However, energetically the most favorable form of (*1*) consists in the all-*trans* zigzag-shaped conformation (*1a*). The conversion of (*1a*) to (*1b*) requires formally a *trans-cis* isomerization of the $\Delta^{7,8)}$ double bond and a rotation about the 8,9-single bond.

For the energy demand of the *cis-trans* isomerization of simple olefines, 251–264 kJ (60–63 kcal) is calculated. In highly resonance-stabilized merocyanines like (*1*), the conjugation of donor and acceptor substituents (here the dialkylamino and

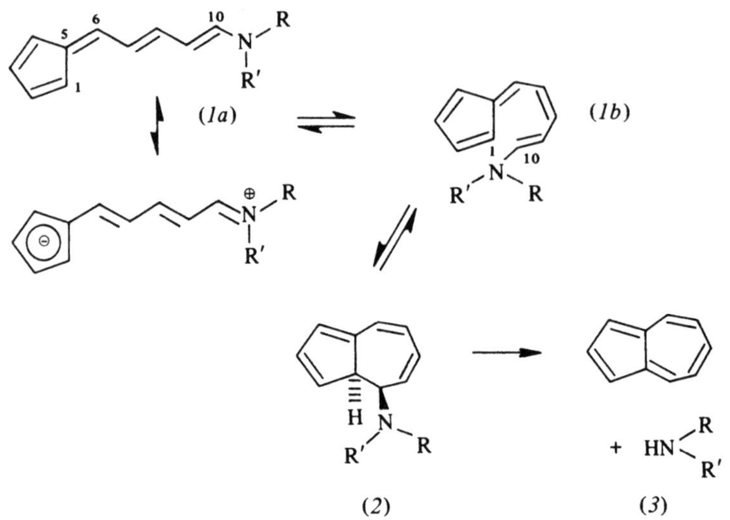

the cyclopentadiene group) across the unsaturated system effects a partial equilibration of the alternate bond orders as visualized by resonance formulas and proved by spectroscopic studies of the ^{1}H-NMR spectra. The real decrease of the bond orders of the double bonds (and some double-bonding of the single bonds) in merocyanines is reflected by the lowered barrier of energy for rotation about the double bonds. In 6-dimethylaminofulvene, the parent compound of (*1*), and in 6-dimethylamino-6-methylfulvene, for the rotation about the $\Delta^{5,6}$ double bond, activation energies of 92.5 kJ (22.1 kcal) and 68.6 kJ (16.4 kcal) respectively were measured[5]. Starting from (*1a*), the adequate form (*1b*) should be reached in a thermal equilibrium quite likely on heating.

In the cyclization step itself a σ-bond is formed utilizing two π-electrons of (*1*) at the expense of some resonance energy. If the ring-closure is a reversible and disrotatory process as we can suppose, a very smooth but irreversible *trans* elimination of dialkylamine in (*2*) can take place, leading immediately to azulene (*3*). Presumably the ring-closure step requires the highest activation energy in the thermolysis and may be the rate-determining process.

B. The Electrocyclic Ring-Closure of 1-Aminohexatrienes

The fact that merocyaninelike but very different 1-dialkylamino-1.3.5-hexatrienes (*4*) also form benzene derivatives smoothly under evolution of dialkylamine by thermolysis extends the general validity of the considerations mentioned even to [π 6 *s*] cyclizations.

The fundamental condition for the ring-closure consists here again in the chance of an effective overlap of the orbitals at the atoms 1 and 6 requiring the adequate conformation and configuration (*4*) of the trienamine. As we can imagine, steric interactions between the substituents in the chain play an important role. Bulky residues at positions 2, 3, 4, 5, or 6, provided an effective acceptor substituent is located at position 6 or 4, may shift the preceding conformational and configurational equilibria in favor of structure (*4*). Hence a rate-enhancement of the reaction to (*6*) or a lower thermolysis temperature was observed in such events.

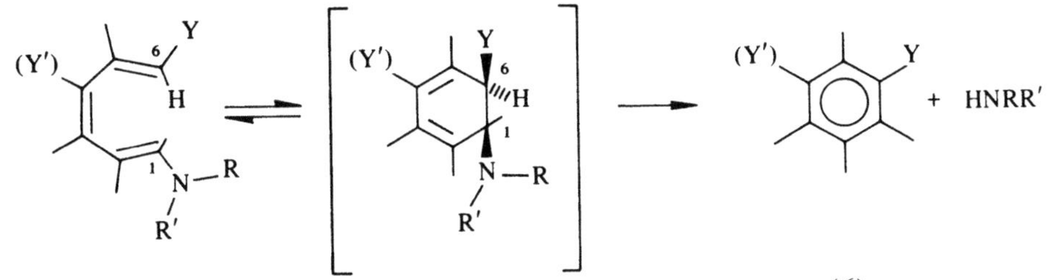

(*4*) (*5*) (*6*)

On the other hand, all factors impeding the most effective orbital overlap in the transition state slow down the reaction rate, or let the thermolysis temperature rise, or prevent the cyclization from taking place at all. Such factors may be: bulky substituents at position 1, which in addition enforce the less favorable Z-configuration of the dialkylamino group requiring *cis* elimination in the following step; steric overcrowding at positions 1 and 6, and finally all structural features producing an angle strain during ring formation.

Kinetic studies have previously been made only in a single case: the thermolysis of 7-dimethylamino-1-phenyl-2,4,6-heptatrien-1-one, a benzoyl-dimethylamino-hexatriene, to benzophenone, reveals a strong first-order reaction. The findings for activation enthalpy [$\Delta H^{\ddagger}$ 84.5 kJ/mol (20.2 kcal)] and activation entropy [$\Delta S^{\ddagger}$ −97.5 J/grad. mol (−23.3 cal/grad mol)] in o-dichlorobenzene as solvent[6] were consistent with the postulated high degree of order for an intramolecular reaction in the transition state. For electrocyclic reactions of alkylated hexatrienes to the corresponding cyclohexadienes, in comparison, $\Delta H^{\ddagger}$ 109–121 kJ (26–29 kcal) and $\Delta S^{\ddagger}$ −29—−50 J/grad. mol (−7—−12 cal/grad. mol)[7]. The negligible dependence of the rate on polarity of the solvent even seems to disprove a polar substitution mechanism and is evidence for an electrocyclic mode of the reaction. This is supported by the observation that only a very weak susceptibility for electronic effects

Rate constants of thermolysis of 7-dimethylamino-1-phenyl-2,4,6-heptatrien-1-one[6]

(°C)	$k_1 \cdot 10^5$ (s^{-1})	Solvent
84	2.69 ± 0.10	o-Dichlorobenzene
89.5	3.49 ± 0.30	o-Dichlorobenzene
100	10.10 ± 0.70	o-Dichlorobenzene
110	17.80 ± 1.20	o-Dichlorobenzene
89.9	3.37 ± 0.20	Mesitylene
94.8	3.98 ± 0.29	Mesitylene
100.8	8.05 ± 0.78	Mesitylene
110	17.65 ± 0.56	Mesitylene

of substituents in the various syntheses could be found unless they stabilize or destabilize the intermediate (*5*) especially.

An essential widening of the scope of the cyclization reaction under discussion consists in the observation that formal 1-aminohexatrienes in which one double bond of (*4*) is part of an aromatic or heteroaromatic system also cyclize smoothly. In other words, the method offers a way for benzannelation of numerous aromatics and heteroaromatics by a simple two-step reaction mostly performed as a one-pot procedure.

In addition one or two carbon atoms in the chain of (*4*) may be replaced by nitrogen, finally forming a six-membered heteroaromatic ring compound.

It is very important for preparative performances that such real and formal cyclizable 1-aminohexatrienes are easy to obtain by aldol- or Knoevenagel-like condensations of various vinamidinium salts (*7*) with a broad spectrum of C—H acid methyl and methylene compounds, as we will see in a following section.

C. MO Calculations as an Aid

In more complex compounds implying the formal 1-aminohexatriene pattern (*4*), the question often arise whether a cyclization will run readily or with difficulty. Also, in some instances with benzannelation a ring-closure is possible in two different directions.

An estimation therefore of the expected readiness in comparison with corresponding examples and the prediction of the preferred way of cyclization should be of great interest. Supposing that the rate-determining step in the ring-closure reaction is the endergonic formation of the intermediate (*5*) from the educt (*4*), and applying the Hammond principle, the transition state will be productlike in structure and energy. This should be correct in most instances, except in those few examples, perhaps of substituted (*4*), that need a still higher activation energy for configurational and conformational transformations.

In the simplest case, *i.e.* in going from (*4*) to (*5*), we see that two π-electrons from the hexatriene are used to form the new σ-bond, leaving a butadiene system in (*5*). The localization approximation concept appropriate for a simple Hückel MO treatment should work well here, yielding useful semiquantitative information[8)].

We define a localization energy $L_{r,s}$, as the π-bonding energy required to isolate two electrons, at positions r and s, from the remainder of the π-network. Such isolation results in a new π-system of two less atoms and two less electrons. The π-energy of the original system is $n\alpha + M\beta$, that of the localized system is $(n-2)\alpha + M_{r,s}\beta$ and the localization energy therefore is given by:

$$L_{r,s} = M - M_{r,s}$$

$L_{r,s}$ is a positive quantity and is a measure of the π-energy change between the educt and the ring-closed product, the intermediate not being isolable. If all σ-bond changes are treated as being effectively constant, even the change of effects of donor and acceptor substituents, we can anticipate a correlation between calculated localization energies $L_{r,s}$ and activation energies of the cyclizations. This simple model of course neglects all steric effects of substituents and steric interactions in the transition state.

Some experimentally treated systems with their π-network (dark lines) are summarized with the π-energy of educt system, M, that of the product system, $M_{r,s}$, and the localization energy $L_{r,s}$ (the units α and β are omitted), in the following figures.

Fig. 1

	M	$M_{r,s}$	$L_{r,s}$
I.	Benzenering formations		
	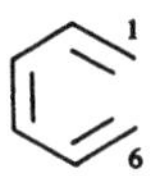		
a)	6.9879	4.4721	2.5158
	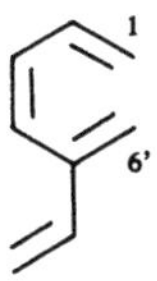	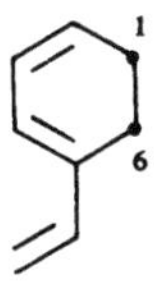	
b)	9.4093	6.9879	2.4214
	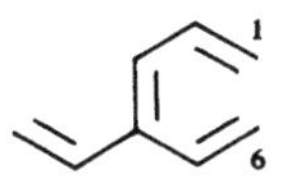	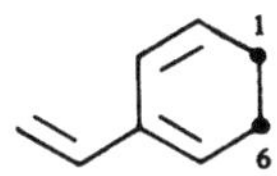	
c)	9.4459	6.8990	2.5469
	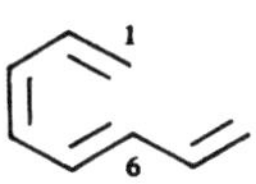	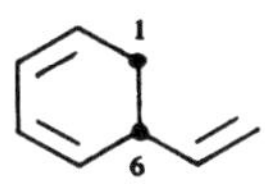	
d)	9.5175	6.4721	3.0454
	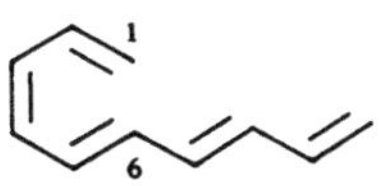	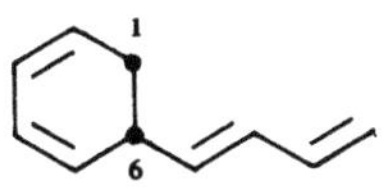	
e)	12.0534	8.9443	3.1091
II.	Naphthalene		
	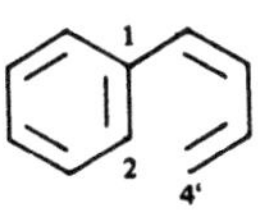	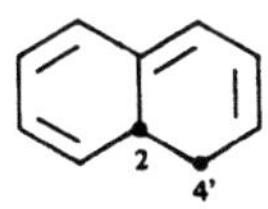	
	12.9321	9.5175	3.4146

Fig. 1 (continued)

	M	$M_{r,s}$	$L_{r,s}$
III.	Azulene		
	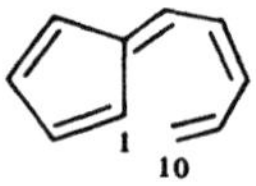	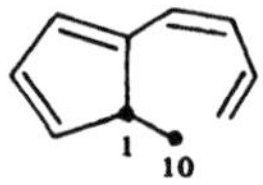	
	12.6130	9.5175	3.0955
IV.	Benzannelation of naphthalene: phenanthrene anthracene		
		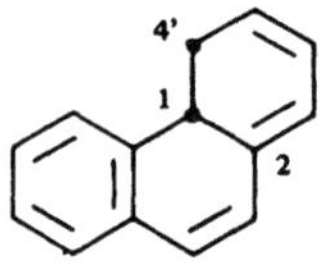	
	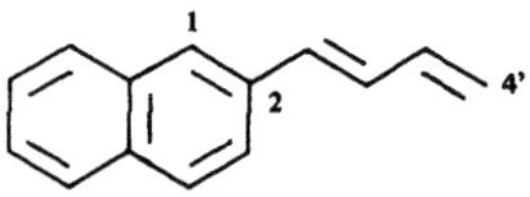	15.4590	3.1618
		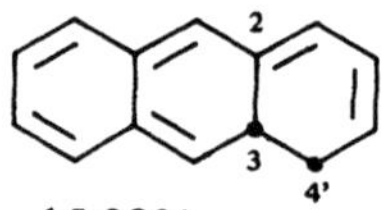	
a)	18.6208	15.0801	3.5407
	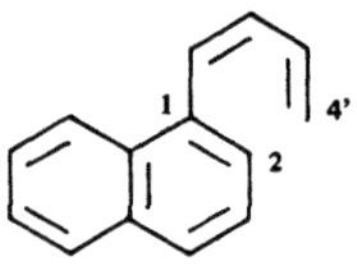	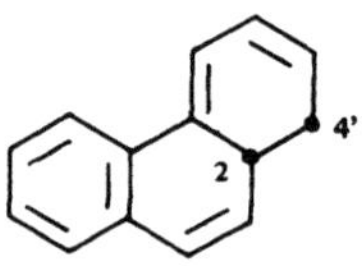	
b)	18.6447	15.3771	3.2675
V.	Benzannelation of azulene:		
	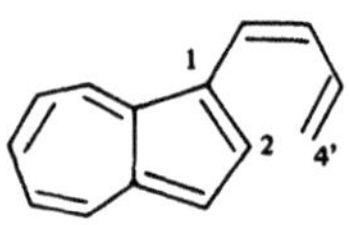	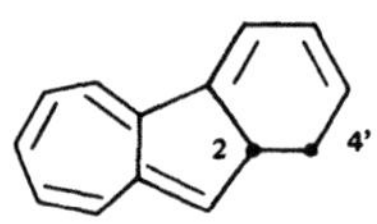	
a)	18.3464	15.0046	3.3418
	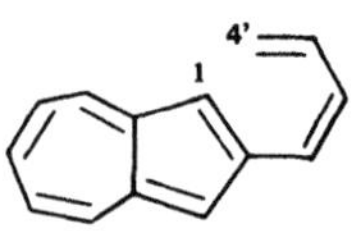	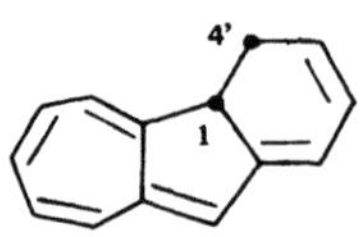	
b)	18.3130	15.1420 Benz(a)azulene	3.1710

Fig. 1 (continued)

M | **$M_{r,s}$** | **$L_{r,s}$**

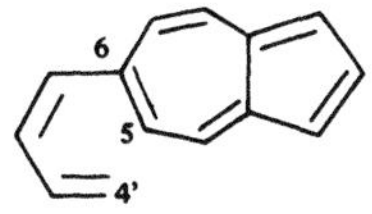

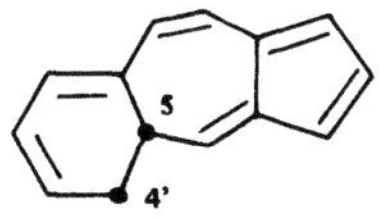

c) 18.3389 | 15.0013 | 3.3376

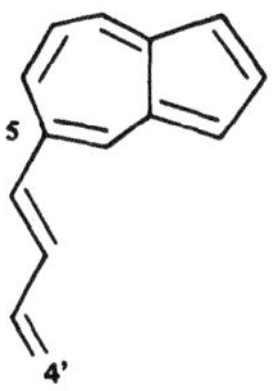

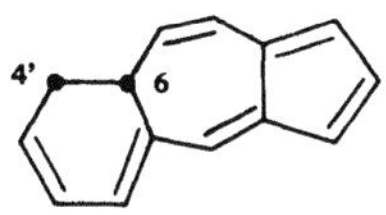

15.0796
Benz(f)azulene | 3.2380

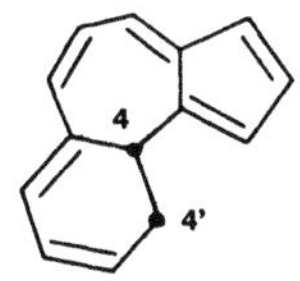

d) 18.3176 | 3.1503

15.1673
Benz(e)azulene

VI. Benzannelation of phenanthrene:

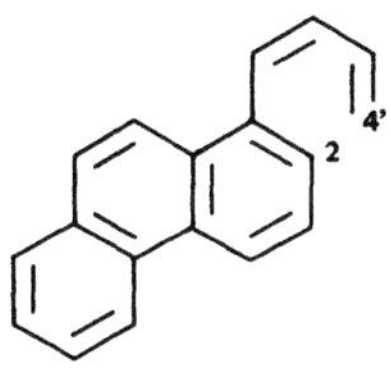

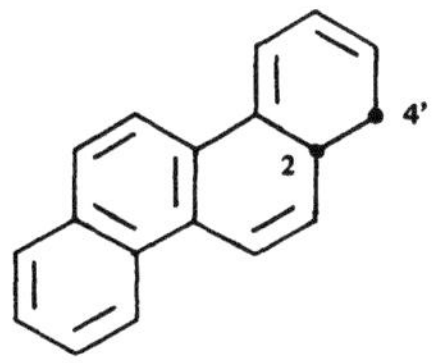

a) 24.4072 | 21.0922 | 3.3150

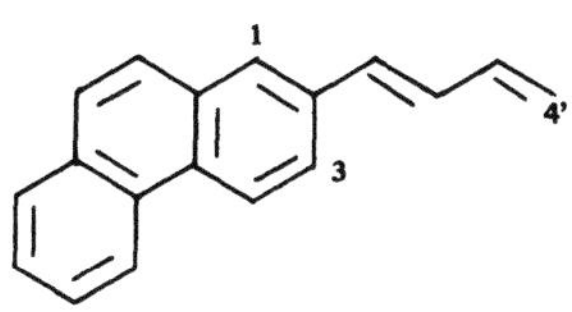

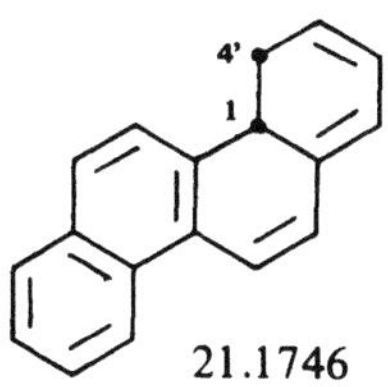

21.1746
Chrysene | 3.2094

b) 24.3840

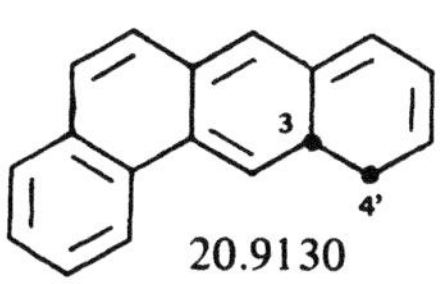

20.9130 | 3.4710

Fig. 1 (continued)

	M	$M_{r,s}$	$L_{r,s}$
c)	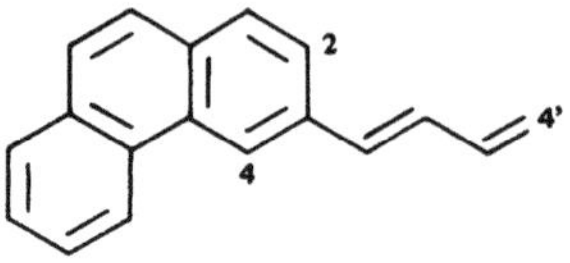 24.3886	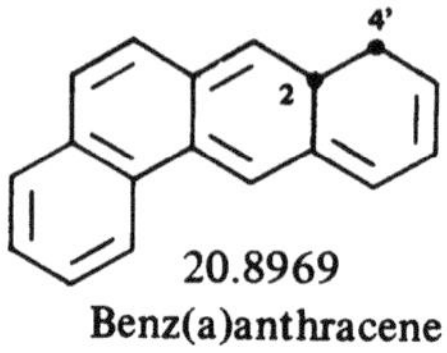 20.8969 Benz(a)anthracene	3.4917
		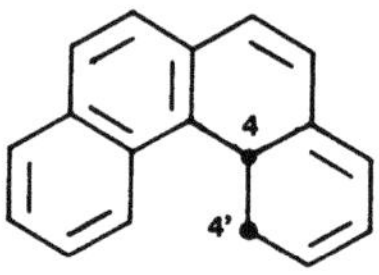 21.1483	3.2403
d)	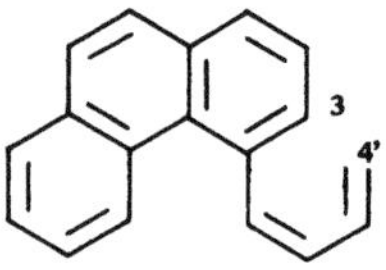 24.4011	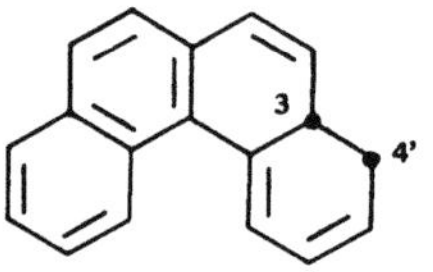 21.0975 Benz(c)phenanthrene	3.3036
e)	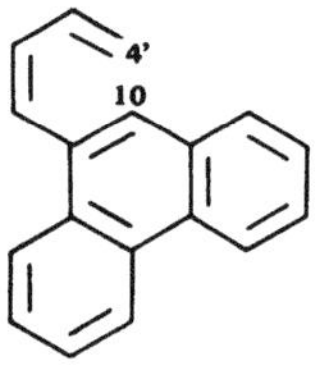 24.4094	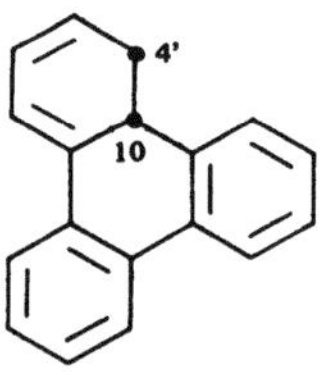 21.3324 Triphenylene	3.0770

VII. Benzannelation of benz(c)phenanthrene:

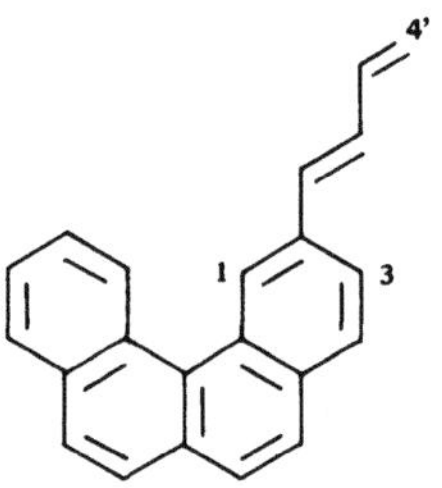 30.1270	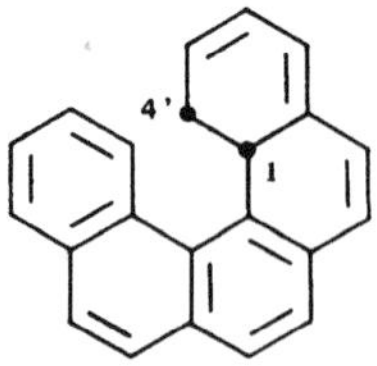 26.9166	3.2104

Fig. 1 (continued)

	M	$M_{r,s}$	$L_{r,s}$
		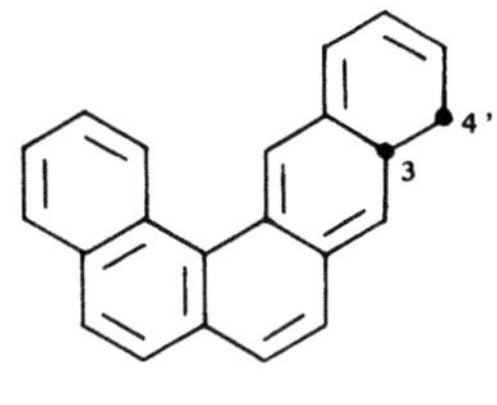	
		26.6244	3.5026
VIII.	Benzannelation of biphenylene:		
	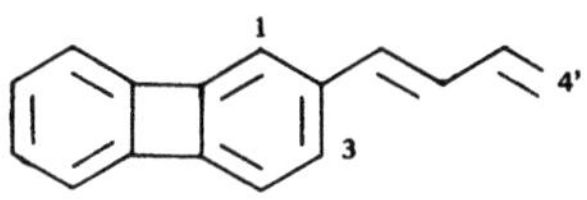	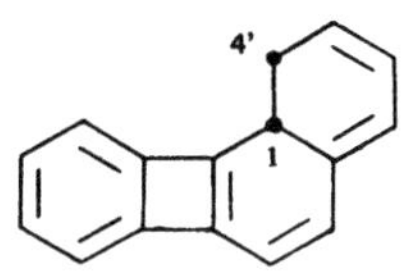	
	21.4565	17.9936	3.4629
		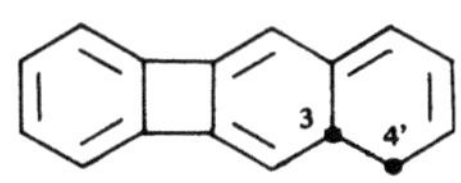	
		18.2042	3.2523
IX.	Pyrene synthesis:		
	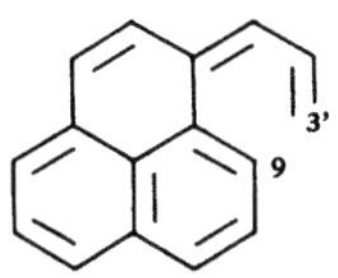	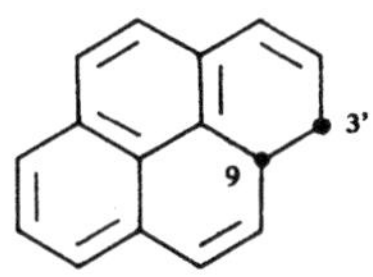	
	21.6595	18.5395	3.1200
X.	Fluoranthene syntheses:		
	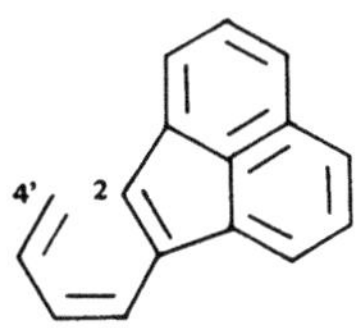	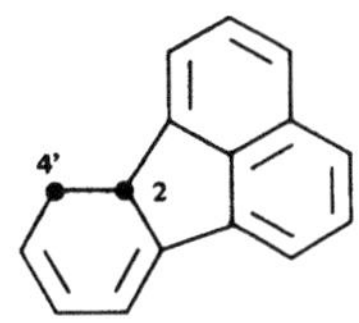	
a)	21.5998	18.6446	2.9552
	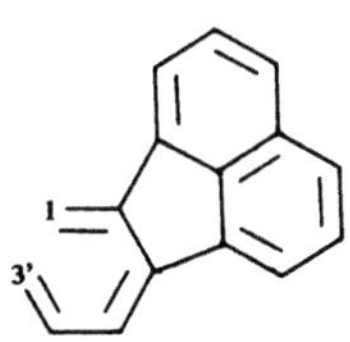	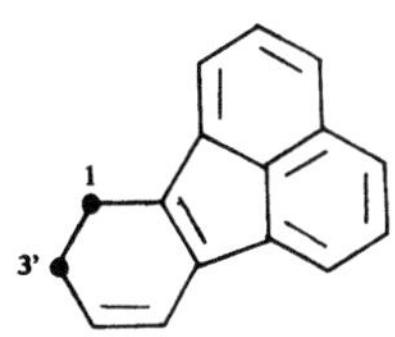	
b)	21.5088	19.0787	2.4301

Fig. 1 (continued)

	M	$M_{r,s}$	$L_{r,s}$
	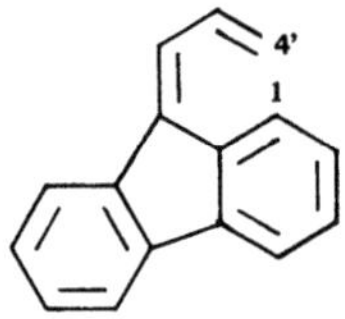	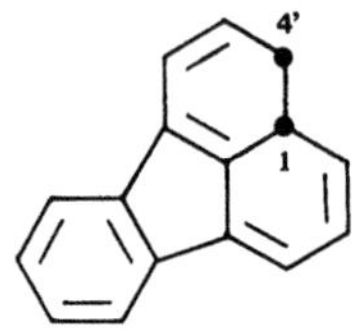	
c)	21.7667	18.3596	3.4071
XI.	Benzfluoranthenes:		
	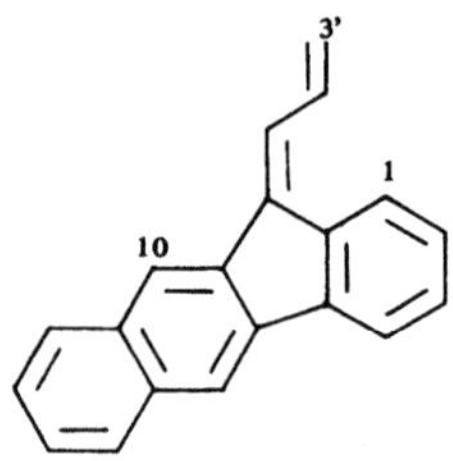	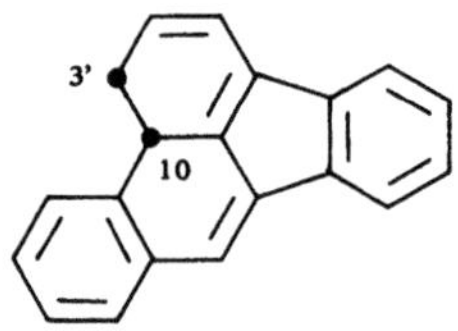	
		24.2923	3.1605
a)	27.4528	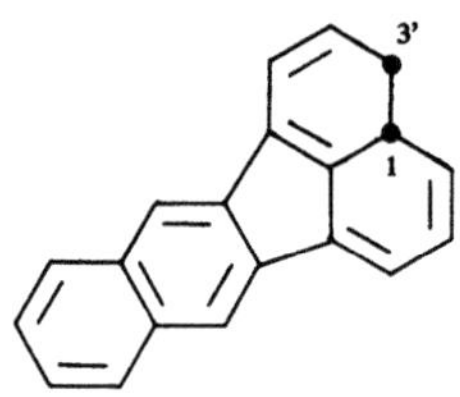	
		24.0443	3.4086
	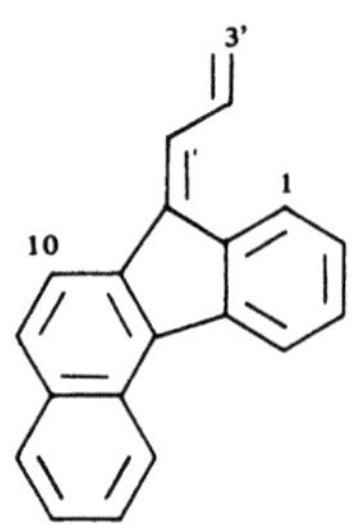	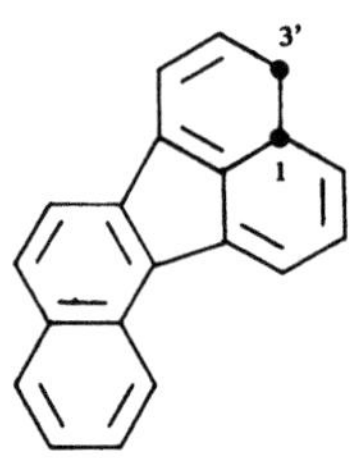	
		24.0771	3.4031
b)	27.4802	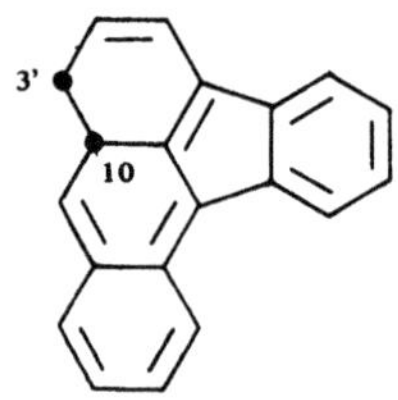	
		23.9803	3.4999

In many of the examples cited with the potentiality of alternate cyclization directions, the chemist will obviously be able to find out the energetically most profitable way simply by writing down the Kékulé structure for the intermediate without any calculation. In a similar manner the easily evaluated Brown's[9] "annelation energy" allows some predictions and shows the general energetic preference of angular over linear katacondensation in polycyclic aromatic ring formations. But more subtle details are lost by the latter method, which also fails for pericondensed systems. For example, the formation of phenanthrene derivatives should succeed more easily starting from the 2-substituted naphthalene derivative than from the 1-substituted one (IVa and b). The considerable difference of activation energies for the cyclization of 1- and 2-substituted azulene derivatives to benz(a)azulene (Va, b), corroborated plainly by experiment, could also only be seen by the localization calculation.

Using a frontier orbital conception the educt should also show some product-developing properties. Indeed, not only do the connecting atoms in all our examples (Fig. 1) possess the same sign in the HOMO as is required for a disrotatory process, but also the magnitude of their coefficients are in a proper correlation with the calculated $L_{r,s}$ values.

III. Synthesis of Benzenoid and Nonbenzenoid Aromatic Ring Systems

A. Reaction Conditions, Materials

If we construct a hexatriene we can perform this in a fictitious manner by means of a substitution reaction in two ways: methyl anion by pentadienyl cation ($C_1 + C_5$ combination) or allyl anion by allyl cation ($C_3 + C_3$ combination). In each case the one partner provides two π-electrons, the other partner four π-electrons to the six-π-electron system hexatriene. Such an imaginary reaction can indeed be realized simply by an aldol- or Knoevenagel-like condensation of a "vinamidinium salt"[10] (*7*) or their vinylogue (*8*) with a corresponding activated (acid) methyl of methylene derivative by means of a base giving cyclizable aminohexatrienes of type (*4*). We can interpret the vinamidinium salts, the 3-dimethylaminoallylidene-N-dimethylammonium perchlorates or tetrafluoroborates (*7*) used, and their vinylogues, the 5-dimethylaminopenta-2,4-dienylidene-N-dimethylammonium salts (*8*), as masked allyl or pentadienyl cations. The carbanions needed are then generated from the C–H acid compounds by added base.

In a typical experiment we solve equimolar amounts of (*7*) or (*8*) and the C–H acid component in a polar, highly boiling solvent under a nitrogen atmosphere, usually quinoline – about 25 ml for a 10 mmol scale preparation – and then add the required equivalents of sodium methoxide in methanol drop by drop, while stirring. The sodium methoxide solution should be prepared by reacting clean sodium with methanol to a 2.5–3.5 *N* titer. Shortly after base addition one observes a coloration, usually yellow or red, which rapidly intensifies, indicating the (*4*) formed. A slow stream of nitrogen is led over the reaction mixture to remove the liberated dimethylamine, the titration of which allows us to watch the course of the reaction. To enhance the rate of the condensation it often becomes necessary to heat slightly to 40–80 °C.

In normal preparations the (*4*) formed is usually not isolated. The end of the condensation step is announced by the cleavage of nearly the theoretical amount of amine. The temperature of the mixture is then raised until a new evolution of dimethylamine sets in. During the thermolysis the color of (*4*) slowly fades and the second equivalent of amine is split off. Finally, the frequently dark reaction solution is worked up in a usual manner, *e.g.*, by distilling off the quinoline in vacuo, or by dissolving the quinoline in dilute mineral acid when it is a small-scale preparation or when the product is too volatile for a separation by distillation. Further purification can be achieved by sublimation, crystalization, and column chromatography on alumina or silica gel.

Consequently most examples performed so far have been outlined in a two-step, one-pot procedure. Several intermediate aminohexatrienes (*4*) are so thermolabile that isolation fails or becomes too wasteful.

For individual preparation and isolation of (*4*) if desired, it is advantageous to use ethanol or pyridine as solvent. The isolation can be recommended if a separation of cyclized product from unconverted starting material becomes difficult or laborious. Because of the high polarity of the colored (*4*), a chromatographic purification and separation from starting material succeeds easily.

In our syntheses the vinamidinium salts (*7*) play an important role as very valuable synthons. A small collection of salts used is given in Table 1. The representatives (4 b–m) render it possible to introduce substitutents in the product at definite positions, assuming they are stable under the reaction conditions applied. As derivatives of malondialdehydes the vinamidinium salts can be prepared from these or related compounds, *e.g.*, 3-chloroacraldehydes, 3-chlorovinylketones[27] and mono- or bisacetals of 1,3-dicarbonyl compounds. But the most effective method, adaptable also to larger-scale preparations consists in the use of the Vilsmeier-Haack-Arnold formylation reagent[25], formed by the action of phosgene or phosphoryl chloride on dimethylformamide. Treatment of acetaldehyde acetals and homologues[12] or the corresponding enolethers[13] with this reagent gives 3-dimethylaminoacraldehydes in high yields, which in turn by alkylation by means of dimethylsulfate or triethyloxonium tetrafluoroborate to intermediate 3-alkoxyallylidene ammonium salts can be converted with dimethylamine to (*7 a–e*)[14, 16]. More efficiently, malonic acids can be formylated directly, in a one-step procedure, to the salts (*7 b–d*)[15, 17, 18]. Likewise, the formylation of chloro- and aryl acetic acids leads to the corresponding salts, *i.e.*, (*7g*) and (*7i*)[15, 19, 20], whereas acetophenone gives 3-chloro-3-phenylallylidene dimethylammonium chloride, whose treatment with dimethylamine finally leads to (*7j*)[15, 21, 26].

The direct formylation of cyanoacetic acid fails, but the acid reacts with dimethylformamide diethylacetal by splitting of carbon dioxide to form 3-dimethylaminoacrylonitrile[23], which is very susceptible to formylation, yielding (*7f*)[17].

For the preparation of salts of type (*7 j–m*), also, the corresponding dimethylaminovinyl ketones can be used, and they are transformed as described for the 3-dimethylaminoacraldehydes. The condensation of dimethylformamide acetals with active methyl or methylene ketones, or Claisen condensation of them with ethylformate, followed by treatment with dimethylammonium chloride[28], presents a rational way to this prestage.

The introduction of a protected amino group succeeds finally with the salt (*7 h*) which could be obtained by formylation of glycocol hydrochloride[22].

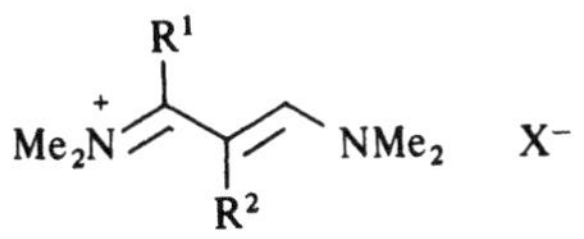

(7)

Table 1. Vinamidinium salts

(7)	R^1	R^2	X	mp. (C°)	Ref.
a	H	H	ClO_4	120	11, 12, 24)
b	H	Me	ClO_4	91–2	12–17)
c	H	Et	ClO_4	96–8	12–15, 17, 18)
d	H	n-Bu	ClO_4	112	12, 17, 18)
e	H	OMe	ClO_4	90–1	15, 22)
f	H	CN	ClO_4	142–3	17)
g	H	Cl	ClO_4	123–4	15, 20)
h	H	$N{=}CH{-}NMe_2$	ClO_4	139	22)
i	H	Ph	ClO_4	195–7	15, 19)
j	Ph	H	ClO_4	138–9	15, 21, 26)
k	Ph	Ph	BF_4	140–2	
l	Me	H	ClO_4	154–5	15)
m	Et	H	ClO_4	143	15)
n	H	NO_2	ClO_4	154–5 (dec.)	15, 17)

Me = CH_3 Et = C_2H_5 Ph = C_6H_5.
Bu = C_4H_9.

The salt (*7 n*) was obtained directly by nitration of (*7 a*) in a 74% yield[15, 17]. This shows the meneidic character of the resonance-stabilized vinamidinium system[10]. For the preparation of the vinylogues of vinamidinium salts, the pentamethinium salts (*8*), there also exist numerous methods, but compared to (*7*) none possesses such a variability for introducing different substituents and is so easy to perform. Table 2 gives a small selection of salts (*8*) used in the following syntheses. The ring-opening reaction of certain pyridinium salts by sec. amines, a long-known reaction[29] represents an approach to (*8*). Only a limited number of (*8*), however, are accessible, with fair yields, in such a way[30]. On the other hand, methyl and methylene ketones condense readily with salts (*7*) in the presence of base to give 5-dimethylaminopentadienones, which can be transformed by one of the methods mentioned to salts (*8*)[6].

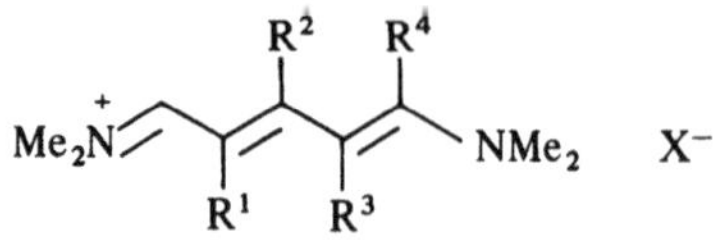

(8)

Table 2. Vinylogues of vinamidinium salts ("pentamethinium salts")

(8)	R^1	R^2	R^3	R^4	X	mp. (C°)	Ref.
a	H	H	H	H	ClO_4	164–5	11, 24, 31)
b	H	Ph	H	H	ClO_4	169	18)
c	H	NMe_2	H	H	ClO_4	184–5	14)
d	H	Me	H	H	ClO_4	205	30)
e	Me	H	H	H	ClO_4	179–81	
f	Ph	H	H	Ph	BF_4	168–71	6)

B. Benzene Nucleus Forming Cyclizations

1. Benzene Ring Formation by C_1 with C_5 Combinations

Arylmethylketones and related hetarylmethylketones (*9*) (the C_1 component) condense rapidly with the salts (*8*) (the C_5 component) to give a high yield of thermo-

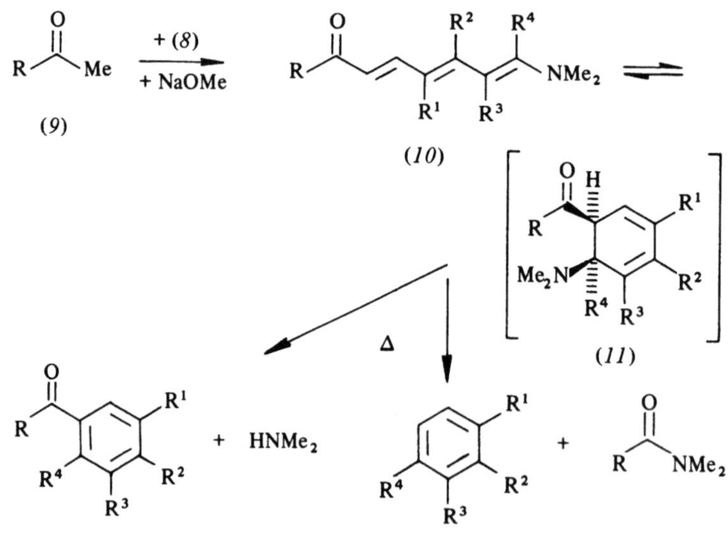

labile aminoheptatrienones (*10*). Some of these slowly begin to decompose, cyclizing already at temperatures below 50°. This condensation and cyclization offers a reaction sequence for converting generally an acetyl group to a benzoyl or substituted benzoyl one[6, 32].

Even double and threefold condensation followed by cyclization can be performed successfully (Table 3, Nos. 10–13). The mixture of arylmethylketone and diarylketone in No. 11 is caused by simultaneously formed mono- and bis-condensation product of type (*10*) of acetone.

The formation of some p-terphenyl (*13*) besides the expected 3-benzoyl-p-terphenyl (*12*) in No. 9 from the highly overcrowed intermediate (*11*) may find an explanation in the *cis*-elimination of dimethylbenzamide.

Earlier it was observed that the treatment of pyrylium salts (*14*) with enamines of cycloalkanones (*15*) in boiling acetonitrile results in the formation of substituted cycloalkenobenzophenones (*17*) if the pyrrolidino residue in (*15*) is used[33]. However, with enamines of morpholine preferentially the cycloalkenobenzenes (*18*) could be isolated. The heptatrienones (*16*), although not isolated, were supposed to be intermediates consistently with our experience, and a disrotatory ring-closure with *cis*-elimination of the corresponding benzamide was proposed for the formation of (*18*)[33] (Table 4).

In the same manner as acetophenones some other methyl compounds can also be transformed with salts (*8*) to the corresponding phenyl compounds. 2-Methylchromone (*19*), and likewise 2,6-dimethyl-4-pyrone by double conversion, lead to flavone (*20*) and 2,6-diphenyl-4-pyrone respectively[36]. Even the very weak acid methyl groups in methylphenylsulphone and in 6-methylazulene (*21*) may be smoothly transformed. A 100% excess of (*8*) and added sodium methoxide in this

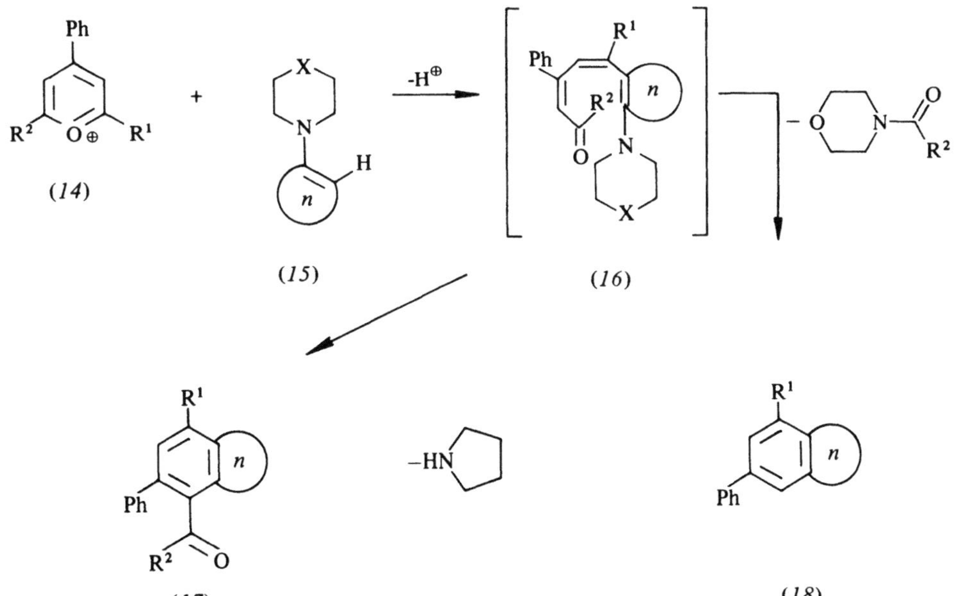

(*14*) (*15*) (*16*) (*17*) (*18*)

Table 3. Conversion of C-acetyl to C-benzoyl

					Reaction conditions			
No.	Product (*12*)	Yield	Ketone (*9*)	(*8*)	(°C)	Time	Heptatrienone (*10*)	Ref.
1	Pivalophenone DNPH mp. 193°	78	2.2-Dimethyl-3-butanone (pinacone)	*a*	150	5 h	Ochre-yellow leaflets (69%) mp. 70–1°	6, 32)
2	Benzophenone mp. 48°	92	Acetophenone	*a*	130	3 h	Brick-red needles (85%) mp. 99–101°	6, 32)
3	4-Methylbenzophenone mp. 57–8°	93	4-Methylacetophenone	*a*	130	3 h	Brown-red needles (90%) mp. 141°	6, 32)
4	4-Methoxybenzophenone mp. 61°	93	4-Methoxyacetophenone	*a*	130	3 h	Red needles (95%) mp. 123°	6, 32)
5	4-Bromobenzophenone mp. 80–1°	93	4-Bromoacetophenone	*a*	130	3 h	Brown-golden needles (85%) mp. 126°	6, 32)
6	4-Benzoylbiphenyl mp. 101–2°	95	Acetophenone	*b*	80–110	1 h	Not isolated	34)
7	4-Dimethylaminobenzo-phenone mp. 90°	70	Acetophenone	*c*	140	12 h	Not isolated	35)
8	3-Methylbenzophenone DPNH mp. 220°	61	Acetophenone	*e*	115	24 h	Not isolated	35)
9	2-Benzoyl-p-terphenyl mp. 123–4° (*12*)	62	Acetophenone	*f*	170–210	24 h	Not isolated	32)
	p-terphenyl (*13*) mp. 208°	20						
10	Benzophenone mp. 48° DPNH mp. 238°	50	Acetone	*a*	140	4 h	Dark-red powder (74%) mp. 280° (dec.)	35)

Table 3. (continued)

No.	Product (*12*)	Yield	Ketone (*9*)	(*8*)	Reaction conditions (°C)	Time	Heptatrienone (*10*)	Ref.
11	4-Acetylbiphenyl mp. 120°	26	Acetone	*b*	140	7 h	Mixture, not isolated	35)
	Di-4-biphenyl ketone mp. 232°	31						
12	4.4′-Bisdimethylamino-benzophenone mp. 174°	41	Acetone	*c*	140–160	12 h	Not isolated	35)
13	1.3.5-Tribenzoylbenzene mp. 120–1°	37	1.3.5-Triacetylbenzene	*a*	115	24 h	Not isolated	6, 32)
14	2-Benzoylthiophene mp. 56°	89	2-Acetylthiophene	*a*	130	3 h	Pink-violet needles (76%) mp. 127° (dec.)	6, 32)
15	4′-Biphenylyl-4-pyridyl-ketone mp. 153–4°	95	4-Acetylpyridine	*b*	80–110	1 h	Not isolated	34)
16	Benzoylferrocene mp. 107–8°	89	Acetylferrocene	*a*	130–170	2 h	Not isolated	34)

Table 4. Benzene derivatives from pyrylium salts and enamines[33)]

Product (*17*)/(*18*)	Yield %	$R^1 = R^2$	n	X
4-Benzoyl-5,7-diphenylindane mp. 127–8°	64.5	Ph	5	–
4-Benzoyl-7-methyl-5-phenylindane mp. 138.5–140.5°	17	Ph	5	–
5-Benzoyl-6,8-diphenyltetraline mp. 137–8°	56	Ph	6	–
2,3-Hexamethylene-4,6-diphenyl-benzophenone mp. 126–7°	63.2	Ph	8	–
2,3-Decamethylene-4,6-diphenyl-benzophenone mp. 186–7°	19.2	Ph	12	–
4-Benzoyl-5,7-diphenylindane mp. 127–8°	26	Ph	5	–O–
4,6-Diphenylindane mp. 101–2°	7.7			
5,7-Diphenyltetraline mp. 88–9°	49.2	Ph	6	–O–
7-(4-Methoxyphenyl)-5-phenyl-tetraline mp. 101–2°	63.5	4-$MeOC_6H_4$	6	–O–
1,2-Decamethylene-3,5-diphenyl-benzene mp. 84.5–85.5°	63.5	Ph	12	–O–

instance, and longer condensation time as well, enhance the yield considerable. The anion of the strongly acidic nitromethane, on the other hand, shows only weakly nucleophilic properties and forms an sodium salt scarcely soluble in organic solvents. The reaction with nitromethane[34)] is therefore performed in boiling pyridine by addition of triethylamine. Added sodium methoxide leads to the formation of some corresponding azoxy compound as by-product.

Table 5. Conversion of methyl to phenyl

				Reaction conditions	
Product	Yield	Starting compound	(*8*)	(°C)	Time
Diphenylsulphone mp. 126–7°	35	Methylphenyl-Sulphone	*a*	200	5 h
Nitrobenzene	73	Nitromethane	*a*	115	2 h
4-Nitrobiphenyl mp. 111–3°	66	Nitromethane	*a*	115	2 h
4,4'-Diphenylazoxy-benzene mp. 209–11°	15				

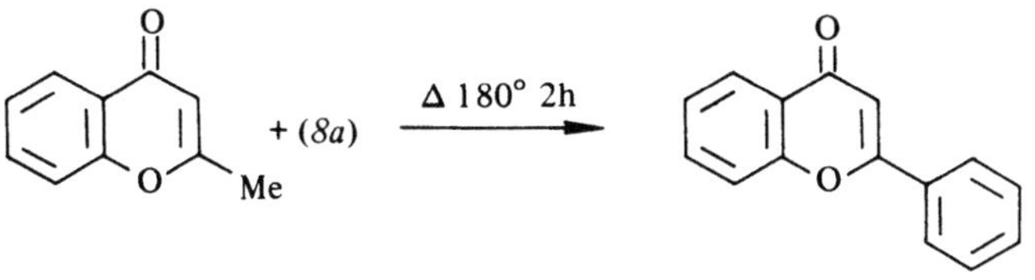

mp. 99° (89%)

(*19*) (*20*)

In analogy:

2,6-dimethyl-4-pyrone + (*8 a*) $\xrightarrow{\Delta\ 180\ ^\circ C\ 3\ h}$ 2,6-diphenyl-4-pyrone mp. 135° (71%)

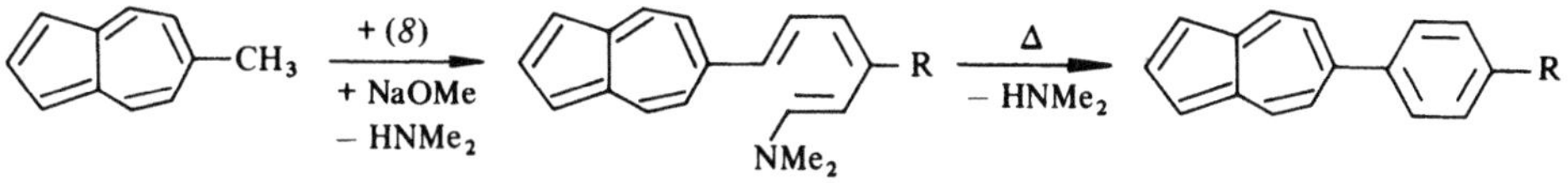

(*21*) (*22*)

6-phenylazulene from (*21*) + (*8 a*) 71%, 200° 3 h mp. 158–9° blue leaflets

6-(4-biphenylyl)azulene from (*21*) + (8 *b*) 57%, 150° 8 h mp. 261° blue needles

6-(4-dimethylaminophenyl)azulene mp. 234–5° dark-green needles from (*21*) + (*8 c*) 54%, 160° 16 h

2. Benzene Ring Formation by C_3 with C_3 Combinations

The second but synthetically much more important approach to benzene ring forming methods, the $C_3 + C_3$ combination, uses as electrophiles the salt (*7*) and stabilized allyl anions generated from diethylglutaconate or glutacononitrile (*23*) and from substituted allyl cyanides (crotonitriles) (*24*) (Table 6) during the reaction. Starting with the corresponding ketones, the nitriles (*24 b–i*) – sometimes mixtures of their double-bond isomers – are simply prepared by Knoevenagel condensation with cyanoacetic acid, followed by decarboxylation[38].

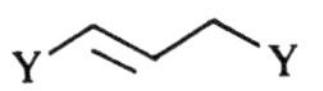

(*23*)

(*23*) a : Y = COOEt diethylglutaconate
b : Y = CN glutacononitrile[37]

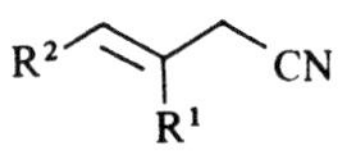

(*24*)

Table 6. Substituted allylcyanides

(*24*)	R^1	R^2		Ref.
a	H	Ph	Styrylacetonitrile	
b	Ph	H	3-Phenylcrotonitrile (80%)	38)
c	$-(CH_2)_3-$		Cyclopentenylacetonitrile (80%)	38)
d	$-(CH_2)_4-$		Cyclohexenylacetonitrile (80%)	38)
e	$-(CH_2)_5-$		Cycloheptenylacetonitrile (81%) bp. 107°/12 torr.	
f	$-(CH_2)_{10}-$		Cyclododecenylacetonitrile (42%) bp. 115–118°/0.1 torr.	
g	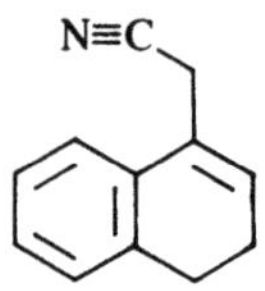		3,4-Dihydronaphthalene-1- (70%) acetonitrile bp. 103–4°/0.1 torr.	
h	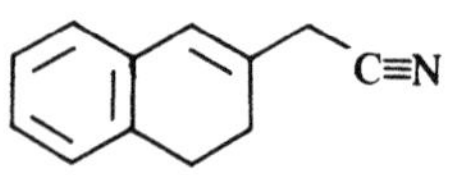		3,4-Dihydronaphthalene-2- (68%) acetonitrile bp. 100°/0.1 torr.	
i	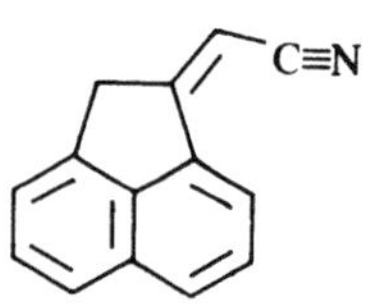		1-Cyanomethyleneacenaphthene (63%) mp. 128–9°, light-yellow needles	

The catalogue of vinamidinium salts (*7*) used also can be extended to mono- and bicyclic derivatives (*7 o–s*), which we prepare in the manner mentioned (Section III.A.) from corresponding ketones as outlined here for the acenaphthylene compound (*7 s*)[16].

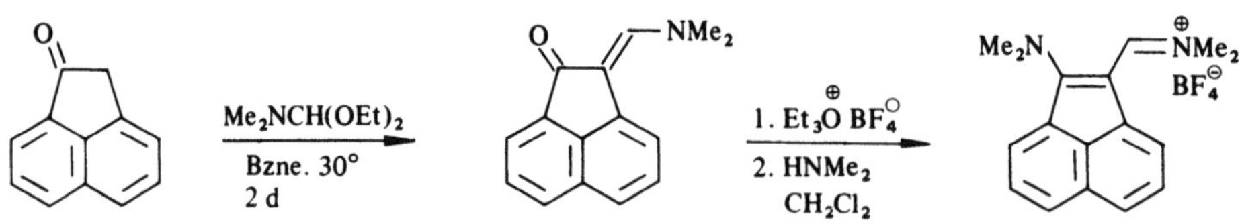

(*25*) (*26*) (*7s*)

81% mp. 98°
Yellow needles

67% mp. 190°
Red-orange needles

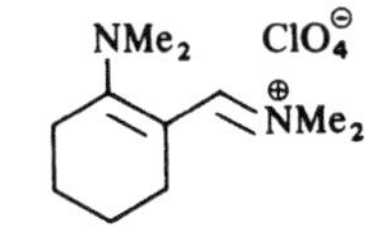

(*7o*)[16)]
mp. 88–89.5° (72%)
Colourless neddles

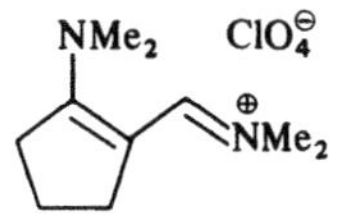

(*7p*)[35)]
mp. 76° (60%)
Colourless needles

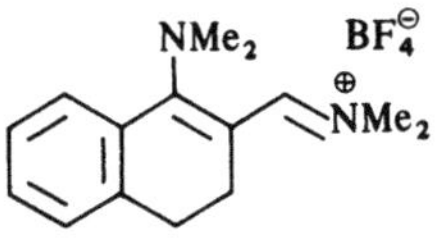

(*7q*)[16)]
mp. 87–90° (79%)
Light-yellow needles

mp. 104–5° (65%)
Yellow needles

(*7r*)[36)]

A surprising range of very different combinations of (*7*) and (*24*) is therefore possible, according to the following reaction scheme:

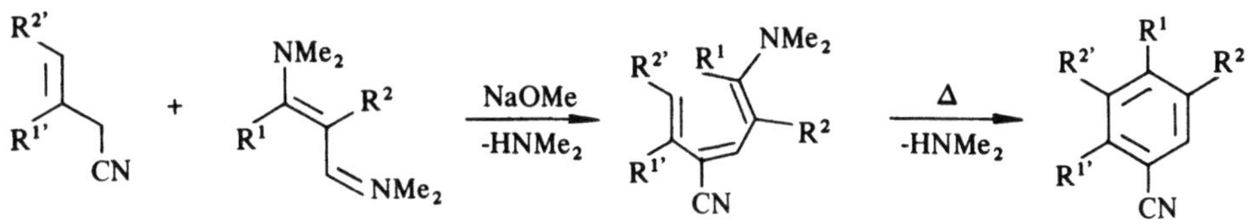

(*23, 24*) (*7*) (*27*)

But this has an essential restriction: some of the used vinamidinium salts (*7*) are sensitive and may be decomposed by treatment with strong base, *e.g.*, sodium methoxide. Especially such salts possessing methyl or methylene groups at odd-numbered, electron-dificient positions of the vinamidinium system tend toward complex self-condensations: a proton abstraction to a diene amine may occur, followed by attack of the unaltered ion:

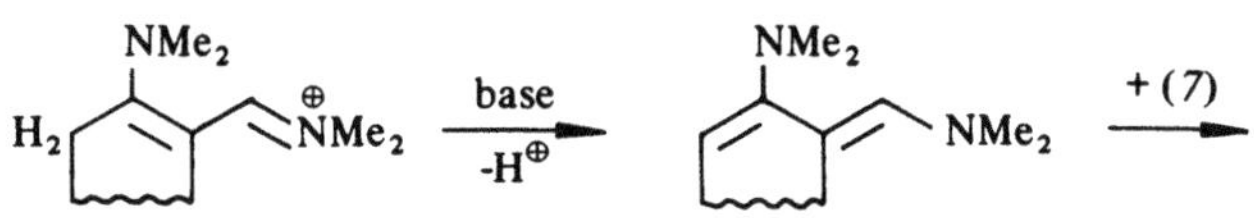

Such salts as (*7 l–m*), (*7 o–p*), and (*7 r*) react successfully only with components of higher acidity that form resonance-stabilized, nucleophilic anions of weak basicity, *e.g.*, (*23 a, b*), (*24 a*), (*24 h*), and (*24 i*).

The nitriles (*24 a–h*), no matter what double-bond isomer is used, are attacked exclusively at carbon atom α of the nitrile group, with the sole exception of (*24 i*). On the other hand, the vinamidinium salts (*7*) always react with their least-substituted iminium carbon position.

All listed compounds gave correct analyses, when tested by ^{1}H NMR spectra. Some nitriles were saponified by potassium hydroxide in diglycol or methylcellosolve

Table 7.

No.	Compound	Yield	Components	Reaction conditions (°C)	Time
	Benzenes				
1	Diethyl-5-nitro-isophthalate mp. 82°	82	(*23a*) + (*7n*)	100	2 h
2	Dimethyl-5-chloroisophthalate[1]) mp. 72–3°; acid: mp. 279–281°	78	(*23a*) + (*7g*)	150	2 h
3	Trimesic acid trinitrile mp. 261–3° (subl.)	85	(*23b*) + (*7f*)	130	8 h
4	5-Phenylisophthalonitrile mp. 122–4°	93 44	(*23b*) + (*7i*) (*24a*) + (*7f*)	70 170	4 h 3 h
5	2,5-Diphenylbenzonitrile mp. 106–7°	62	(*24b*) + (*7i*)	200	3 h
6	3,4-Diphenylbenzonitrile mp. 111–112.5°	55	(*24a*) + (*7j*)	220	40 h
7	3,5-Diphenylbenzonitrile mp. 109–110°	98	(*24a*) + (*7i*)	170	3 h
8	3,4,5-Triphenylbenzonitrile mp. 156–7°	31	(*24a*) + (*7k*)	220	8 h
	Indanes				
9	4,6-Indanedicarbonitrile mp. 134–5°	77	(*23b*) + (*7p*)	170	6 h
10	6-Methoxy-4-indanecarbonitrile oil, acid: mp. 160–2°	75	(*24c*) + (*7e*)	180	6 h
11	6-Phenyl-4-indanecarbonitrile oil, acid: mp. 193–6°	78	(*24c*) + (*7i*)	180	6 h
	Tetralines				
12	7-Methyl-5-tetralincarbonitrile bp. 88–9°/0.1 torr.; amide: mp. 216°, acid: mp. 167–8°	71	(*24d*) + (*7b*)	180	6 h
13	7-Methoxy-5-tetralincarbonitrile oil; acid: mp. 163–5°	74	(*24d*) + (*7e*)	180	6 h
14	7-Phenyl-5-tetralincarbonitrile mp. 113–4°	76	(*24d*) + (*7i*)	180	6 h
	Benzocycloheptene				
15	3-Methyl-6,7,8,9-tetrahydro-5H-benzocycloheptene-1-carbonitrile bp. 95–102°/0.1 torr.; mp. 69° amide: mp. 192–200°	62	(*24e*) + (*7b*)	180	6 h
	Benzocyclododecene				
16	3-methyl-5,6,7,8,9,11,12,13,14-decahydrobenzocyclodecene-1-carbonitrile mp. 92–3°	89	(*24f*) + (*7b*)	170	2 h

Table 7. (continued)

No.	Compound	Yield	Components	Reaction conditions (°C)	Time
	... *9,10-Dihydrophenanthrenes*				
17	... 2,4-Diethyldicarboxylate oil, acid: mp. 275°	96	(*23a*) + (*7q*)	130	4 h
18	... 2,4-Dicarbonitrile mp. 144–5°	92	(*23b*) + (*7q*)	70	2 h
19	... 1,3-Dicarbonitrile mp. 151–2°	87 88	(*23b*) + (*7r*) (*24h*) + (*7f*)	140 140	4 h 1.5 h
20	... 4-Carbonitrile mp. 97°	60	(*24g*) + (*7a*)	180	7.5 h
21	... 2-Methyl-4-carbonitrile mp. 85–6°	65	(*24g*) + (*7b*)	160	6 h
22	... 2-Methoxy-4-carbonitrile mp. 87°	77	(*24g*) + (*7e*)	180	0.5 h
23	... 2-Phenyl-4-carbonitrile mp. 145°	79	(*24g*) + (*7i*)	160	3.5 h
24	... 1-Carbonitrile mp. 81°	83	(*24h*) + (*7a*)	175	3 h
25	... 3-Methyl-1-carbonitrile mp. 83°	91	(*24h*) + (*7b*)	160	2 h
26	... 3-Methoxy-1-carbonitrile mp. 116°	93	(*24h*) + (*7e*)	160	1.3 h
27	... 3-Phenyl-1-carbonitrile mp. 114°	90	(*24h*) + (*7i*)	160	1.3 h
	... *Fluoranthenes*				
28	... 7,9-Diethyldicarboxylate mp. 112–113° yellow needles	93	(*23a*) + (*7s*)	90	3 h
29	... 7,9-Dicarbonitrile mp. 234–5°, pale yellow needles	92	(*23b*) + (*7s*)	160	6 h
30	... 7-Carbonitrile mp. 167–7.5°, light-yellow needles	90	(*24i*) + (*7a*)	150	18 h
31	... 9-Chloro-7-carbonitrile mp. 177–8°, yellow needles	58	(*24i*) + (*7g*)	180	8 h
32	... 9-Methyl-7-carbonitrile mp. 153–4°, light-yellow needles	87	(*24i*) + (*7b*)	120	4 h
33	... 9-Methoxy-7-carbonitrile mp. 174–5°, orange-yellow needles	90	(*24i*) + (*7e*)	90	2 h
34	... 9-Phenyl-7-carbonitrile mp. 157–8°, yellow needles	78	(*24i*) + (*7i*)	100	5 h

[1]) A replacement of the ethyl ester group by methyl takes place, owing to methoxide.

(160°, 18–20 h) and the corresponding carboxylic acids decarboxylized in boiling quinoline, in the presence of copper powder, to known hydrocarbons.

Although Table 7 is by no means exhaustive, it gives a good impression of the versatility of the method. 9,10-Dihydrophenanthrenes, for instance, may be formed using the bicyclic allylcyanides, (*24g*) or (*24h*) and open-chain vinamidinium salts (*7a–n*) or open-chain allyl derivatives, *e.g.*, (*23a*) or (*23b*), and the bicyclic vinamidinium salts (*7q*) or (*7r*). Thus, 9,10-dihydrophenanthrene-1,3-dicarbonitrile (Table 7, No. 19) could be synthesized in two ways:

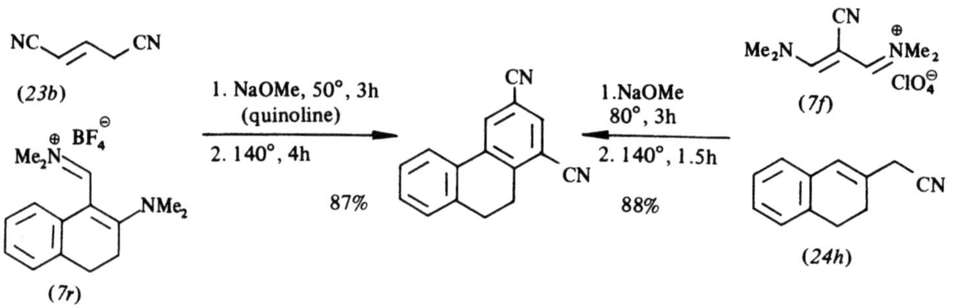

More complex hydrocarbons are simply to approach by combination of two cyclic components, *e.g.*, benz(c)phenanthrene (*29*), dibenzo(c,g)phenanthrene (pentahelicene) (*31*), and benz(c)chrysene (*33*):

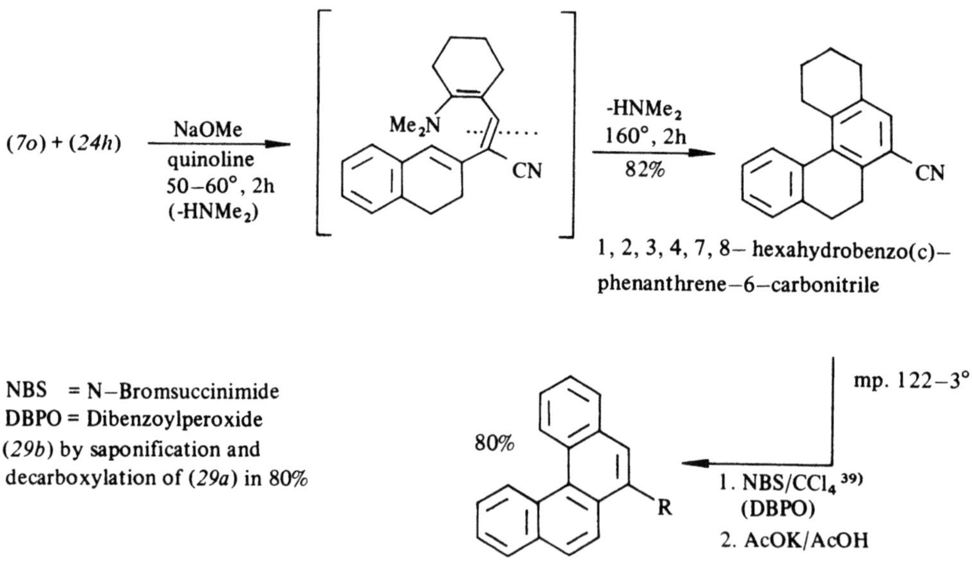

(*29*) a: R = CN mp. 124–6°
b: R = H mp. 67–8° [40]

Benzo(c)phenanthrene

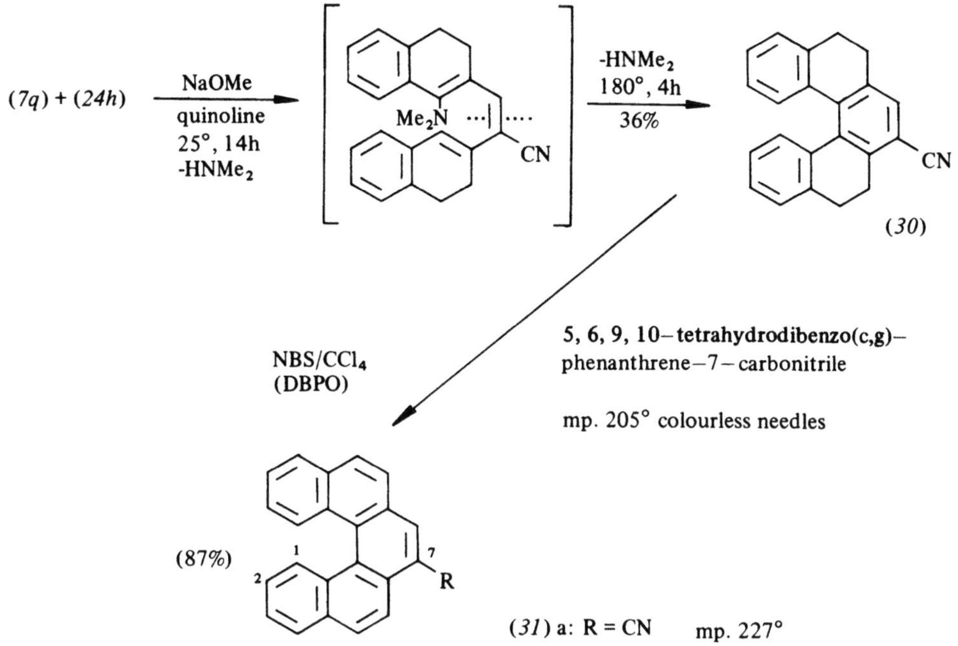

5, 6, 9, 10–tetrahydrodibenzo(c,g)–phenanthrene–7–carbonitrile

mp. 205° colourless needles

Dibenzo(c, g)phenanthrene

(*31*) a: R = CN mp. 227°

b: R = H mp. 175–7° [41]

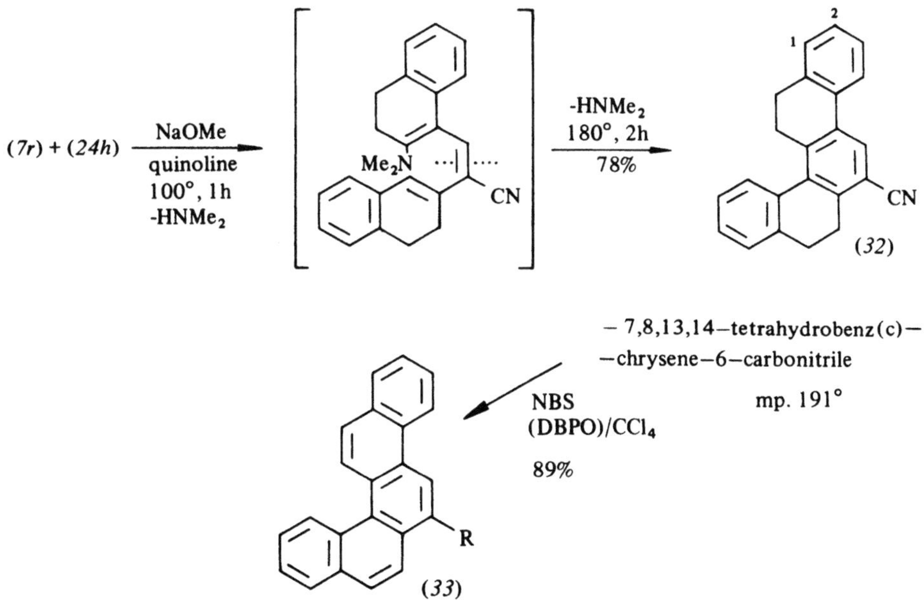

– 7,8,13,14–tetrahydrobenz(c)––chrysene–6–carbonitrile

mp. 191°

a: R = CN mp. 183°

b: R = H mp. 126–7° [41, 42]

Benzo(c)chrysene

The intermediate hexatrienes – formulas in brackets – in all three examples can cyclize only from the less favorable Z-form, relative to the dimethylamino group. Nevertheless, the most sterically overcrowded hexatriene forms the tetrahydropentahelicene (*30*) once in a 36% yield.

Appropriate nitro-allyl derivatives for condensation to hexatrienes are usually not easily available, with the exception of nitromethyl-1-cyclohexene and nitromethyl-1-cycloheptene (*34*)[43]. Like nitromethane, (*34*) does not condense with vinamidinium salts (*7*) and sodium methoxide, but yields insoluble sodium salts.

A successful condensation in a slightly exothermic reaction can be achieved with (*34*) and 3,3-diethoxy-1-dimethylaminopropenes (*35 a–c*)[6, 44] to the yellow nitrohexatrienes (*36*) which cyclize smoothly to form cycloalkenonitrobenzenes (*37*) on heating in quinoline:

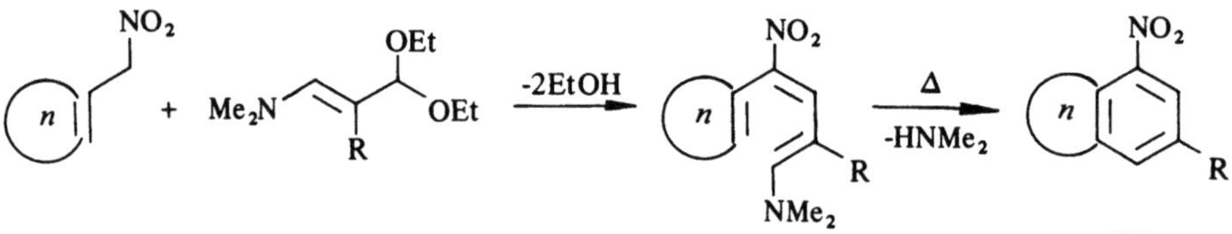

(*34*) (*35*) (*36*) (*37*)

a: R = H bp. 96°/10 torr.[6]
b: R = Me bp. 48°/0.1 torr.[44]
c: R = Ph bp. 105–110°/0.1 torr[44]

Table 8. Cycloalkenonitrobenzenes

(*37*)	Yield	R	n	(*36*)	Reaction conditions (°C)	Time
5-Nitrotetraline bp. 85°/0.1 torr. mp. 33–4° [45]	72	H	6	Yellow plates (76%), mp. 123° (dec.)	150	0.8 h
7-Methyl-5-nitrotetraline bp. 93–5°/0.1 torr.	73	Me	6	Not isolated	100	5 h
5-Nitro-7-phenyltetraline mp. 108–9°	71	Ph	6	Not isolated	80	3.5 h
1-Nitro-5-benzo-6,7,8,9-tetrahydrocycloheptene bp. 93–5°/0.1 torr. mp. 48°	62	H	7	Yellow oil (73%)	140	3 h

1,2,3,4-Tetrahydroisoquinolines (*39*) are also accessible in moderate yields using the aza-analogues (*38 a, b*) of cyclohexenylacetonitrile (*24 d*). A mixture of the double-bond isomers (*38 a, b*) could simply be obtained in 74% yield by Knoevenagel condensation of N-methyl- or N-ethoxycarbonyl-4-piperidinone with cyanoacetic acid.

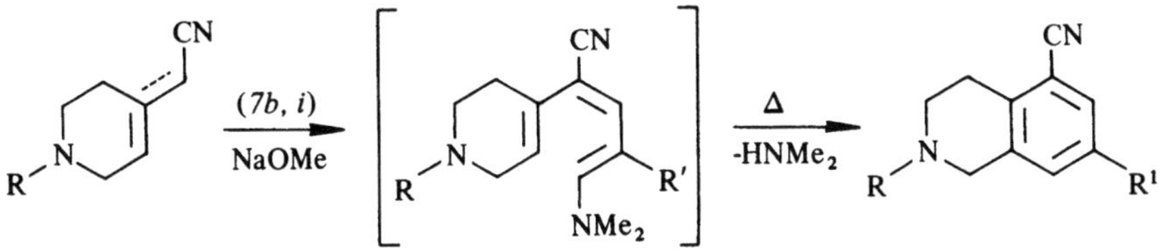

(38) (39)

a: R = Me, bp. 116–117°/10 torr.
b: R = CO_2Et, bp. 108–110°/0.05 torr.

Table 9. 1,2,3,4-Tetrahydroisoquinoline-5-carbonitriles (*39*)

R	R′	yield %	mp.	Reaction conditions (°C)	Time
Me	Me	58	73–74°	180	5 h
CO_2Et	Me	29	102°	150	5 h
CO_2Et	Ph	70.4	127	160	4 h

3,5,5-Trimethylcyclohex-2-enone(isophorone) readily condenses with salts (*7*) in the presence of sodium methoxide to orange-red heptatrienones (*40*)[6], which begin at 90° to cyclize to 3,3-dimethyl-1-tetralones (*41*), losing dimethylamine:

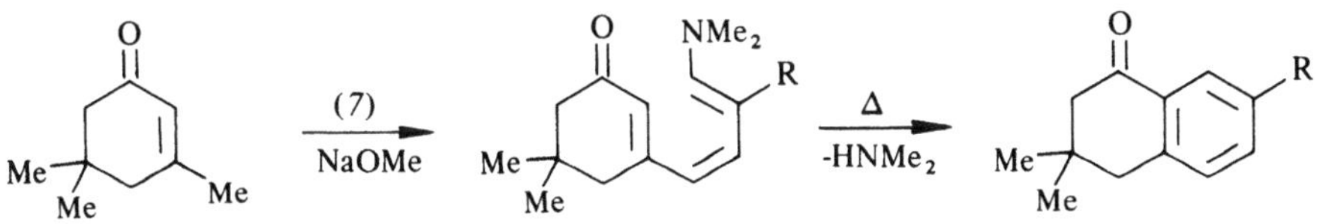

(40) (41)

a) R = OMe
b) R = N=CH–NMe_2

(*41 a*) R = OMe: 7-methoxy-3,3-dimethyl-1-tetralone (84%), oil, bp. 107–108°/0.3 torr. (110°, 4 h)

(*41 b*) R = N=CH–NMe_2: 7-dimethylaminomethyleneamino-3,3-dimethyl-1-tetralone (82%)[32] mp. 99° (subl.) (140°, 2 h), saponification with ethanolic potassium hydroxide gives 7-amino-3,3-dimethyl-1-tetralone (75%) mp. 156–157°

Likewise the reactive methyl groups in 2-methylchromone (*19*) and 2,6-dimethyl-4-pyrone are also susceptible to condensation with vinamidinium salts (*7*), as are their vinylogues (compare Section B. 1.). The colored condensation products (*42*) smoothly cyclize to xanthones (*43*):

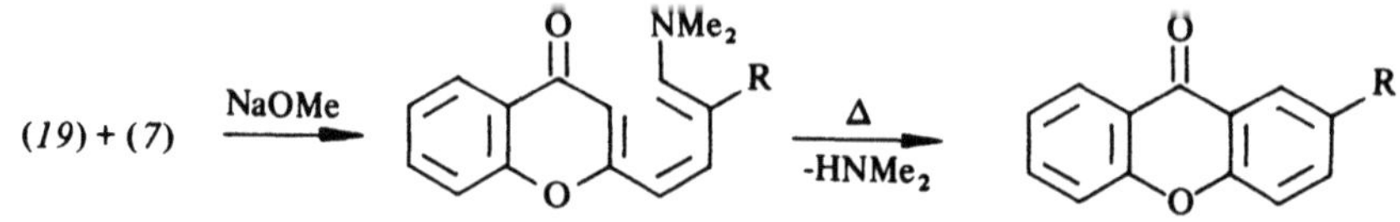

(42) (43)

R = OMe, Ph

Table 10. Xanthones (*43*)

2-methoxyxanthone (95%) mp. 129–130°	(*19*) + (*7e*)		170°, 6 h
2-phenylxanthone (97%) mp. 157–158°	(*19*) + (*7i*)		170°, 6 h
2,7-dimethylxanthone (80%) mp. 141–142°	2,6-dimethyl-4-pyrone + (*7b*) (double condensation)		190°, 8 h

3. Azulene Ring Formation by C_3 with C_7 Combinations

The reaction under consideration, namely the Ziegler-Hafner azulene synthesis, regarded as the prototype of electrocyclic ring-closure with elimination, consists in a (π 10 *s*)-cyclization of the fulvenoid decapentaene (*1*), which was formed by condensation of sodium cyclopentadienide with the vinylogous vinamidinium salt (*8a*). The synthesis of (*1*) therefore consists of a C_5 with C_5 combination. Cyclizable decapentaenes (*45*), likewise, are also obtained in a C_3 with C_7 combination by condensation of the readily available cyclic heptamethinium salt (*44*)[46], with stabilized allyl anions generated from (*23a, b*) or (*24a*) by addition of sodium methoxide to the reaction mixture. Without intermediate isolation the violet-red (*45*) cyclize smoothly to the azulenes (*46*)[47] on heating:

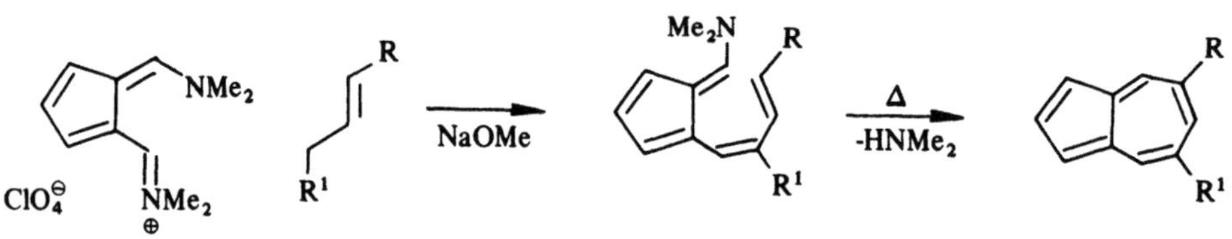

(44) (23) or (24) (45) (46)

In the usual manner the salt (*44*) can also be condensed with the reactive nitriles (*24h*) and (*24i*) and finally cyclized to the azulenes (*48*), (*49*), and (*50*). Starting with (*24h*), one obtains passing over the fulvene derivative (*47*), the dihydro-naphtho-azulene (*48*) as the main product (45%), together with a small amount (6%) of the azulene (*49*). Bromosuccinimide in carbon tetrachloride dehydrogenates (*48*) to (*49*) in 89% yield, without any substitution by bromine.

Table 11. Azulenes (*46*)

Azulene-/UV (CH_2Cl_2)	R	R'	Yield %	Components	Reaction conditions (°C)	Time
5,7-Diethyldicarboxylate dark-violet needles mp. 82–83°	CO_2Et	CO_2Et	50	(*44*) + (*23a*)	100	2 h
λ_{max} (log ϵ) = 662 (2.00), 604 (2.37), 556 (2.39), 387 (4.12), 368 (3.96), 322 (3.98), 293 (4.91), 282 nm (4.67)						
5,7-Dicarbonitrile dark-blue needles mp. 201–202°	CN	CN	53	(*44*) + (*23b*)	160	6 h
λ_{max} (log ϵ) = 636 (2.40), 605 (2.36), 582 (2.43), 386 (4.13), 367 (3.90), 320 (4.01), 289 (4.92), 281 nm (4.71)						
7-Phenyl-5-carbonitrile blue-green needles mp. 87–88°	Ph	CN	42	(*44*) + (*24a*)	180	6 h
λ_{max} (log ϵ) = 635 (2.70), 602 (2.68), 582 (2.70), 382 (4.19), 365 (4.03), 285 nm (4.81)						

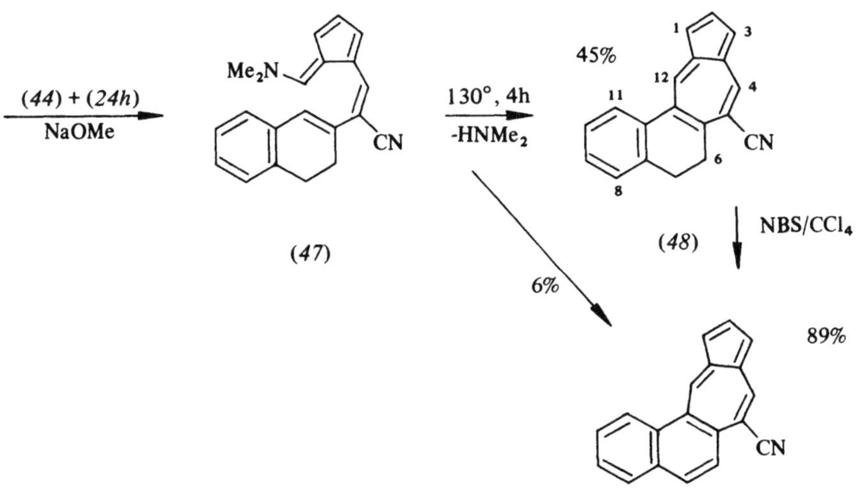

(*49*)

6,7-dihydronaphtho(1,2-f)azulene-5-carbonitrile (*48*)
green-blue needles, mp. 117–118°
λ_{max} (log ϵ) = 685 (2.35), 623 (2.66), 592 (2.62), 575 (2.62), 398 (4.35), 378 (4.03), 333 (4.19), 302 (4.82), 291 (4.65), 254 nm (4.37)

naphtho(1,2-f)azulene-5-carbonitrile (*49*)
green needles, mp. 188–189°
λ_{max} (log ϵ) = 689 (2.14), 621 (2.52), 576 (2.36), 421 (3.56), 400 (3.81), 383 (3.84), 350 (4.65), 343 (4.66), 313 (4.70), 242 nm (4.65)

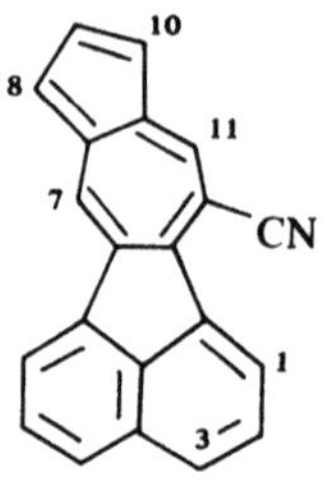

(*50*)

azuleno(5,6-a)acenaphthylene-12-carbonitrile (*50*), (130°, 7 h) 54%, green needles, mp. 187–188°
λ_{max} (log ϵ) = 690 (2.89), 658 (2.87), 456 (3.21), 430 (3.65), 410 (3.85), 379 (4.74), 362 (4.75), 342 (4.67), 275 nm (4.45)

Likewise the weakly acid 3,5,5-trimethylcyclohex-2-enone (isophorone) and the more active 2-methylchromone (*19*) react with (*44*) to give the corresponding azulenes (*51*) and (*52*).

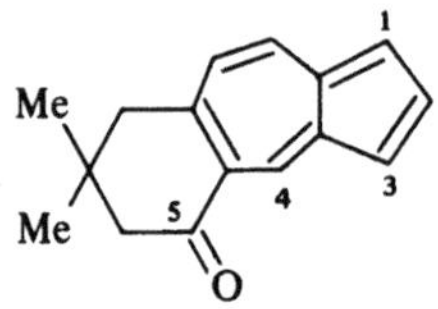

(*51*)

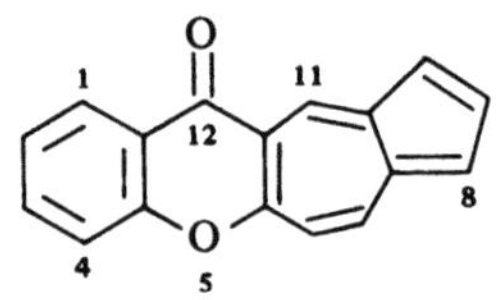

(*52*)

7,7-dimethyl-5,6,7,8-tetrahydrobenz(f)azulene-5-one (*51*): (160–170°, 24 h; 15%) blue-violet needles, mp. 68–69°
λ_{max} (log ϵ) = 644 (2.23), 587 (2.60), 565 (2.56), 545 (2.63), 394 (4.04), 381 (3.96), 301 (3.97), 288 nm (4.69)

azuleno(6,5-b)chromen-12-one (*52*): (200°, 20 h; 43%) red-brown needles, mp. 211°
λ_{max} (log ϵ) = 612 (1.79), 519 (2.41), 487 (2.45), 404 (3.91), 354 (4.19), 324 (4.81), 314 nm (4.70)

C. Polycyclic Aromatics by Benzannelation

The most exciting and perhaps the most versatile application of the electrocyclic ring-closure elimination method may be the benzannelation reaction. The ready availability of the starting compounds and in general the high yields make this new mode of formation of long- and well-known aromatics the synthesis of choice, especially if an aimed substitution is desired in the compound.

The condensation of vinamidinium salts (*7*) with reactive derivatives of different methylarenes, preferably with arylacetonitriles in quinoline by addition of the equivalent amount of sodium methoxide leads in the usual manner to derivatives of 1-aryl-4-dimethylaminobuta-1,3-dienes. These compounds, often not isolated, cyclize on heating. As we have already discussed (Section I. B.), such formal aminohexatrienes, in which one double bond is part of an aromatic system, need a greater activation energy for cyclization than comparable open chain aminohexatrienes. An additional extra energy gap by the localization of two π-electrons in the cyclization step is here associated with the loss of some aromatic resonance of the connected aromatic ring system. For our purposes, a good estimation of this energy amount in comparisons may be arrived at by means of the calculated HMO localization energies $L_{r,s}$ (Section C., Fig. 1).

1. Naphthalenes

Simple benzannelation of benzene leads to naphthalene. Related to our synthesis, a great number of benzene derivatives with an acidified methylene group that are ready for condensation with (*7*), *e.g.*, desoxybenzoin (activated by carbonyl), benzylphenylsulphones (activated by sulphonyl), esters of phenylacetic acids, and best-suited, the phenylacetonitriles (activated by cyano) can easily be transformed to arylaminobutadienes (*54*) and cyclized to naphthalenes (*56*):

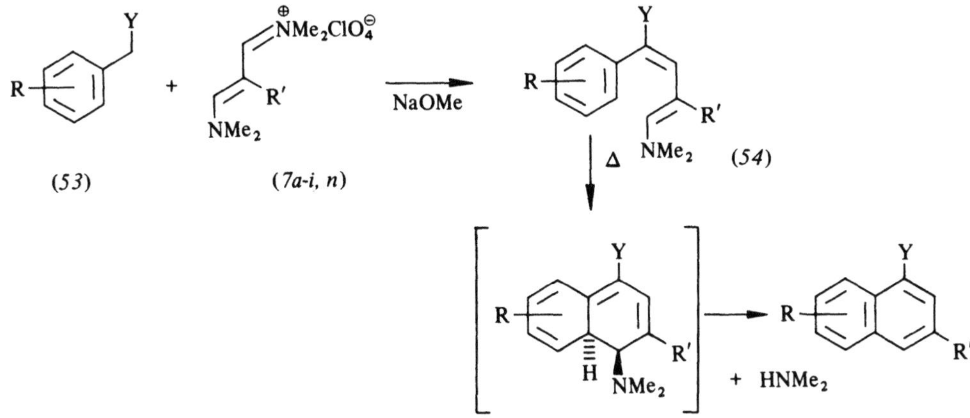

(*53*) (*7a-i, n*) (*54*) (*55*) (*56*)

On steric grounds only arylaminobutadienes (*54*) generated from vinamidinium salts (*7a–i, n*) that are derived from malondialdehydes can be cyclized successfully.

Peculiarily, the compounds (*54*) with Y = NO_2, obtained in good yields by condensation of phenylnitromethanes with 3,3-diethoxypropenes (*35*) fail to cyclize on heating.

Starting with *meta*-substituted phenylacetonitriles (*53*), a ring-closure of the intermediate (*54*) in two different directions should be possible, giving rise to a mixture of the 5- and the 7-substituted 1-naphthonitrile (*56a* + *56b*):

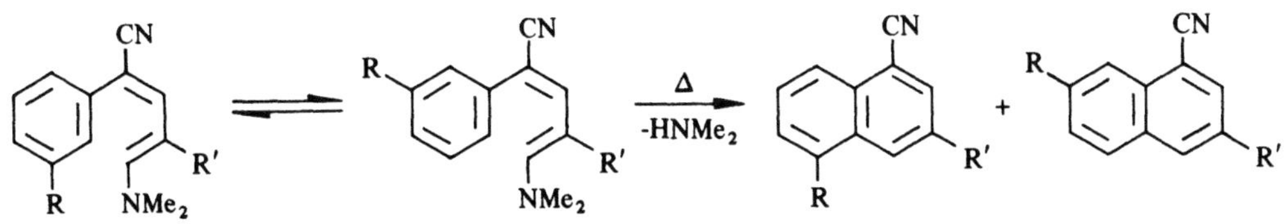

(*54*) (*56a*) (*56b*)

Table 12. Naphthalenes (*56*)

No.	Naphthalene derivative	Yield %	Y	Components	Reaction conditions (°C)	Time
1	1-Benzoyl-3-phenylnaphthalene mp. 98–99°	94	PhCO	Desoxybenzoin + (*7i*)	170	4 h
2	3-Methyl-1-phenylsulphonylnaphthalene, mp. 155–156°	96[1])	$PhSO_2$	Benzylphenylsulphone + (*7b*)	200	3 h
3	3-Butyl-1-ethylnaphthoate oil, acid: mp. 78°	72	CO_2Et	Ethylphenylacetate + (*7d*)	200	10 h
4	3-Ethyl-1-naphthonitrile mp. 46–47°	97	CN	Phenylacetonitrile + (*7c*)	180	3 h
5	3-Phenyl-1-naphthonitrile mp. 128–129°	91	CN	Phenylacetonitrile + (*7i*)	180	1.5 h
6	3,6-Dimethyl-1-naphthonitrile mp. 105	84	CN	4-Tolylacetonitrile + (*7b*)	200	1 h
7	6-Methoxy-3-methyl-1-naphtonitrile, mp. 99°	91	CN	4-Methoxyphenylacetonitrile + (*7b*)	180	6 h
8	3,5,6,7-Tetramethoxy-1-naphthonitrile, mp. 132–133°	70	CN	3,4,5-Trimethoxyphenylacetonitrile + (*7e*)	180	4 h
9	3-Phenyl-5,6,7-trimethoxy-1-naphthonitrile, mp. 129–130°	76	CN	3,4,5-Trimethoxyphenylacetonitrile + (*7i*)	160	4 h
10	3-Butylnaphthalodinitrile mp. 96–97° (subl.)	96	CN	2-Cyanophenylacetonitrile + (*7d*)	200	3 h
11	3-Methoxynaphthalodinitrile mp. 207–208°	90	CN	2-Cyanophenylacetonitrile + (*7e*)	200	5 h

[1]) Reaction with the double amount of benzylphenylsulphone, based on (*7b*).

The relation to each other of the two isomers formed may depend mainly on the nature the substituent R, as we can expect, and to a minor extent on the substituent R′. Table 13 shows the results of some experiments: the slender, rod-shaped carbonitrile group strongly favors the formation of the 5-isomers, contaminated only with small amounts of the 7-isomers in the case of present alkyl substituent R′. The methoxy- and the chloro-substituents may have a similar steric demand, and they give rise to the formation of considerable portions of the 7-isomers besides the 5-isomers. Finally the most bulky trifluoromethyl group affords the 7-compounds exclusively.

Another way to naphthalenes starts with 2-butene-1,4-dicarbonitrile (β-dihydromucononitrile) (*57*)[44, 48]. With vinamidinium salts (*7*) in pyridine or quinoline (*57*) undergoes very smoothly at room temperature, adding sodium methoxide, a double condensation to the deep-colored decapentaenes (*58*), whereas a monocondensation

Table 13. Naphthalenes from 3-substituted phenylacetonitriles

Naphthalene (*56 a*/*56 b*)	R	R′	Yield %	Components	Reaction conditions (°C)	Time
Naphthalene-1,5-dicarbonitrile (pure): mp. 261–263° (subl.)	CN	H	59	3-Cyanophenyl-acetonitrile + (*7a*)	210	24 h
3-Methoxynaphthalene-1,5-di-carbonitrile (pure): mp. 223–224°	CN	OMe	96	3-Cyanophenyl-acetonitrile + (*7e*)	200	4 h
3-Methylnaphthalene- 1,5-dicarbonitrile / 1,7-dicarbonitrile } mixture (10:1) mp. 183–198°	CN	Me	83	3-Cyanophenyl-acetonitrile + (*7b*)	180	3 h
3-Ethylnaphthalene- 1,5-dicarbonitrile / 1,7-dicarbonitrile } mixture (20:1) mp. 134–135°	CN	Et	95	3-Cyanophenyl-acetonitrile + (*7c*)	170	3 h
5-Chloro- / 7-chloro- } 3-methyl-1-naphthonitrile mixture (3:1) mp. 98–113°	Cl	Me	95	3-Chlorophenyl-acetonitrile + (*7b*)	180	2 h
3,5-Dichloro- / 3,7-dichloro- } 1-naphthonitrile mixture (3:2) mp. 134–137°	Cl	Cl	90	3-Chlorophenyl-acetonitrile + (*7g*)	180	6 h
5-Methoxy-3-methyl- / 7-methoxy-3-methyl- } 1-naphthonitrile mixture (2:1) mp. 59–80°	OMe	Me	83	3-Methoxyphenyl-acetonitrile + (*7b*)	170	3 h
3-Methyl-7-trifluoromethyl-1-naphthonitrile (pure): mp. 102–3°	CF_3	Me	85	3-Trifluoromethyl-phenylacetonitrile + (*7b*)	180	8 h
3-Methoxy-7-trifluoromethyl-1-naphthonitrile (pure): mp. 98–99°	CF_3	OMe	82	3-Trifluoromethyl-phenylacetonitrile + (*7e*)	180	10 h

does not succeed, also varying reaction conditions and the relation of (*57*) to (*7*). Only by twofold cyclization of (*58*) can the naphthalenes (*60*) be formed. The first ring-closure of (*58*) to the benzene derivative (*59*) should afford with $L_{1,6}$ = 3.109 a much smaller activation energy than the second one with $L_{2,4'}$ = 3.415 (Figs. 1, *1e* and II). Heating (*58*) splits off two equivalents of dimethylamine, as expected. Indeed each portion of dimethylamine begins to evolve at a different temperature indicating a step-by-step cyclization. Performing the reaction sequence with (*57*) and (*7a*) we succeeded in isolating the corresponding intermediates, the decapentaene (*58*) and the benzene (*59*). Directly related to (*54*), the benzenes (*59*) can cyclize in two directions, which also leads to a mixture of the 1,5- and 1,7-dinitriles (*60a, 60b*). By column chromatography on alumina the isomeric dinitriles – the symmetric (*60a*) with no dipole moment migrating more rapidly on the column – could be separated[44].

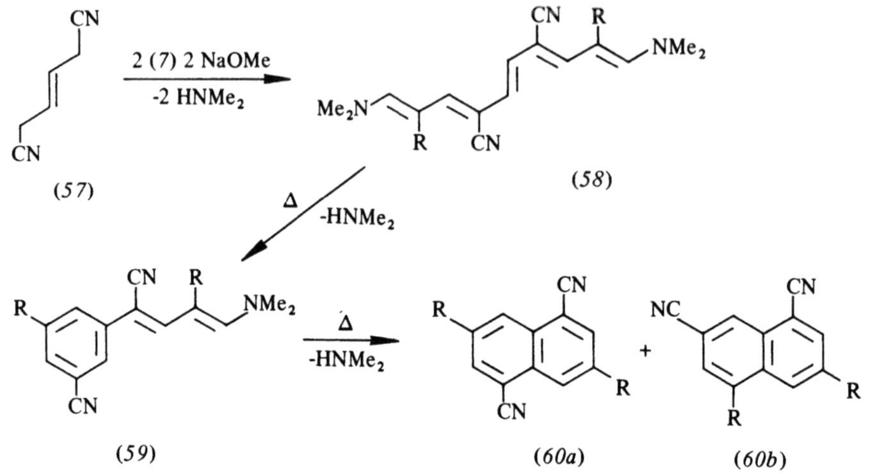

Table 14. Naphthalene dicarbonitriles from 2-butene-1,4-dicarbonitrile

Naphthalene-	Yield %	R	Components		Reaction conditions step 1	Reaction conditions step 2
a) -1,5-Dicarbonitrile, mp. 261–63°	34[1])	H	(57)	(7a)	190°, 16 h	280–300°, 3 h
b) -1,7-Dicarbonitrile	0	H				
a) -3,7-Dimethyl-1,5-dicarbonitrile mp. 276–277°	39	Me	(57)	(7b)	120–30°, 3 h	160°, 3 h
b) -3,5-Dimethyl-1,7-dicarbonitrile mp. 213–214°	10	Me				
a) -3,7-Diethyl-1,5-dicarbonitrile mp. 169°	51	Et	(57)	(7c)	120–30°, 2 h	160°, 3 h
b) -3,5-Diethyl-1,7-dicarbonitrile mp. 176°	8.5	Et				
a) -3,7-Dimethoxy-1,5-dicarbonitrile mp. 291–293°	42	OMe	(57)	(7e)	140°, 3 h	170°, 3 h
b) -3,5-Dimethoxy-1,7-dicarbonitrile mp. 269–271°	22	OMe				
a) -3,7-Diphenyl-1,5-dicarbonitrile mp. 256°	62	Ph	(57)	(7i)	120°, 1 h	160°, 1.5 h
b) -3,5-Diphenyl-1,7-dicarbonitrile	0	Ph				

[1]) On direct heating of isolated (59) without solvent.

2. Mesobenzanthrones

The conversion of anthrone (61) to mesobenzanthrones (63) resembles the preceding naphthalene syntheses: a modified benzene derivative becomes benzannelated. Anthrone (*61*) condenses with (*7e, i*) to the deep orange-red enaminoketones (*62*), but oxygen must be strictly excluded for acceptable yields. The cyclization takes place at 170–190°, giving the yellow mesobenzanthrones (*63a, b*)[34].

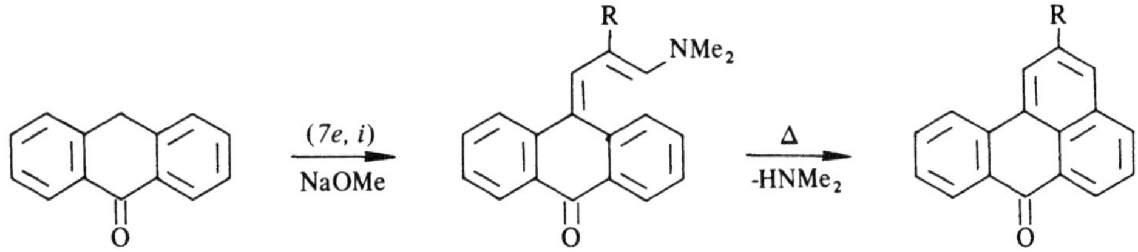

(*61*) (*62*) (*63*)

a : R = OMe
b : R = Ph

2-methoxymesobenzanthrone (*63a*), mp. 171–172° (subl.) (60%)/170–190°, 7 h
2-phenylmesobenzanthrone (*63b*), mp. 196–197° (AcOH) (87%)/170–180°, 4 h

3. Phenanthrenes

The easily available naphthalene acetonitriles are suitable starting components for the benzannelation of naphthalene[32, 34, 44]. The naphthalene-1-acetonitrile (*64*) produces phenanthrene-4-carbonitriles (*66*). The direction of cyclization in (*65*) is structurally fixed. It turns out the nitrile group in (*66*) to be sterically strongly screened. Basic hydrolysis under very hard conditions succeeds only to the stage of the corresponding amide.

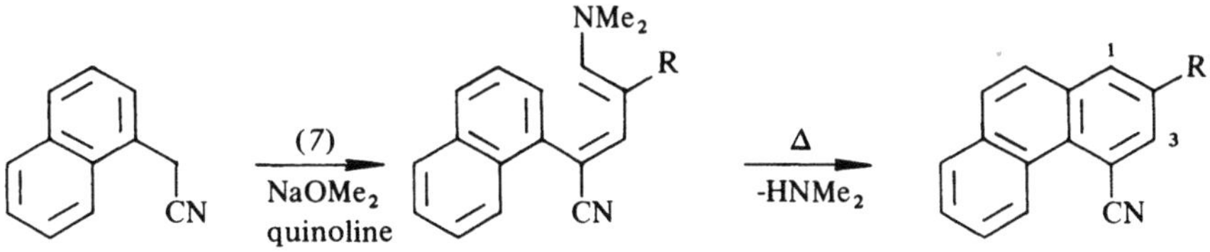

(*64*) (*65*) (*66*)

R	Phenanthrene-4-carbonitrile (*66*)	Yield %	Components	Reaction conditions (°C)	Time
a) H	--, mp. 105–107°	91	(*64*) + (*7a*)	200–220	40 h
b) OMe	2-Methoxy--, mp. 155.5°	95	(*64*) + (*7e*)	180–200	2.5 h
c) Ph	2-Phenyl--, mp. 169–171°	90	(*64*) + (*7i*)	160	3 h

Analogously, acenaphthene-5-acetonitrile (*68*) leads to acephenanthrene-10-carbonitriles (*69*). The starting (*68*) is obtained in a three-step sequence from acenaphthene-5-carbaldehyde (*67*).

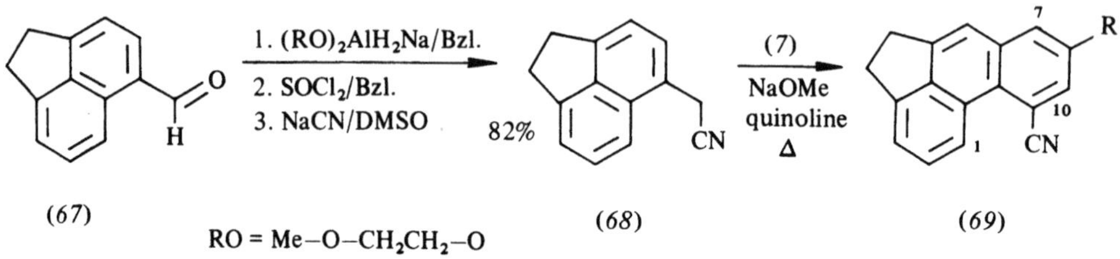

R	Acephenanthrene-10-carbonitrile (*69*)	Yield %	Components	Reaction conditions (°C)	Time
a) H	---, mp. 158°	57	(*68*) + (*7a*)	200–220	40 h
b) Me	8-Methyl--, mp. 174–175°	58	(*68*) + (*7b*)	170–200	3 h
c) OMe	8-Methoxy--, mp. 170–171°	85	(*68*) + (*7e*)	180–200	5 h

The naphthalene-2-acetonitrile (*70*) may serve as the source for many synthetically valuable phenanthrene-1-carbonitriles (*72*). The nitrile group in (*72*) can be converted to other functional groups or be smoothly eliminated in high yield by hydrolyzing it with potassium hydroxide in diethyleneglycole, followed by decarboxylation of the acid in boiling quinoline, in the presence of copper powder.

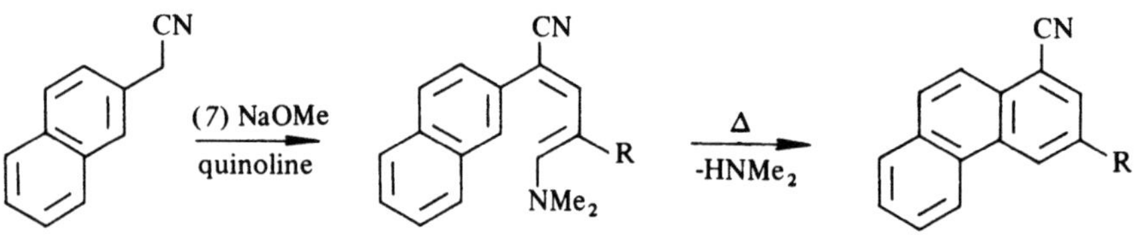

R	Phenanthrene-1-carbonitrile (*72*)	Yield %	Components	Reaction conditions (°C)	Time
a) H	----, mp. 125° (instab.)/130° (stab.)	98	(*70*) + (*7a*)	220	4 h
b) Me	3-methyl--, mp. 142°	95	(*70*) + (*7b*)	170–190	1 h
c) Et	3-ethyl---, mp. 83–84°	96	(*70*) + (*7c*)	140–160	3 h
d) OMe	3-methoxy--, mp. 153°	96	(*70*) + (*7e*)	180–200	3 h
e) Cl	3-chloro--, mp. 174–175°	91	(*70*) + (*7g*)	200	4 h
f) Ph	3-phenyl--, mp. 143–144°	90	(*70*) + (*7i*)	160	3 h

The intermediate (*71*) should formally be able to cyclize in two directions. As outlined in Section II. C, the activation energies for the single-ring closures should be selected according to calculated localization energies. Angular annelation to phenanthrenes with $L_{1,4'} = 3.162$, and linear annelation to anthracenes with $L_{3,4'} = 3.541$ (Figs. 1, IVa) are so different in energy-demand that formation of the phenanthrenes clearly prevails. Another fact which is evident from the tables: a voluminous substituent R (R′) in the side chain of (*54*), (*58*), (*66*), and (*71*) strongly facilitates the cyclization and lowers the thermolysis temperature considerably.

To gain a linear annelation to an anthracene by blocking up position 1 in (*71*) with a methyl group, we synthesized the aminonitriles (*74a, b*) by condensation of (*7b*), or (*7i*), with 1-methylnaphthalene-2-acetonitrile (*73*). The latter was obtained from 1-methyl-2-naphthoic acid[50] in three steps to give an overall yield of 60%. All attempts to enforce the cyclization of (*74a*) or (*74b*), however, were defeated. An uncontrolled decomposition of (*74a, b*) to tarry products takes place above 300°[44].

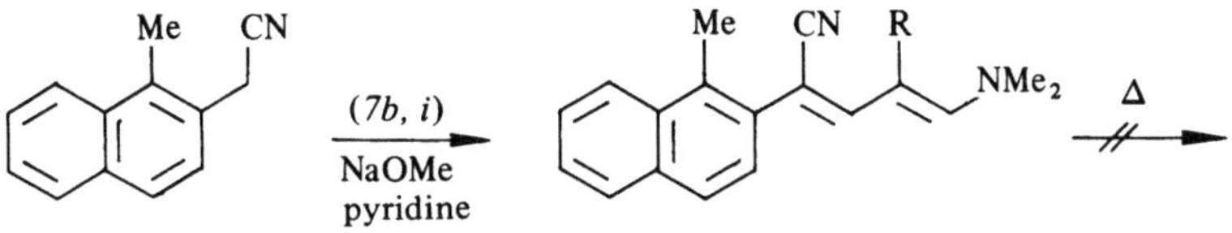

(*73*)

mp. 75–76°

(*74*) a : R = Me, mp. 124–125° (73%)
b : R = Ph, mp. 175–178° (95%)
Yellow needles

Phenanthrenes can also be generated by double condensation and thermolysis of benzene derivatives. Benzene-1,3-diacetonitrile (*75a*) and benzene-1,4-diacetonitrile react with (*7b*) to give not very soluble, brick-red aminonitriles (*76*) and (*78*), yielding the phenanthrenes (*77*) and (*79*). The cyclization doubtless proceeds first to a naphthalene and then to the phenanthrene, but no differentiation between the two cyclization steps could be observed. 3,6-Dimethylphenanthrene (*80*)[51], a valuable starting material for the synthesis of helicenes[3b, c] was obtained by hydrolysis and decarboxylation of (*79*). Benzene-1,2-diacetonitrile failed to give an ascertainable double-condensation product with (*7*).

4. Chrysenes

Three tetracyclic systems, triphenylene, benz(c)phenanthrene and chrysene can be derived by angular benzannelation of phenanthrene and hence these hydrocarbons may be synthesized from corresponding phenanthreneacetonitriles. The phenanthrene-1-acetonitrile (*81*) used for the preparation of the chrysenes (*82*) was obtained from phenanthrene-1-carbonitrile (*72a*) in a sequence of conventional steps: hydrolysis to the acid (84%) by KOH in triglycol, reduction to the carbinol (82%) by sodium dihydrido-bis(methoxyethoxy)aluminate, conversion by thionyl chloride in benzene to the chloromethyl derivative (98%), and finally reaction of the latter with sodium cyanide in DMSO to (*81*) (94%).

Since (*72a*) was obtained by the benzannelation reaction of naphthalene-2-acetonitrile, the above synthesis may serve as an example of the twofold application of the benzannelation principle.

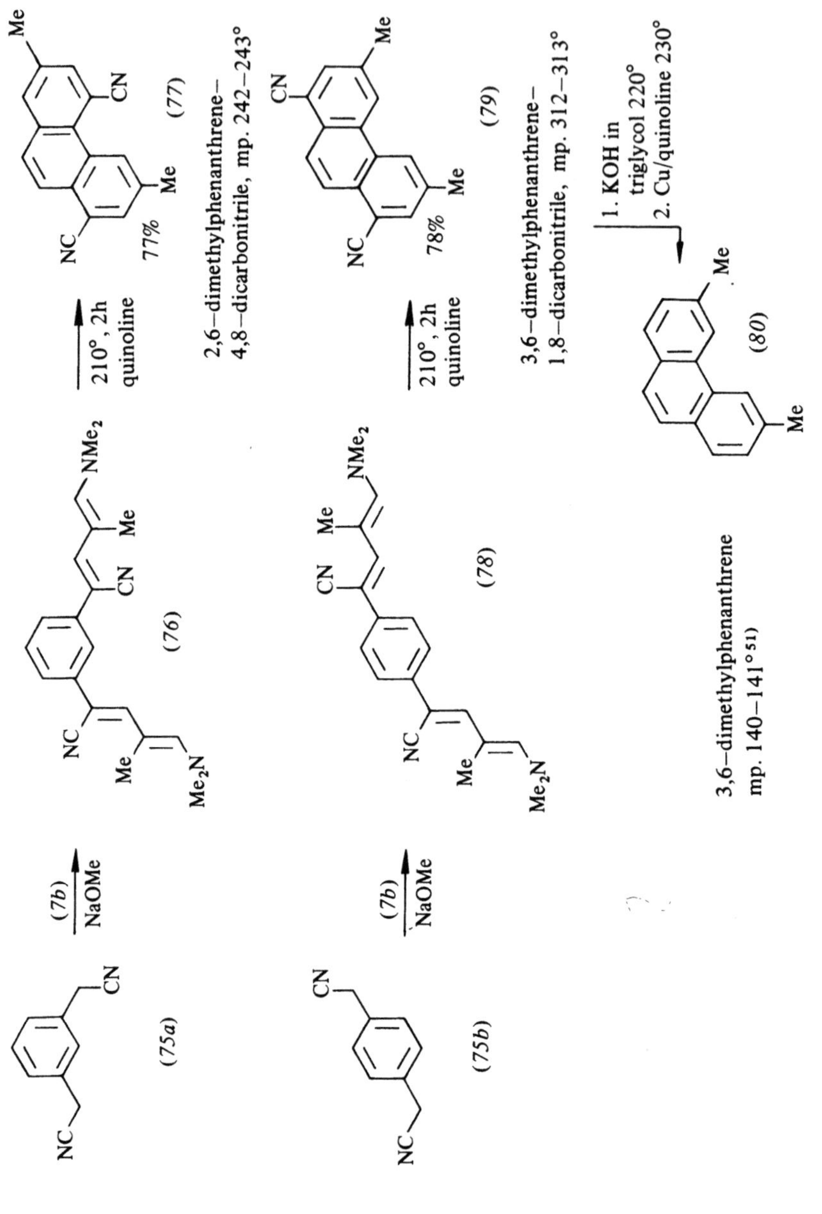
NC
CN
(75a)
(7b)
NaOMe
NC
CN
Me
NMe2
Me2N
Me
(76)
210°, 2h
quinoline
NC
CN
Me
Me
77%
(77)
2,6–dimethylphenanthrene–
4,8–dicarbonitrile, mp. 242–243°
NC
CN
(75b)
(7b)
NaOMe
NC
CN
Me
Me
NMe2
Me2N
(78)
210°, 2h
quinoline
NC
CN
Me
Me
78%
(79)
3,6–dimethylphenanthrene–
1,8–dicarbonitrile, mp. 312–313°
1. KOH in
triglycol 220°
2. Cu/quinoline 230°
Me
Me
(80)
3,6–dimethylphenanthrene
mp. 140–141° 51)

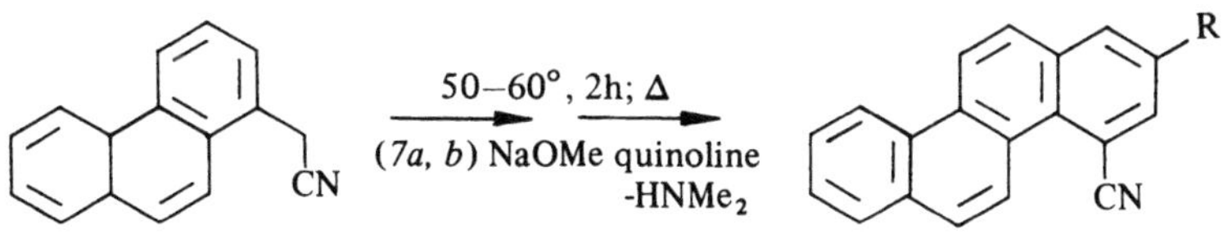

(*81*) mp. 128–129° (*82*) a: R = H,
b: R = Me

(*82a*): chrysene-4-carbonitrile, mp. 151–152° needles (70%) (210°, 80 h)

(*82b*): 2-methylchrysene-4-carbonitrile, mp. 208–209° plates, (86%) (160–180°, 3.5 h)

Double condensation of naphthalene-2,6-diacetonitrile (*83*)[52] with (*7a, b, i*) leads to the hardly soluble, red-brown enaminonitriles (*84*) which we isolated for intermediate purification. Their thermolysis gives the chrysene-1,7-dicarbonitriles (*85a–c*) in high yields[44]. Hydrolysis followed by decarboxylation in the presence of copper powder affords chrysene, mp. 253° (74%)[53] from (*85a*) and 3,9-dimethylchrysene, mp. 244°, plates (50%) from (*85b*).

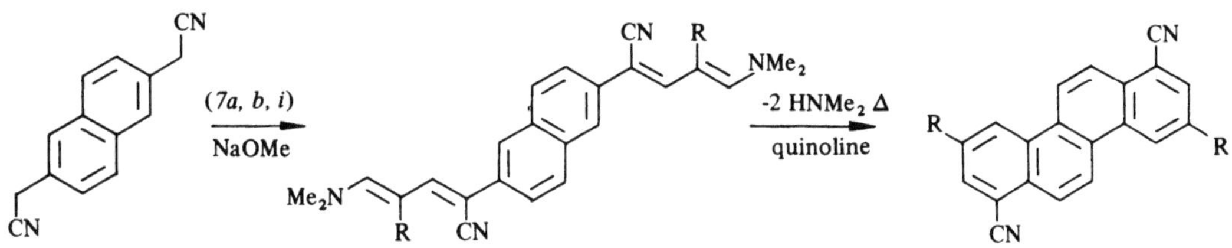

(*83*) (*84*) (*85*)

a): R = H (92%)
b): R = Me (91%)
c): R = Ph

(*85a*): chrysene-1,7-dicarbonitrile, mp. > 360° (dec.), (98%) (190–210°, 12 h)

(*85b*): 3,9-dimethylchrysene-1,7-dicarbonitrile, mp. > 360° (dec.), (88%), (170°, 2 h)

(*85c*): 3,9-diphenylchrysene-1,7-dicarbonitrile, mp. > 360° (dec.), [70%, without isolation of (*84c*)], (130°, 4 h)

5. Triphenylenes

Phenanthrene-9-carbonitrile, prepared by bromination of phenanthrene to the 9-bromo derivative[54] and its transformation by cuprous cyanide in dimethylformamide[55], served as starting material for the needed phenanthrene-9-acetonitrile (*86*). By the same sequence of steps as when converting phenanthrene-1-carbonitrile (*72a*) to (*81*) the 9-carbonitrile was converted to (*86*), giving an overall yield of 60%. The standard procedure leads very smoothly to the triphenylenes (*87*):

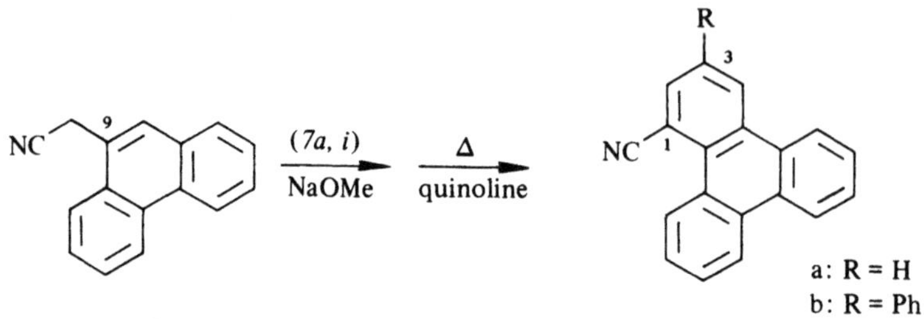

(86) (87)

triphenylene-1-carbonitrile, mp. 128–129° (90%) (180–200°, 4 h)

3-phenyltriphenylene-1-carbonitrile, mp. 162–163° (91%) (150°, 3 h)

6. Benzo(c)phenanthrenes

One pathway to benzo(c)phenanthrene by the cyclization-elimination method has been outlined already (Section III., B. 2.) and another we will see later. A synthesis by benzannelation should start with phenanthrene-4-acetonitrile or with phenanthrene-3-acetonitrile (*88*) as the one component. For steric reasons the preparation of the first has only poor chances, but (*88*) is readily available: 3-methylphenanthrene, mp. 62–63°[56], obtained by hydrolysis and decarboxylation from the nitrile (*72 b*) (Section III., C. 3.) in an overall yield of 92% was side-chain brominated by N-bromosuccinimide in CCl_4, and the 3-bromomethylphenanthrene, mp. 116–117° (67%), converted to (*88*), mp. 82–83° (91%).

Thermolysis of the isolated aminonitrile (*89*), mp. 169° (85%) obtained by condensing (*88*) with (*7 b*) in pyridine, gives rise to the formation of a mixture (83%) of two different isomeric nitriles in nearly equal amounts, the expected 2-methylbenzo(c)phenanthrene-4-carbonitrile (*90*) and the 9-methylbenzo(a)anthracene-11-carbonitrile (*91*).

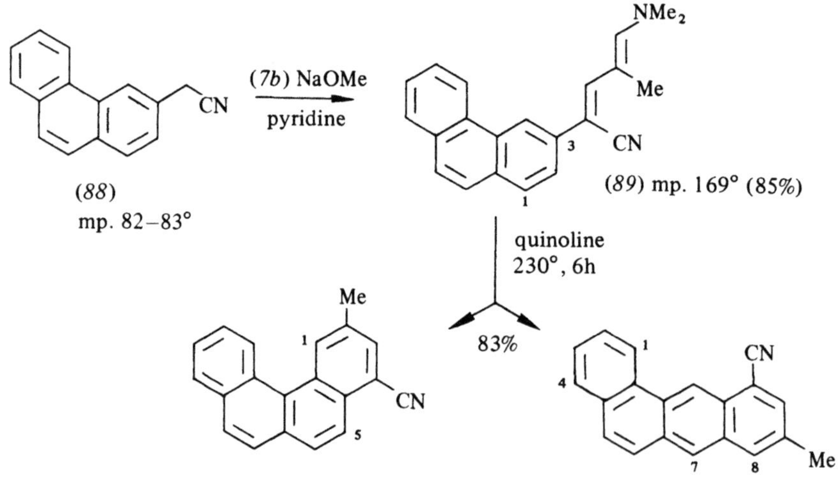

(90) (91)

The structure of (*90*) and (*91*), ascertained by their ^1H NMR spectra can be evidenced simply by transformation to the known hydrocarbons, 2-methylbenzo(c)-phenanthrene, mp. 81–82°[42] and 9-methylbenzo(a)anthracene, mp. 150–151°[57].

For the observed angular benzannelation of the naphthalene (*71*) to the phenanthrene (*72*), one calculates an energy profit of 0.379 β over the linear benzannelation to anthracene. It turns out by computation that the angular ring-closure of (*89*) to (*90*) should be favored above the linear one to (*91*) still by 0.251 β, a planar or nearly planar arrangement of the involved π-system provided. An inspection shows that some strong restraint exists against reaching the ready but sterically very overcrowded conformation of (*89*) forming (*90*), so that the linear benzannelation to (*91*) may successfully compete against the angular ring-closure.

In a four sequence, starting with 3,6-dimethylnaphthalene-1-carbonitrile (*56/6*) (Section III., C. 1.; Table 12), we prepared the naphthalene-2,7-diacetonitrile (*92*), mp. 114°, which undergoes smoothly a double condensation with (*7b*) to the symmetric aminonitrile (*93*), mp. 218° (dec.) (96%).

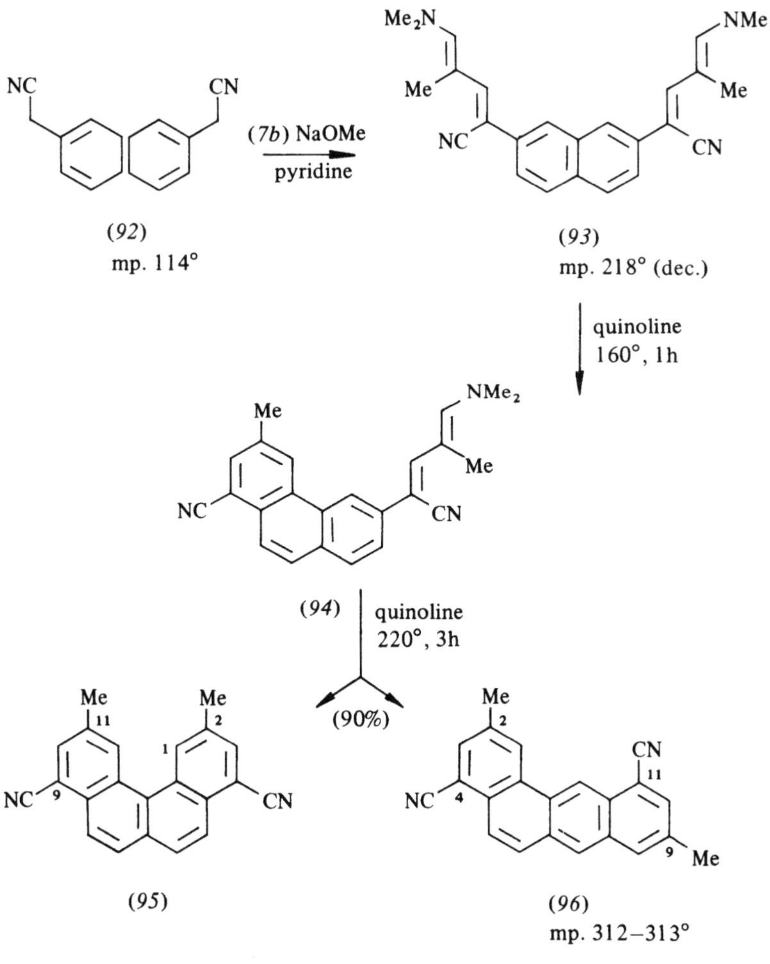

In thermolysing (*93*) one observes, by the evolution of dimethylamine, the first cyclization step at 160°, no doubt to the intermediate phenanthrene (*94*), clearly differentiated from the second one at 220°. The latter cyclization leads like the above (*89*) to a (1 : 1)-mixture (90%) of two isomeric dinitriles, the 2,11-dimethylbenzo(c)phenanthrene-4,9-dicarbonitrile (*95*) and the 2,9-dimethylbenzo(a)-anthracene-4,11-dicarbonitrile (*96*), mp. 312–313°.

A better chromatographic separation was achieved after hydrolysis and decarboxylation to the hydrocarbons, the 2,11-dimethylbenzo(c)phenanthrene, mp. 129° (only one resonance signal for methyl in the ^{1}H NMR spectrum) and the 2,9-dimethylbenzo(a)anthracene, mp. 162°[58].

By calculation the angular benzannelation of the yellow benz(c)phenanthrene derivative (*98*), from condensation of 11-methylbenzo(c)phenanthrene-2-acetonitrile with (*7b*) obtained, should favored by 0.292 β above the linear annelation.

But the sterical hindrance for angular cyclization to a dibenzo(c,g)phenanthrene is found to be so much increased here that thermolysis, achieved only under very strong conditions (240°, 12 h), leads exclusively to the linear annelated 2,11-dimethyldibenzo(b,g)phenanthrene-13-carbonitrile (*99*), yellow needles mp. 243–244° (66%). Conversion of it by hydrolysis and decarboxylation gives 2,11-dimethyldibenzo(b,g)phenanthrene, mp. 155–156°.

2,11-Dimethylbenzo(c)phenanthrene, readily accessible by a new synthesis, as described later (Section E. 1), side-chain brominated by N-bromosuccinimide and converted to (*97*), was used as starting compound in the above synthesis.

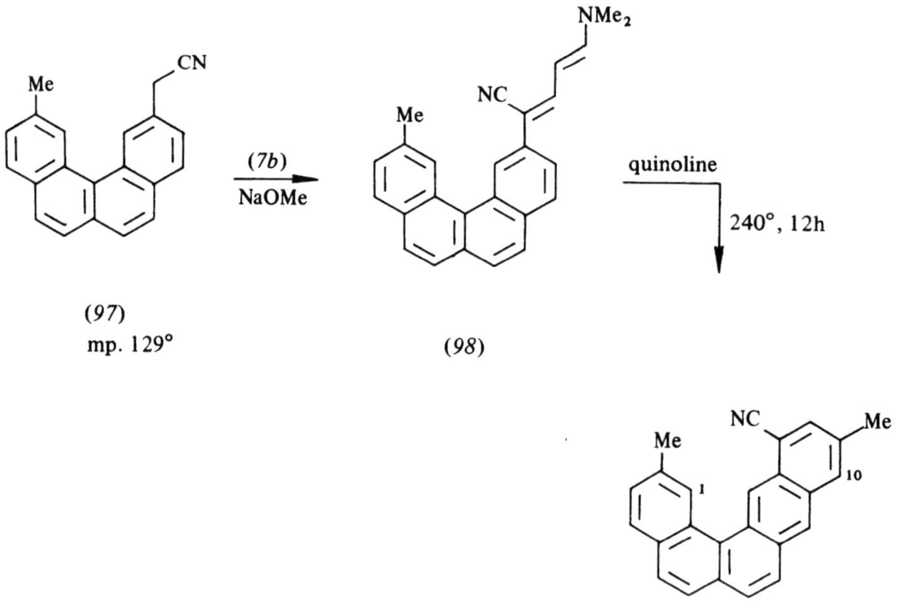

(*97*)
mp. 129°

(*98*)

(*99*)
mp. 243–244° (66%)
Yellow needles

7. Benzo(b)biphenylenes

In a conventional seven-step synthesis beginning with biphenylene, biphenylene-2-acetonitrile (*100*), mp. 105–107°, was prepared. It condenses readily with (*7*) to the orange-yellow aminonitriles (*101*) (not isolated). In accordance with the MO calculation the ring-closure to the benz(b)biphenylene-6-carbonitriles (*102*) is strongly favored, theoretically by 0.211 β (Section II. C., Fig. 1 VIII), over the isomeric benzo(a)biphenylenes of which no traces could be detected in the reaction products[59]. By hydrolysis and decarboxylation (*102a*) was converted to benzo(b)-biphenylene, mp. 236–237°, Ref. 242–243°[60].

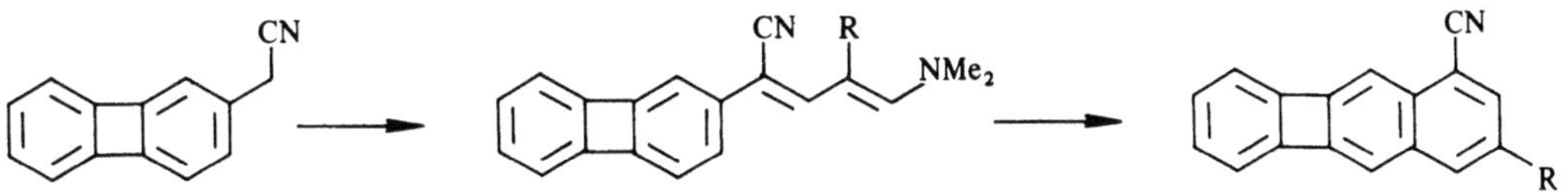

(*100*)
mp. 105–107°

(*101*)

(*102*)

R	---Benzo(b)biphenylene-6-carbonitrile	Yield	Components	Reaction conditions (°C)	Time
a) H	----, mp. 184–185°, pale-yellow needles	93	(*101*) + (*7a*)	205	3 h
b) Et	8-Ethyl---, mp. 169–170°, pale-yellow plates	90	(*101*) + (*7c*)	145	4 h
c) OMe	8-Methoxy---, mp. 216–217°, orange-yellow needles	89	(*101*) + (*7e*)	145	5 h

8. Pyrenes

The tricyclic hydrocarbon phenalene (*103*)[61] can readily be transformed to one of the strongly resonance-stabilized species: the phenalene radical, the phenaleniate anion and the phenalenium cation. Therefore its acidity is quite sufficient, without any additional activating substituent, to generate the extremely oxygen-sensitive, orange phenaleniate anion by means of sodium methoxide. The latter condenses very smoothly with vinamidinium salts (*7*) and their vinylogues (*8*) to deeply colored derivatives (*104*) of the unknown, cross-conjugated 1-methylenephenalene, named "phenafulvene"[62]. We prepared them some time ago[15] in an attempt to test the theoretical prediction that the introduction of electron-donating substituents at exocyclic positions should stabilize the system.

Whereas (*104b*) is thermostable up to 200°, a slow decomposition to pyrene and dimethylamine takes place on heating (*104a*) to the melting point. This was the first example of six-ring closure by thermolysis which we observed, and the starting-point of the whole story suggesting this recent review. In parallel to the known azulene synthesis we originally explained the cyclization by an intramolecular electrophilic-nucleophilic attack in the highly polar (*104a*)[15].

To prepare pyrenes (*106*), especially the otherwise only difficultly accessible 2-substituted ones, the reaction may be performed more conveniently in the usual

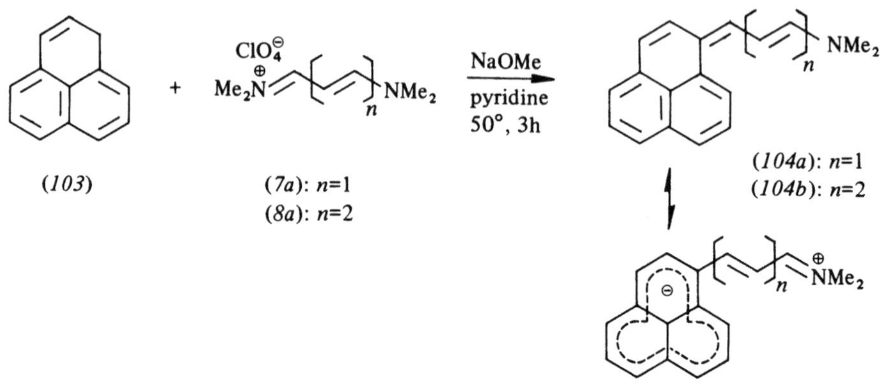

(*104a*): 1-(3-dimethylaminoallylidene)phenalene (91%), mp. 150–151° (dec.)
λ_{max} (log ϵ) in MeCN: 496 (4.53), 278 nm (4.23)

(*104b*): 1-(5-dimethylaminopentadiene-2,4-ylidene)phenalene (72%), mp. 132–134°
λ_{max} (log ϵ) in MeCN: 517 (4.76), 382 (3.86), 331 nm (4.23)

manner in quinoline and without isolation of the intermediate of the sensible enaminophenafulvenes (*104*). The cyclization of (*104*) seems to correspond to the naphthalene phenanthrene formation of (*65*) to (*66*), but in comparison to the latter the calculated localization energy should be diminished by 0.1475 β. In good correlation with that, the thermolysis temperature was found to be exceedingly low.

Using a mixture of the protomeric phenylphenalenes[61] in a condensation-cyclization reaction with (*7a*), one receives a (1 : 1) mixture of 1-phenylpyrene (*106o*) and 4-phenylpyrene (*109*) in 94% yield[15]. The uniform phenylphenaleniate ion (*107*) generated during the reaction shows a partition of the charge between positions 1, 3, 4, 6, 7, and 9. Enaminophenafulvenes arise only by attack of (*7a*) on positions 3, 4,

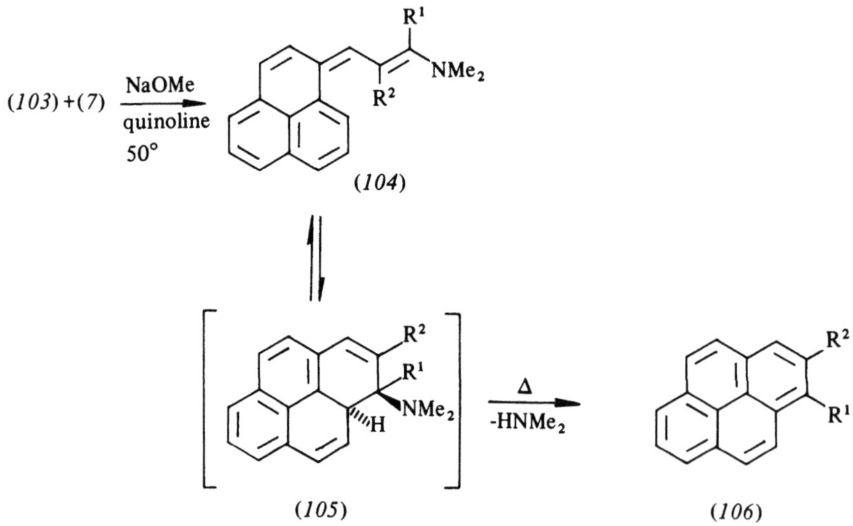

6, 7, and 9 of (*107*), but a reaction on the latter, sterically strongly screened position seems unlikely. Only the remaining four, isomeric enaminophenafulvenes, *e.g.*, (*108a*) and (*108b*), can cyclize, leading to (*106o*) and (*109*):

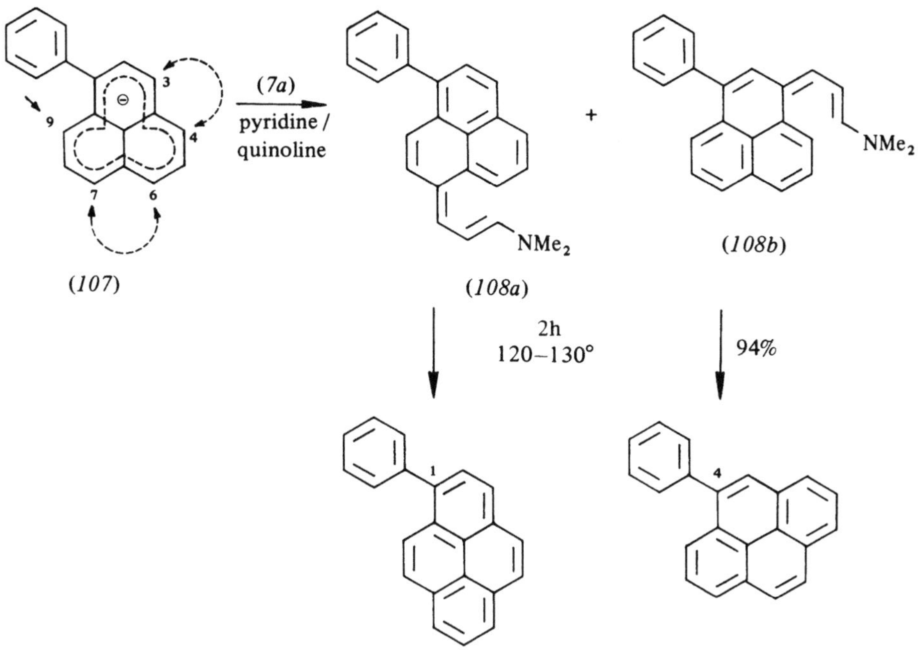

(*107*) (*108a*) (*108b*)

(*106o*) mp. 83° [63)]

(*109*) mp. 130–131° (Ref. [64)] 132–133°)

The instable 1,2-dihydro-5H-cyclopenta(c,d)phenalene (*110*), like phenalene, condenses with (*7a*) to a dark-red enamine (*111*) and provides uniformly the 1,2-dihydrocyclopenta(c,d)pyrene (*112*) in moderate yield (25%) on heating[65)]:

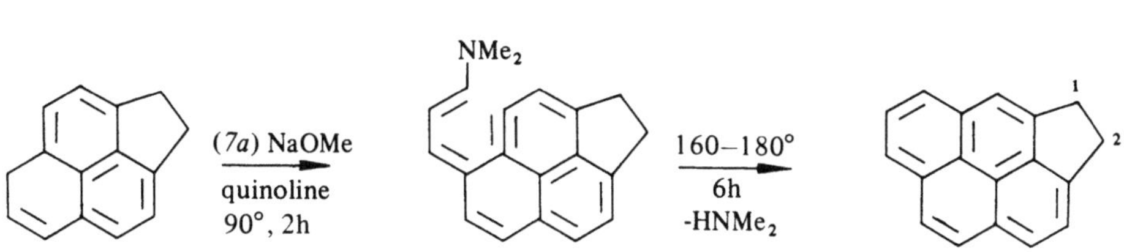

(*110*) mp. 115° (dec.)

(*111*)

(*112*) mp. 131°

The 7H-benz(d,e)anthracene (*113*) and the 6H-benzo(c,d)pyrene (*117*), both much weaker acids than phenalene, form the polydentate ions (*114*) and (*118*) and condense like (*107*) with (*7*) to mixtures of isomeric enamines. From the reaction of

Table 15. Pyrenes (*106*)[15]

(*106*)	----Pyrene	Yield %	R^1	R^2	Components (*103*) +	Reaction conditions (°C)	Time
a	----, mp. 155°	93	H	H	(*7a*)	150–160	2 h
b	2-Methyl--, mp. 142–143°	79	H	Me	(*7b*)	60–70	2 h
c	2-Ethyl--, mp. 68–71°	97	H	Et	(*7c*)	60–70	2 h
d	2-Methoxy-, mp. 105–106°	63	H	OMe	(*7e*)	60–70	2 h
e	2-Chloro--, mp. 143°	38	H	Cl	(*7g*)	130–140	2 h
f	2-Nitro--, mp. 192–194°	30	H	NO_2	(*7n*)	60–70	2 h
g	2-Phenyl--, mp. 166°	74	H	Ph	(*7i*)	100	2 h
h	2-(p-Methoxyphenyl)-, mp. 157–158°	80	H	$4\text{-}C_6H_4OMe$	[1]	80	2 h
i	2-(p-Chlorophenyl)--, mp. 165–166°	87	H	$4\text{-}C_6H_4\text{-}Cl$	[1]	90	2 h
j	2-(p-Bromophenyl)--, mp. 150–151°	79	H	$4\text{-}C_6H_4\text{-}Br$	[1]	90	2 h
k	2-(α-Naphthyl)---, mp. 112–113°	83	H	$\alpha\text{-}C_{10}H_7$	[1]	80	2 h
l	2-(β-Naphthyl)---, mp. 210–211°	81	H	$\beta\text{-}C_{10}H_7$	[1]	80	2 h
m	1-Methyl---, mp. 66–68°	88	Me	H	(*7l*)	170	2 h
n	1-Ethyl---, mp. 94°	87	Et	H	(*7m*)	150	2 h
o	1-Phenyl---, mp. 83°	67	Ph	H	(*7j*)	190–200	2 h
p	7,8,9,10-Tetrahydro-benz(a)---, mp. 112°	83	$-(CH_2)_4-$		(*7o*)	60	2 h

[1]) The vinamidinium salts needed were prepared from the corresponding arylacetic acids by formylation with dimethylformamide phosphoryl chloride, like (*7i*)[15].

(*113*) with (*7a*) we isolated in 82% yield a (1 : 2) mixture of benzo(a)pyrene (*115a*) and benzo(e)pyrene (*116 a*), likewise with (*7b*) in 60% yield the corresponding 2-methyl derivatives (*115 b*) and (*116 b*). As expected, from (*117*) and (*7a*) results in 33% yield a mixture of anthanthrene [= dibenzo(d,e,f–m,n,o)chrysene] (*119*) and benzo(g,h,i)perylene (*120*), also in a 1 : 2 proportion[15].

Readily accessible also are derivatives of benzo(a)pyrene, *e.g.*, (*123*) applying the usual benzannelation method using pyrene-1-acetonitrile (*121 d*). The latter was prepared starting from pyrene-1-carbaldehyde[67] (*121 a*) by reduction with sodium bismethoxyethoxy-dihydroaluminate in benzene to the carbinol (*121 b*), this by thionyl chloride to (*121 c*), and finally with sodium cyanide in DMSO leading to (*121 d*).

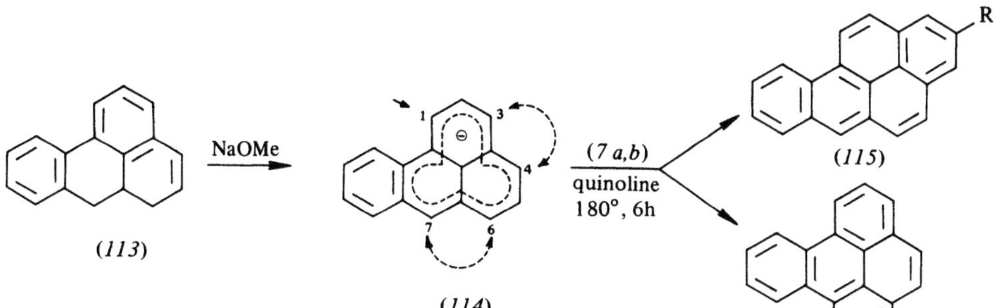

(113) (114) (115) (116)

(115a): R = H, mp. 179°;
(116a): R = H, mp. 178–179°
(115b): R = Me, mp. 176–177° [66]

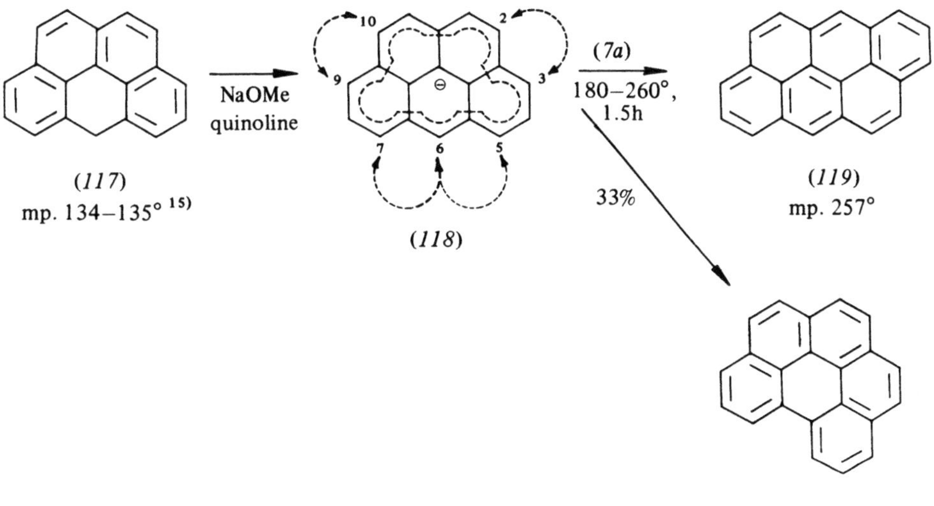

(117) mp. 134–135° [15]

(118)

(119) mp. 257°

(120) mp. 273°

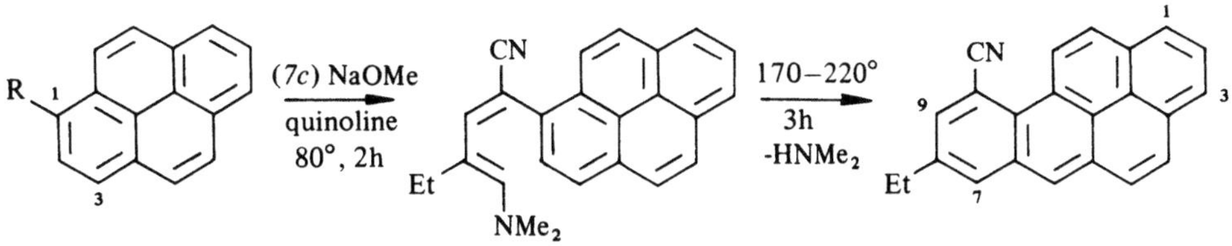

(121) (122) (123)

a) R = CH=O
b) R = CH_2OH, mp. 121–122° (99%)
c) R = CH_2Cl, mp. 146–147° (96%)
d) R = CH_2CN (98%)
mp. 111–112° (labile form), mp. 121–122° (stable form)

8-ethylbenzo(a)pyrene-10-carbonitrile (89%) mp. 242–243°, yellow plates.

9. Fluoranthenes

In Table 7 (Section III., 2, Nos. 28–34) we have listed some fluoranthenes. They were synthesized starting from the corresponding acenaphthene derivatives, constructing the benzene ring as shown here by formulas:

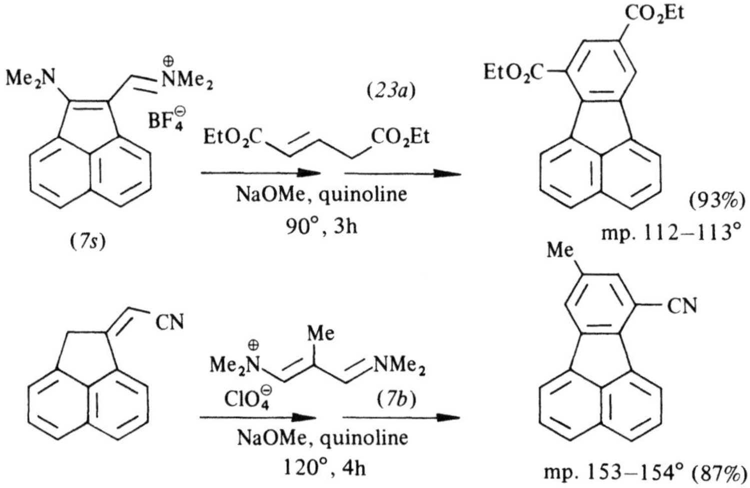

A third way to fluoranthenes (*126*) should be opened by the thermal cyclization of the long-known, stable, colored enaminofulvenes of fluorenes[24], *e.g.* (*125*), which are isomeric with the corresponding derivatives (*104*) of phenalene:

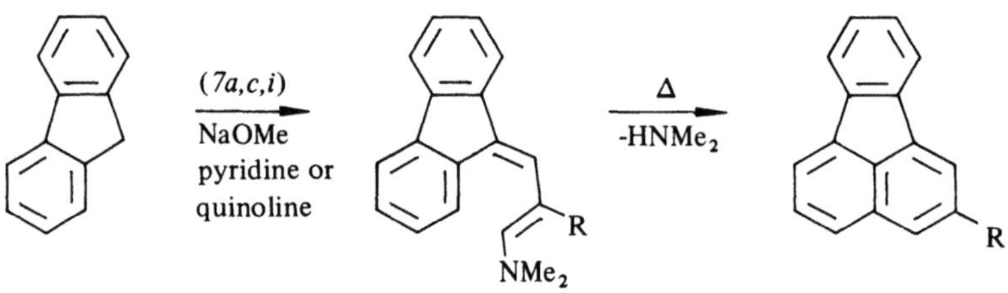

(*124*) (*125*) (*126*)

The cyclization of (*125*) to (*126*) parallel to the formation of naphthalenes (*56*) from (*54*) (Section C., 1.) should afford a rather smaller activation energy than the latter (see Fig. 1, Xc Section 1. C.) in a superficial view. However, the parent compound [(*125a*), R = H] fails to cyclize on heating up to 250°. Instead, sublimation and at higher temperatures uncontrolled decomposition take place. A more detailed inspection shows that the bond angles in (*125*) effect an increased distance between the centers C-1 and C-4′, but an effective orbital overlap of these centers is an essential condition for a ring-closure. The formation of (*126*) therefore involves considerable angle strain. Only factors which depress the thermolysis temperature in other instances also, *e.g.*, bulky substituents or/and variation of the π-system lowering the localization energy, $L_{r,s}$, bring the ring-closure of (*125*) to (*126*) within reach:

(*126 c*): R = Et; *2-ethylfluoranthene*, mp. 108–109° (40%) from (*124*) + (*7 c*), 210°/70 h (intermediate (*125 c*) not isolated)
(*126 i*): R = Ph; *2-phenylfluoranthene*, mp. 80–81° (55%), 200° 2 h; from (*124*) + (*7 i*); (*125 i*): yellow crystals mp. 123° (70%)

In all cases the yields are moderate and the conditions of cyclization relatively hard, but the starting materials are readily available and cheap.

Likewise benzo-annelated fluoranthenes, (*128*), (*130*), and (*132*), are obtained by thermolysis of the corresponding enamino derivatives of benzofluorenes (*127*), (*129*), and (*131*). Whereas one direction of cyclization is open for (*127*), there exist two possibilities of ring-closure direction for (*129*) and (*131*), but only the one was realized which we expected in correlation to the calculated lower localization energies (see Fig. 1, XI., Section II., C.):

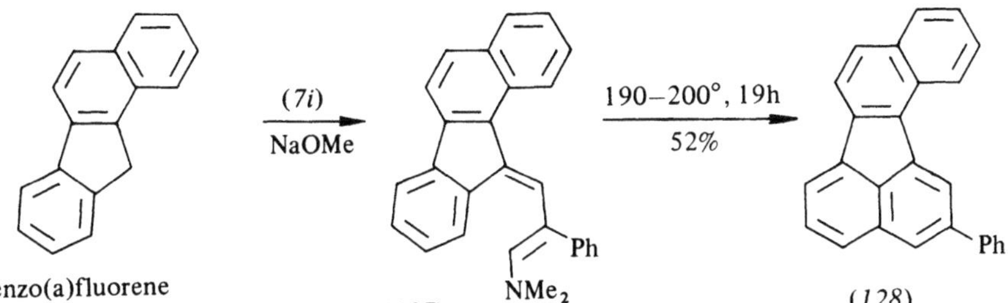

Benzo(a)fluorene (*127*) (*128*)

Yellow needles mp. 158°
2–Phenylbenzo(j)fluoranthene

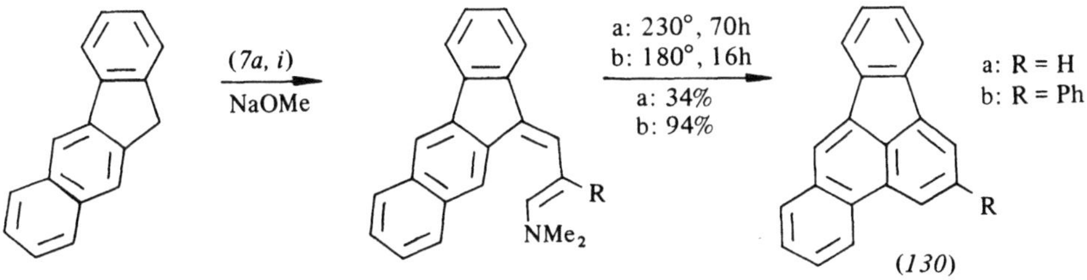

Benzo(b)fluorene (*129*) (*130*)

a: Benzo(e)acephenanthrylene mp. 165° colourless needles
b: 2–Phenyl–, mp. 195°

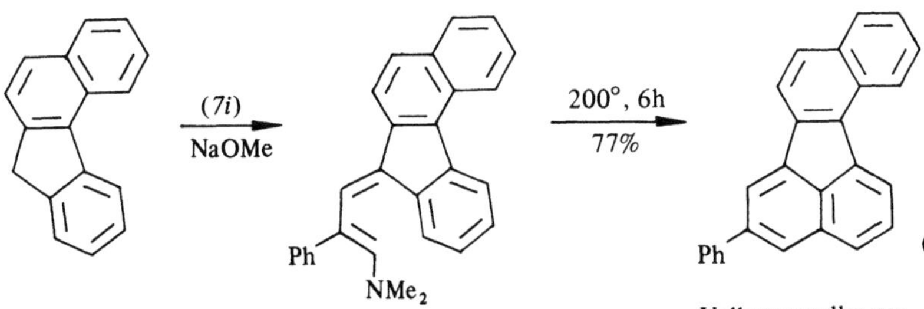

Benzo(c)fluorene (*131*) (*132*)

Yellow needles mp. 172°
5–Phenylbenzo(j)fluoranthene

In analogy to fluorene the more acidic 5H-indeno(1,2-b)pyridine (= 4-azafluorene) (*133*)[6] smoothly condenses with (*7i*) to (*134*) which cyclize uniformly to 6-phenyl-acenaphtho(1,2-b)pyridine (= 2-phenyl-7-azafluoranthene) (*135*)[44]:

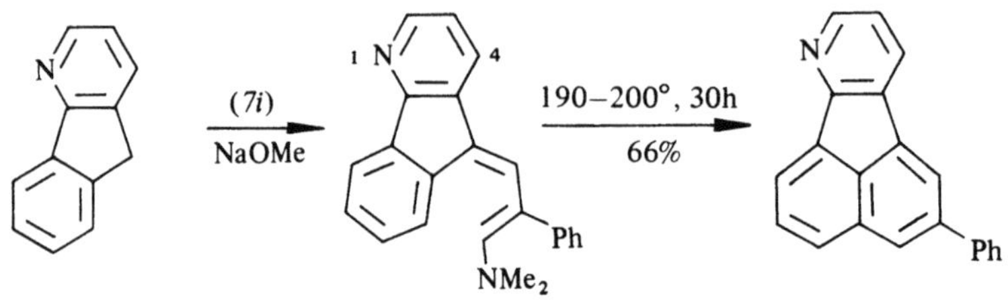

(*133*) (*134*) (*135*) mp. 112°

Benzannelation of (*134*) to the pyridine nucleus seems to be strongly disfavored (small HOMO frontier orbital in position 4 of the pyridin ring?).

Indene itself also gives enaminofulvenes by condensation, but cyclization fails. From the 2,3-diphenylindene (*136*), the 1,2,4-triphenylacenaphthylene (*138*) can be prepared:

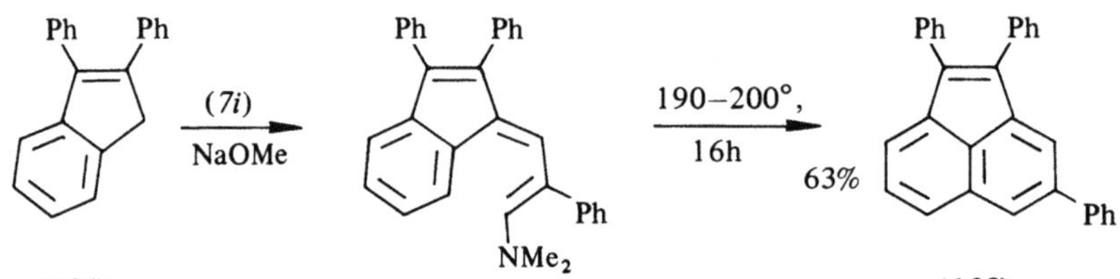

(*136*) (*137*) (*138*)

Orange plates, mp. 199°

Another, instructive example for the high angle strain inherent in combinations of flat hexagonal with pentagonal ring systems is represented by the synthesis of benzo(g,h,i)fluoranthene (*143*). Starting with acenaphthenone (*25*), the cyclization of the derived dienaminoketones (*139*)[6] once more succeeds under rigorous conditions only in the presence of a bulky substituent R, *e.g.*, methyl or phenyl, to the benzo(d,e,f)fluorenones (*140a, b*). Wolff-Kishner reduction of (*140b*) to the fluorene (*141*) and repeated condensation with (*7i*) to (*142*) followed by cyclization leads finally to the symmetrically substituted (*143*) exclusively. The other possible isomer may be disfavored by steric hindrance:[32]

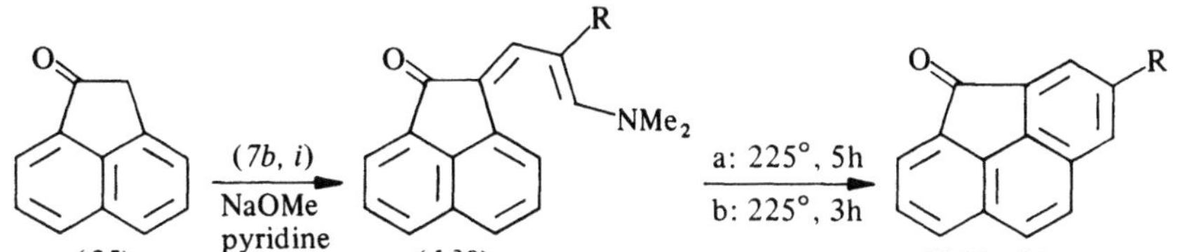

(25) (139) (140a, b)

a: R = Me (87%)
Dark-red needles mp. 139–140°
b: R = Ph (95%)
Red needles mp. 160–161°

(*140 a*): 2-methylcyclopenta(d,e,f)phenanthrene-4-one, (42%) yellow needles, mp. 133–136°
(*140 b*): 2-phenylcyclopenta(d,e,f)phenanthrene-4-one, (96%) yellow needles, mp. 119°

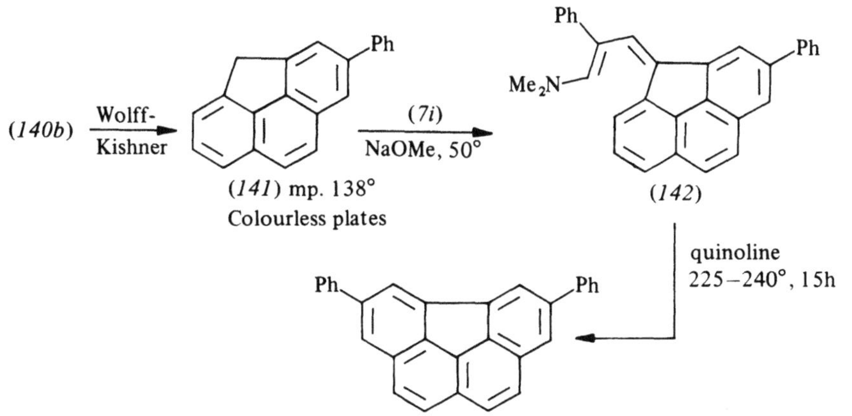

(141) mp. 138°
Colourless plates

(142)

(143) (81%)
Yellow needles, mp. 184°
4,7–diphenylbenzo(g,h,i)–
fluoranthene

D. Azulenes and Azulenoids by Annelation Reactions

1. Benzannelation to Azulenes

Methyl groups in position 4, 6, or 8 of azulene are only just acidic enough to undergo condensation with vinamidinium salts (*7*) and their vinylogues (*8*), in presence of sodium methoxide. With (*7*) the 6-methylazulene (*121 a*) slowly forms the deep-red azulene dienamines (*144*) and, on heating, the benz(f)azulenes (*145*). But the reaction does not run to completion and therefore unconverted (*121 a*) can be recovered at the end[18]. Introduction of an additional acceptor-substituent in the methyl group, like the cyano- or methoxycarbonyl-group (*121 b, c*), strongly enhances the rate of condensation, but the influence on the cyclization step is uncertain. The hitherto unknown 6-azuleneacetonitrile (*121 b*) was obtained by action of cyanogen on sodium 6-methylene-azuleniate[68].

In order to obtain the benz(a)azulenes (*146*) and (*151*), azulene-1-acetonitriles (*146*)[70] and azulene-2-acetonitrile (*149*)[71] are condensed with vinamidinium salts (*7*) – the 1-methyl- and 2-methylazulene are unreactive – to give the corresponding azulene dieneaminonitriles (*147*) and (*150*). In agreement with the theoretical prediction, the latter (*150*) cyclize especially readily and quickly on slight heating. The localization energy linked with the ring-closure of type (*150*) is calculated to be 0.171 β lower than that of type (*147*).

The dienaminonitrile derived from the azulene-5-acetonitrile involving two possibilities of direction of ring-closure should preferentially cyclize, according to the MO theory, to give the benz(e)azulenes. But all attempts to synthesize the azulene-5-acetonitrile testing these predictions have hitherto been unsuccessful.

2. Azulenoids by ($_\pi 10_s$)-Cyclizations

The 4-methyl group of the azulenes (*152*) condense slowly as mentioned with (*7*) to give the red azulene-4-dieneamines (*153*). The latter are related to the dieneaminofulvene (*1*) of the classical Hafner-azulene synthesis, and on heating afford in a ($_\pi 10_s$)-ring-closure the cyclopenta(e,f)heptalenes (*154*)[18, 72], rather than the benz(e)azulenes in a ($_\pi 6_s$)-cyclization.

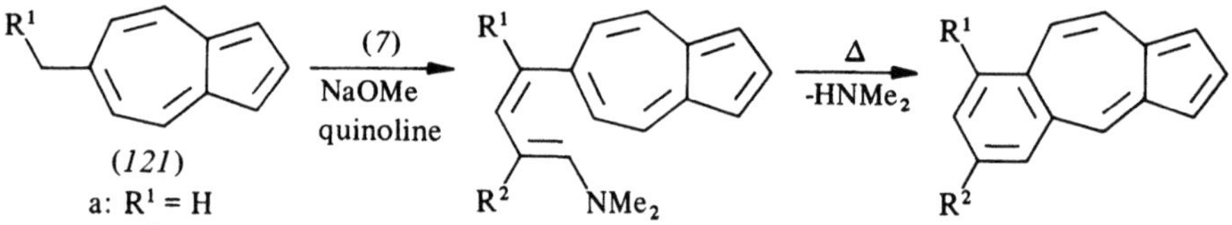

(*121*)
a: R^1 = H
b: R^1 = CN
c: R^1 = CO_2Me

(*144*)

(*145*)

Table 16. Benzazulenes

No.	(*145*) ---Benz(f)azulene	R^1	R^2	Yield %	Reaction conditions: Components	(°C)	Time
1	[1]) ------, mp. 160–161° [69])	H	H	74	(*121 a*) + (*7 a*)	200–210	12 h
2	6-Ethyl--, mp. 91°	H	Et	97	(*121 a*) + (*7 c*)	180–190	14 h
3	6-Benzyl-, mp. 96–97°	H	Ph–CH_2	79	(*121 a*) + (*7**)	180–190	20 h
4	6-Methoxy--, mp. 94–95°	H	OMe	68	(*121 a*) + (*7 e*)	190–200	20 h
5	6-Phenyl--, mp. 146–147°	H	Ph	71	(*121 a*) + (*7 i*)	180	14 h
6	----8-Carbonitrile, mp. 95°	CN	H	37	(*121 b*) + (*7 a*)	180	3 h
7	6-Phenyl--8-carbonitrile, mp. 165–166°	CN	Ph	75	(*121 b*) + (*7 i*)	70	1.5 h
8	6-Methoxy-8-methoxycarbonyl----, mp. 107–108°	CO_2Me	OMe	18	(*121 c*) + (*7 e*)	180	2 h

*) = 2-Benzyl-3-dimethylaminoallylidene-dimethylammonium perchlorate prepared in analogy to (*7 b–d*) by formylation of benzylmalonic acid with DMF-$POCl_3$; mp. 145° (92%)[18].

[1]) λ_{max} (log ϵ) in CH_2Cl_2 = 657 (2.02), 597 (2.51), 547 (2.66), 388 (3.26), 369 (3.57), 351 (3.60), 302 (4.54), 288 nm (4.80).

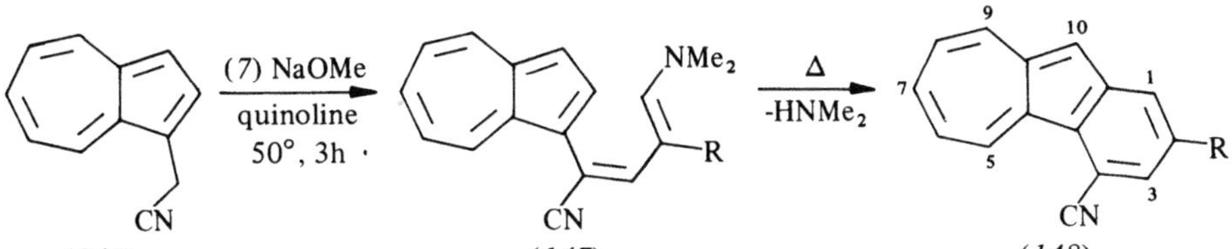

(146) (147) (148)

	(148) --Benz(a)azulene-4-carbonitrile	R	Yield %	Components	Reation conditions (°C)	Time
1	----, mp. 87–88° (subl.)[1])	H	41	(146) + (7a)	200–210	24 h
2	2-Phenyl--, mp. 174–175°	Ph	79	(146) + (7i)	150	18 h

1) λ_{max} (log ϵ) in CH_2Cl_2 = 655 (2.69), 596 (2.78), 555 (2.71), 402 (3.74), 381 (3.89), 362 (3.80), 342 (3.78), 308 nm (4.86)

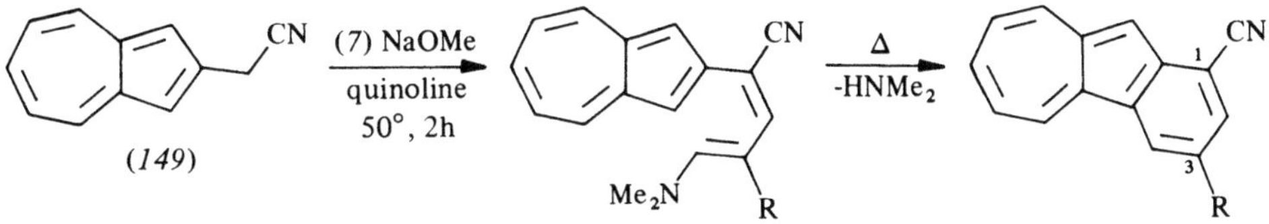

(150) (151)

	(151) --Benz(a)azulene-1-carbonitrile	R	Yield %	Components	Reaction conditions (°C)	Time
1	----, mp. 166–167°	H	77	(149) + (7a)	190	7 h
2	3-Methyl---, mp. 115–116°	Me	84	(149) + (7b)	100	2.5 h
3	3-Phenyl---, mp. 139–140°	Ph	84	(149) + (7i)	100	2.5 h

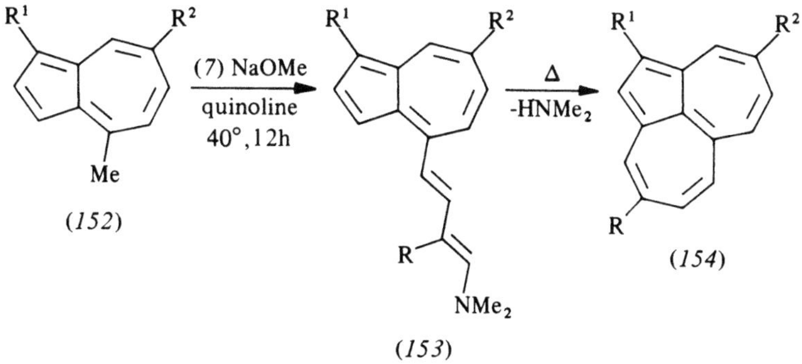

Table 17. Cyclopenta(e,f)heptalenes (*154*)

(*154*)	R	R¹	R²		Yield %¹)	Components	Reaction conditions (°C)	Time
a	H	H	H	Red-violet powder mp. 64–65°	16	4-Methylazulene (*7a*)	190	48 h
b	Ph	H	H	Red-brown powder mp. 87–88°	20	4-Methylazulene (*7i*)	190	48 h
c	Ph	Me	i-Pr	Red-violet crystals mp. 78–79°	62	Guaiazulene (*7i*)	180	24 h

¹) Incomplete conversion, ca. 40%–60%.

This mode of cyclization opened a new access[74] to the stable, tetracyclic peri-condensed, nonbenzenoid aromatic hydrocarbon dicyclopenta(e,f–k,l)heptalene (*159*), also named[73] azupyrene. The 4,5-cyclopentenoazulene (*156*) needed was formed in 28% yield by a Hafner synthesis from the pentamethinium salt (*155*)[75] and sodium cyclopentadienide:

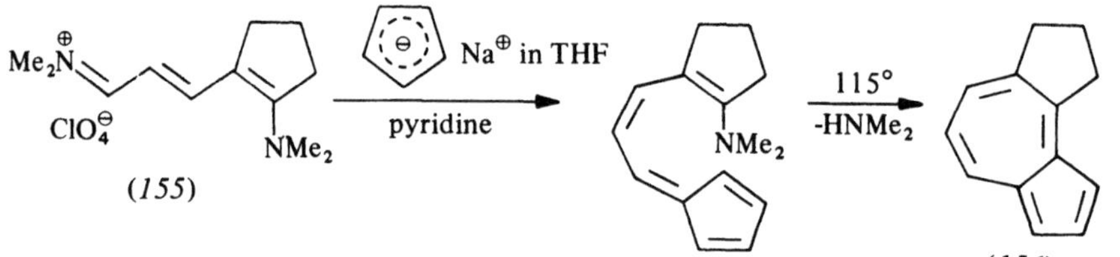

(*155*)

(*156*)
Blue oil (28%)

The azulene (*156*) slowly condenses with (*7a*) to the red-violet azulene dieneamine (*157*) and cyclization occurs at 150–200°, giving in 58% yield by a 70% conversion a mixture of the expected 1,2-dihydrodicyclopenta(e,f–k,l)heptalene (*158*) and the azupyrene (*159*), formed by simultaneous dehydrogenation[74]. Treatment of the mixture with chloranil in benzene leads quantitatively to the pure (*159*), bronze-shining, black plates, mp. 257–259° (subl.).

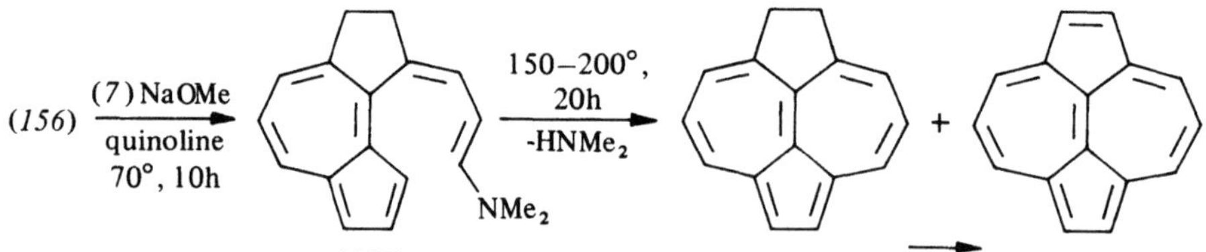

(*157*) (*158*) -2 H (*159*)

(*159*): UV (in CH_2Cl_2), λ_{max} (log ε) = 484 (4.21), 471 (3.50), 461 (3.21), 454 (3.27), 444 (3.17), 432 (2.75), 423 (2.75), 410 (2.89), 397 (2.58), 387 (2.50), 358 (3.64), 344 (4.26), 334 (4.20), 309 (4.28), 299 (4.30), 285 (4.46), 267 (4.99), 254 nm (4.70)

The azupyrene shows a very characteristic UV spectrum, with numerous sharp absorption bands[73, 74]. The high thermodynamic stability of (*159*) is consistent with a marked tendency for its formation, though in minor amounts, provided the structurally appropriate components are brought to reaction[76]. A 1% yield of azupyrene (*159*) is obtained if a solution of 6-dimethylaminovinylfulvene (*160*) in pyridine is treated with iodine in the presence of sodium methoxide and, after refluxing and distilling off the pyridine, the mixture is heated in quinoline up to 230°. The bis-fulvene (*161*) may be an intermediate in the formation of (*159*):

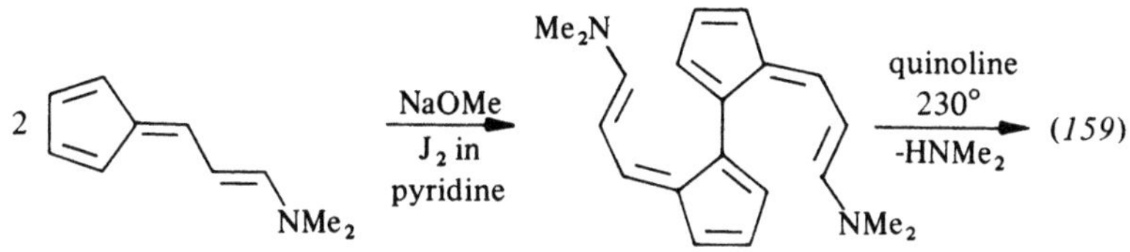

(*160*) (*161*)

Also condensation of the salt (*155*) with (*160*), followed by heating in quinoline, furnishes (*159*) in a 5% yield besides the azulene (*156*) (2.5%). The fulvenoid compound (*162*) may here be the intermediate, and in accordance with this assumption the azulene (*156*) could be generated by a splitting reaction of the azulene dieneamine formed (*157*).

Furthermore, the formation of the azupyrene (*159*) was achieved in a 2.5% yield, besides the azulene (*156*) in 3% yield, by heating a condensation mixture of the stable nonamethinium salt (*164*), cyclopentadiene, and sodium methoxide in quinoline at 230° for 48 hours, which should also create (*162*).

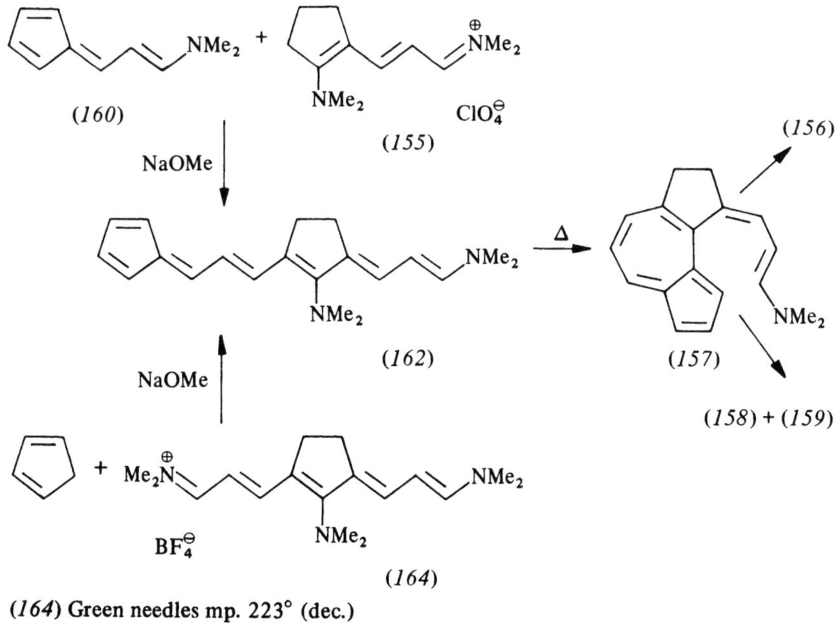

(*164*) Green needles mp. 223° (dec.)

The dyestuff used, salt (*164*), was obtained as shown in the following scheme by double condensation of cyclopentanone with the acetal (*35 a*) to the divinylogous urea (*163*) and conversion of the latter to (*164*):

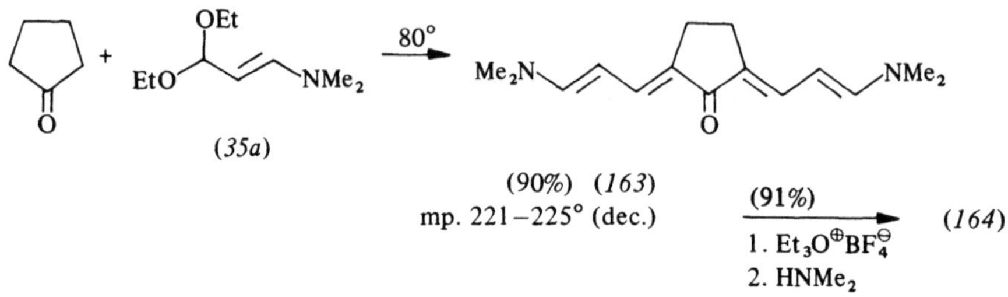

If one employs the benzanalogous indene polymethinium salt (*165*) in the condensation reaction with cyclopentadiene and heats for 20 h up to 230°, the benzo(a)-dicyclopenta(e,f–k,l)heptalene [= benz(a)azupyrene] (*166*) results in a 3.6% yield. The synthesis of the (*165*) used can be seen in the following scheme[6, 74]:

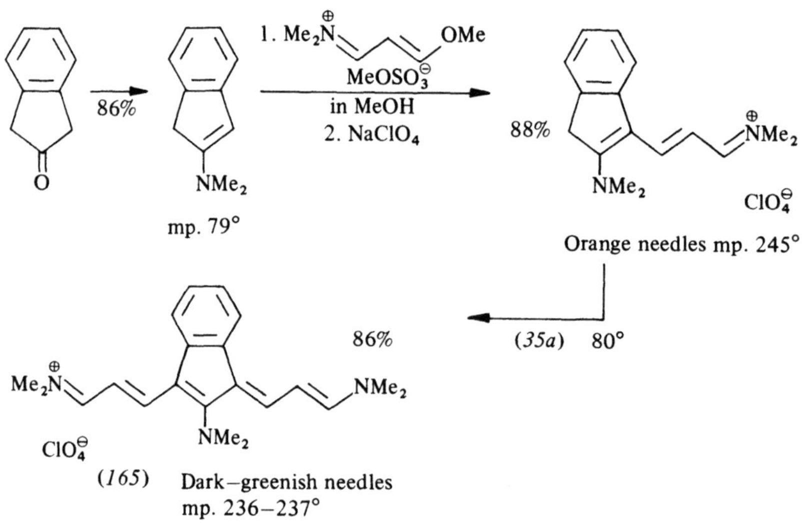

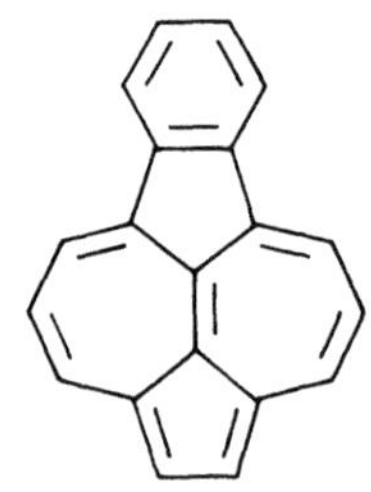

benzo(a)dicyclopenta(e,f–k,l)heptalene dark-brown needles, mp. 198–200°
UV(CH_2Cl_2): λ_{max} (log ε) = 498 (3.84), 485 (3.98), 464 (3.66), 453 (3.54), 422 (3.09), 387 (4.28), 369 (3.94), 348 (4.06), 342 (4.02), 300 (4.82), 291 (4.75), 286 (4.74), 272 (4.78), 263 (4.88), 253 (4.87), 237 nm (4.81)

(*166*)

As mentioned in Section III., B. 3., the use of the cyclic heptamethinium salt (*44*) in condensation-cyclization reaction provides an approach to several azulenoid compounds. But arylacetonitriles usually fail to condense with (*44*) under ordinary reaction conditions. The strongly electrophilic (*44*) reacts in an irreversible manner with methoxide or other bases before the strongly basic anion is created by proton abstraction from the arylacetonitrile. However, the more acidic compounds, *e.g.*, phenalene (*103*)[15], acenaphthenone (*25*), fluorene (*124*), and benzofluorenes condense smoothly with (*44*) by addition of sodium methoxide and can be cyclized thermally to new, polycyclic azulene derivatives[47], as can be visualized by the formulas:

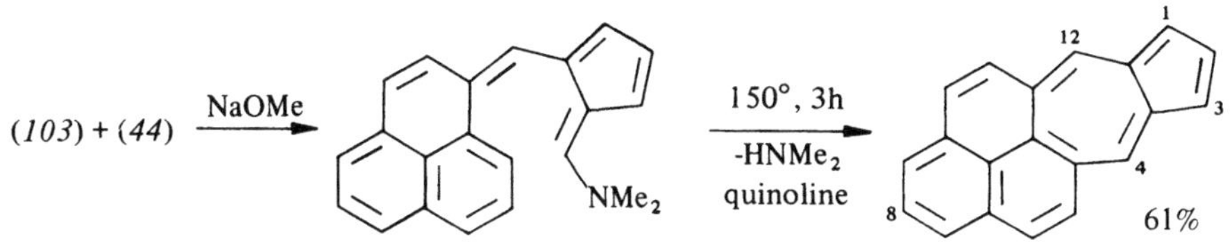

(*167*)
Red violet

(*168*)
Azuleno(5, 6, 7–c, d)phenalene
Green plates, mp. 184°

UV (EtOH): λ_{max} (log ε) = 739 (3.26), 726(2.78), 714 (2.70), 700 (3.11), 674 (3.08), 665 (3.15), 641 (2.90), 631 (2.93), 613 (2.85), 449 (4.97), 441 (4.65), 424 (4.50), 402 (3.96), 397 (3.90), 384 (3.71), 380 (3.72), 360 (3.85), 316 (5.11), 309 (5.06), 247 nm (4.38)

Also a methyl and a phenyl derivative of (*168*) were prepared; analogously, the related 1,2-dihydroazuleno(5,6,7-c,d)cyclopenta(g,h)phenalene (*169*)[65] was also obtained in moderate yield using the 1,2-dihydro-5H-cyclopenta(c,d)phenalene (*115*). Both hydrocarbons, (*168*) and (*169*) are strongly suspected of cancerogenity[77].

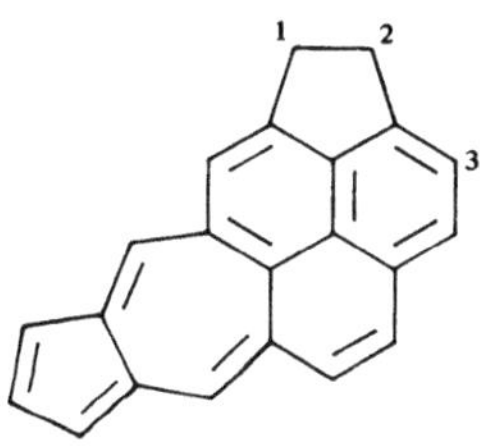

(*169*) Grey-green needles, mp. 154°
(8% yield)

As we accounted for the difficulties mentioned in ring-closures of fluorene derivatives (*125*) to fluoranthenes, the combinations of five- and six-membered carbon ring systems in a planar arrangement are associated with considerable angle strain. This obstacle should be cancelled out essentially in combinations of five- and seven-membered ring systems. The condensation products (*170*) and (*172*) of acenaphthenone and fluorene with (*44*) in the following examples indeed cyclize under relatively smooth conditions by formation of such systems[47].

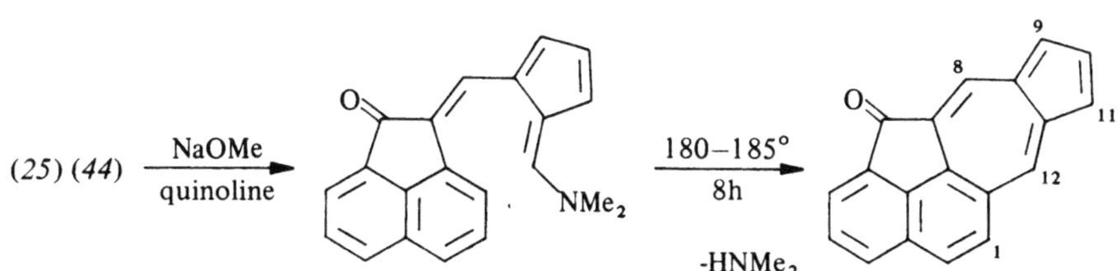

(*170*) Deep-violet solution

(*171*) 73%

(*171*) azuleno(3,6,7-b,c)acenaphthylene-6-one grey-green needles, mp. 172–173°
UV (CH_2Cl_2): λ_{max} (log ϵ) = 690 (2.43), 617 (2.83), 571 (2.91), 358 (4.07), 302 (4.76), 288 nm (4.89)

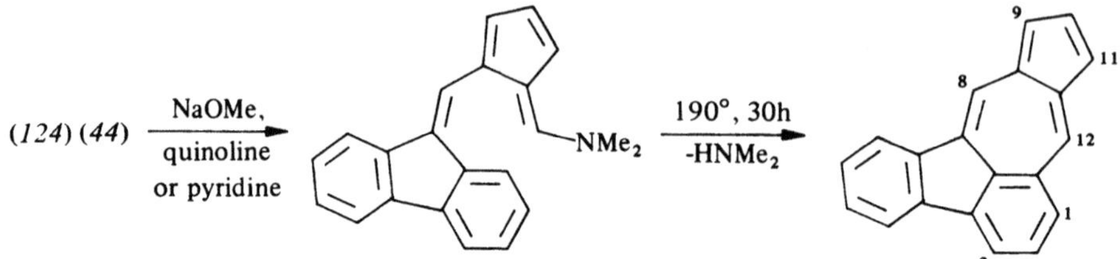

(*172*)
Brick-red needles
mp. 130° (dec.) 48%

(*173*) 41%

(*173*) azuleno(5.6.7-j,k)fluorene, grey-green needles, mp. 144°
UV (CH_2Cl_2): λ_{max} (log ϵ) = 668 (2.47), 604 (2.84), 556 (2.93), 435 (3.97), 415 (3.99), 332 (4.44), 302 nm (4.74)

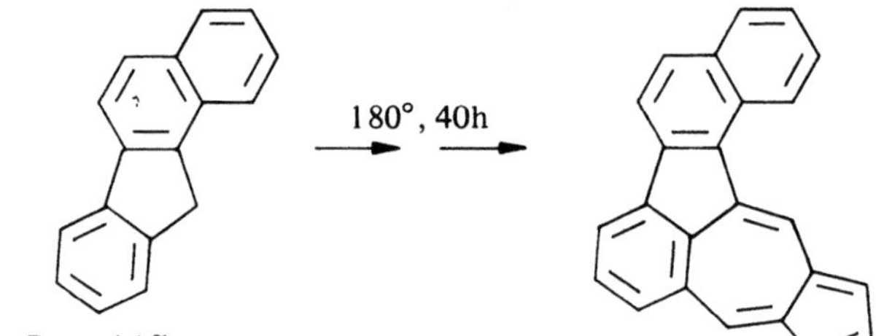

Benzo(a)fluorene

(*173a*) 35% Shiny-black needles mp. 159°

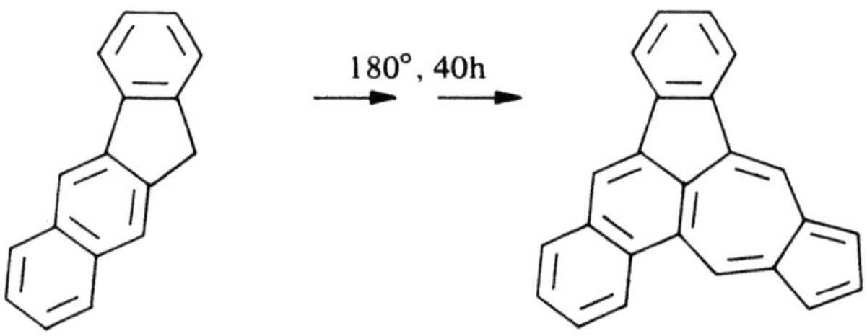

Benzo(b)fluorene

(*173b*) 30% Light-green needles mp. 259°

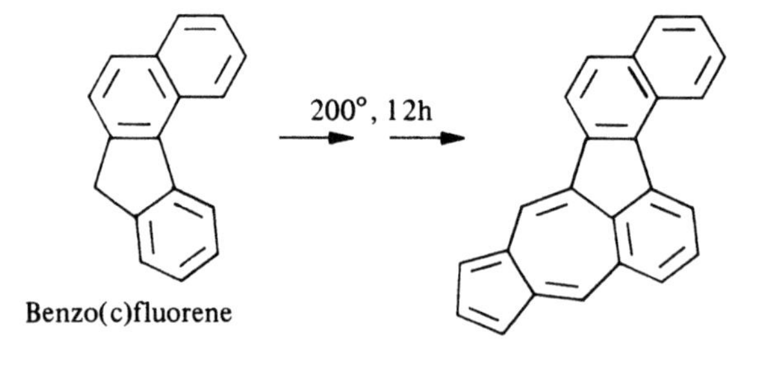

Benzo(c)fluorene

(*173c*) 19% Shiny bronze needles mp. 190°

Starting with the benzofluorenes, in the same manner one obtains the corresponding azuleno(5,6,7-j,k)benzofluorenes (*173 a–c*), according to expectation[47].

In an analogous manner, we succeeded finally in synthesizing the 5-carbonitrile of the kata-condensed tetracyclic azuleno(1,2-f)azulene (*176*) in a 50% yield by condensation of (*44*) with azulene-2-acetonitrile (*149*) to the red-violet aminonitrile (*174*) and smooth thermolysis of it[78]. The hydrocarbon (*176*) is the first known member of the theoretical 12 isomeric azulenoazulenes of which considerable stability and aromatic properties are predicted[79]. Indeed, (*176*) can readily be formylated by dimethylformamide-phosphorylchloride in a typical aromatic-electrophilic substitution to the 1-carbaldehyde (*177*)[78].

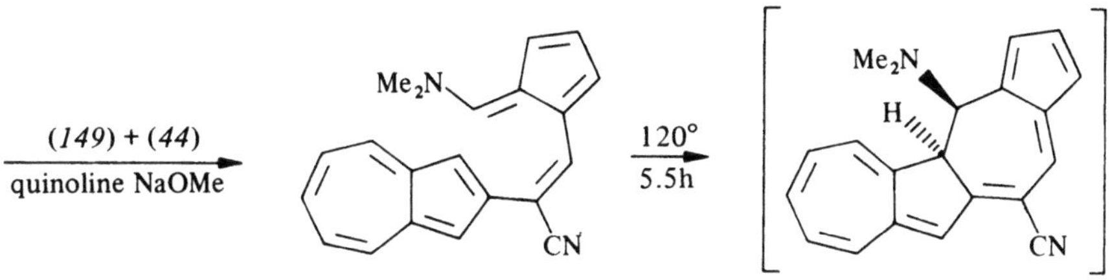

(*174*) (*175*)

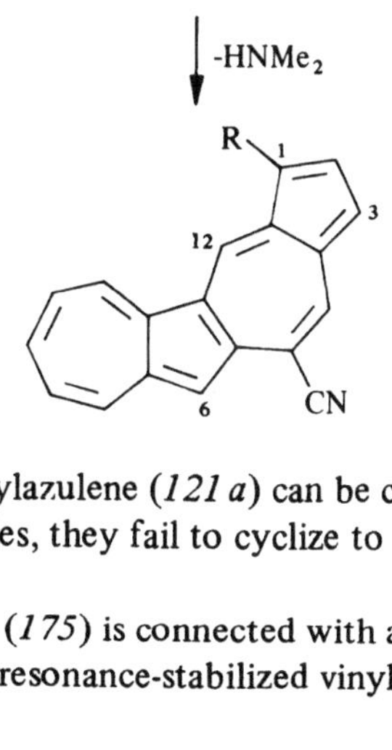

(*176*): azuleno (1,2-f)azulene-5-carbonitrile dark-brown-violet needles, mp. 210–211° (dec.) UV (CH_2Cl_2): λ_{max} (log ϵ) = 666 (2.93), 434 (4.79), 407 (4.63), 362 (4.56), 336 (4.49), 259 nm (4.37)

(*176*): R = H (50%)
(*176*) → (*177*): R = CHO (75%) dark plates, mp. > 340° (dec.)

Although azulene-1-acetonitrile (*146*) and 6-methylazulene (*121 a*) can be condensed with (*44*) to corresponding colored aminonitriles, they fail to cyclize to azulenoazulenes as hoped.

Only the ring-closure of (*174*) to the intermediate (*175*) is connected with a distinctly low localization energy, as visualized by the resonance-stabilized vinylogue sesquifulvalene structure in (*175*).

E. Helicenes and Related Systems by Combination of Thermolytic Ring-Closure with Photodehydrocyclization

Dissolved orange-red 1-dimethylamino-2,2-bis(9-fluorenylidenemethyl)ethylene (*178*) slowly fades out in sunlight, generating strongly fluorescent solutions. Irradiation of (*178*) in tetrahydrofuran by means of a high pressur mercury lamp at 600 W (TQ 718, Hanau-Quarzlampen GmbH) allows one to isolate a yellow hydrocarbon, the benzo(e)-fluoreno(9,1-kl)acephenanthrylene (*180*) in 64% yield[32, 80]. The direct formation of (*180*) by light involves two ring-closures and seems to be a unique property of (*178*). We can split up the synthesis of (*180*) also in two different cyclization steps: first by thermolysis of (*178*) at 180–190°, which leads by the usual electrocyclic process with elimination to 2-(9-fluorenylidenemethyl)fluoranthene (*179*) in 81%

yield, and second by irradiation of a benzene solution of the latter (*179*) in the presence of oxygene and traces of iodine to effect a photodehydrocyclization[3], finally affording (*180*).

The combination of the two cyclization processes, not limited to the example above, opens a new general access to helicenes and related compounds.

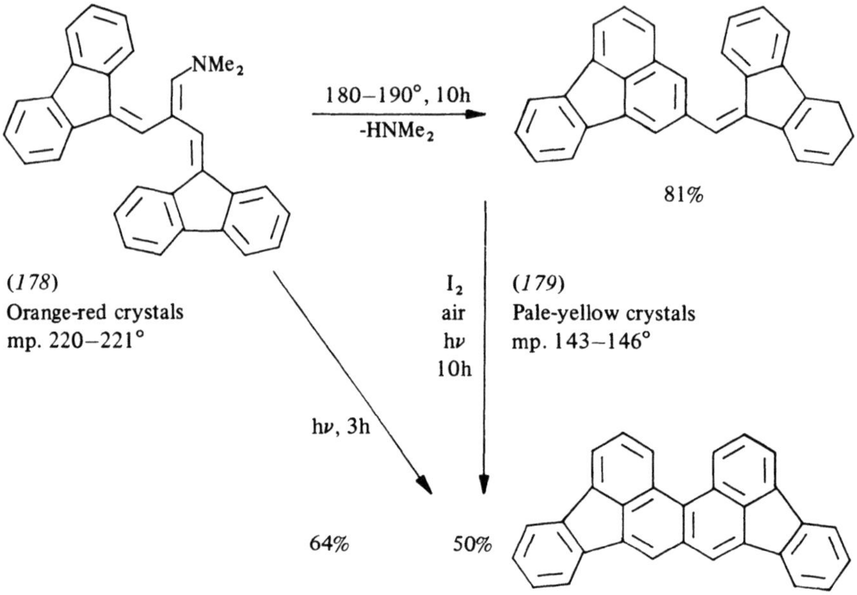

(*178*)
Orange-red crystals
mp. 220–221°

(*179*)
Pale-yellow crystals
mp. 143–146°

(*180*)
Yellow needles
mp. 259–260°

The dicationic vinamidinium salt (*181*), a derivative of triformylmethane, represents a valuable starting compound for our new syntheses. It is readily accessible and has been prepared by formylation of (*7a*)[24], chloro- and bromoacetic acid[20] and malonic acid[20, 80], with dimethylformamide/phosphorylchloride.

Thus the fulvenoid (*178*) is obtained in a 87% yield by the twofold condensation of fluorene with (*181*) in the presence of sodium methoxide[24]. Analogously, condensation of (*178*) succeeds with phenylacetonitriles (*53*), naphthalene-2-acetonitrile (*70*), and phenanthrene-3-acetonitrile in a corresponding manner to the colored, cross-conjugated aminonitriles (*182*), (*185*), and (*188*), as outlined in the following reaction schemes.

1. Benzo(c)phenanthrenes

As mentioned above, the reaction sequence of thermolysis and irradiation generates two new, angular anellated benzene rings. Starting with mononuclear phenylacetonitriles (*53*), by double condensation one finally reaches the tetracyclic benz(c)-phenanthrene-5,8-dicarbonitriles (*184*):

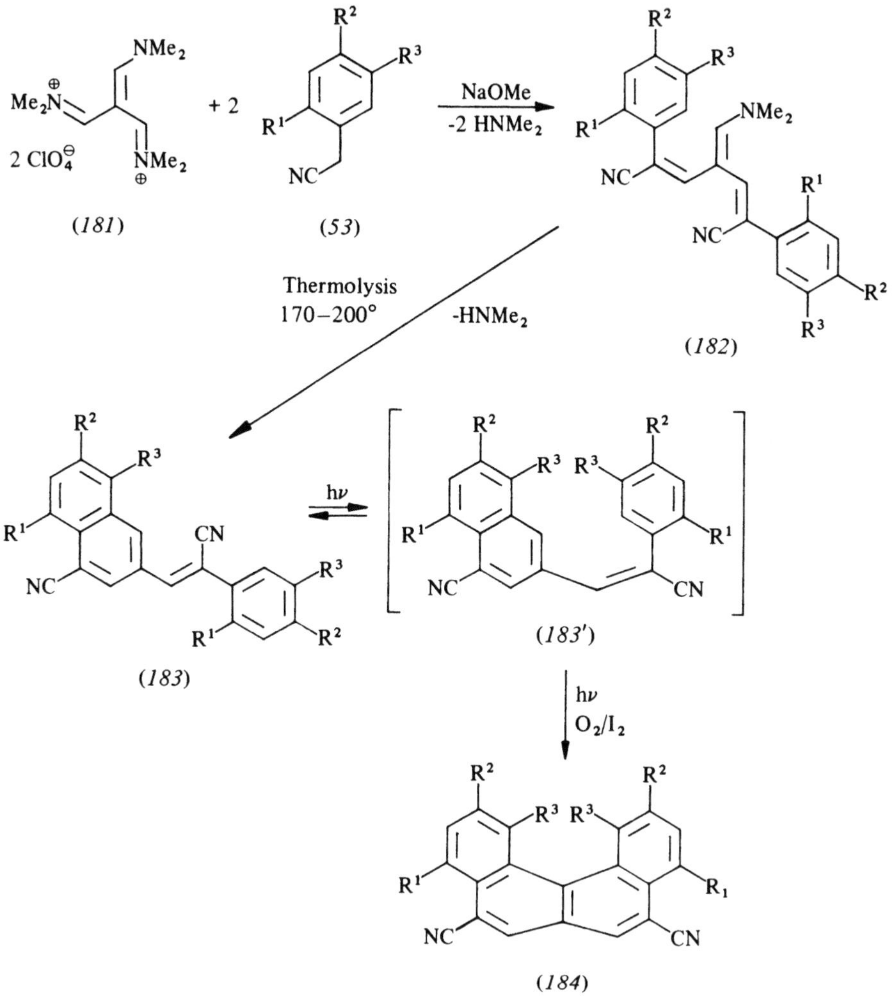

Table 18. Benz(c)phenanthrene-5,8-dicarbonitriles (*184*)

(*184*)		R^1	R^2	R^3	(*182*)	(*183*)
a	--, mp. 331–332° irr. (73%)	H	H	H	mp. 136–137° (67%)	mp. 152–153° 180°, 2 h (98%)
b	2,11-Dimethyl--, mp. 327°, irr. (85%)	H	Me	H	Not isolated	mp. 183° 180°, 5 h (60%)
c	2,11-Dimethoxy--, mp. 319–320°, irr. (76%)	H	OMe	H	Not isolated	mp. 170–171° 170°, 8 h (40%)
d	2,4,9,11-Tetramethyl-, mp. > 350°, irr. (60%)	Me	Me	H	Not isolated	mp. 170–172° 190°, 2 h (20%)
e	1,4,9,12-Tetramethyl-, mp. 317–318°, irr. (60%)	Me	H	Me	Not isolated	mp. 188–189° 200°, 2 h (46%)

We prepared (*184d*) and (*184e*) to study overcrowding of the methyl groups in positions 1 and 12 by the diamagnetic anisotropy effects of the aromatic rings. The 2,11-CH_3 groups in (*148d*) give rise to a singlet at τ = 7.42 ppm in the ^1H NMR spectra – the 4,9-CH_3 are found at τ = 6.75 ppm – whereas the 1,12-CH_3 groups in (*148e*) are shielded to τ = 8.28 ppm.

Pure benz(c)phenanthrene, mp. 67–68° and 2,11-dimethylbenz(c)phenanthrene, mp. 129–130°, can be simply obtained hydrolyzing (*148a*) or (*148b*) by potassium hydroxide in triglycol at 180° and decarboxylation of the dicarboxylic acids in boiling quinoline in the presence of copper powder, giving a 91% or 70% yield, respectively[32, 81]. The sterically screened carbonitrile groups in (*148d*) and (*148e*) cannot be saponified.

2. Phenanthro(3,4-c)phenanthrenes ("Hexahelicenes")[32, 80]

An analogous synthesis is that of the two hexahelicenes (*187a*) and (*187b*), using naphthalene-2-acetonitrile (*70a*) or 7-methylnaphthalene-2-acetonitrile (*70b*) in

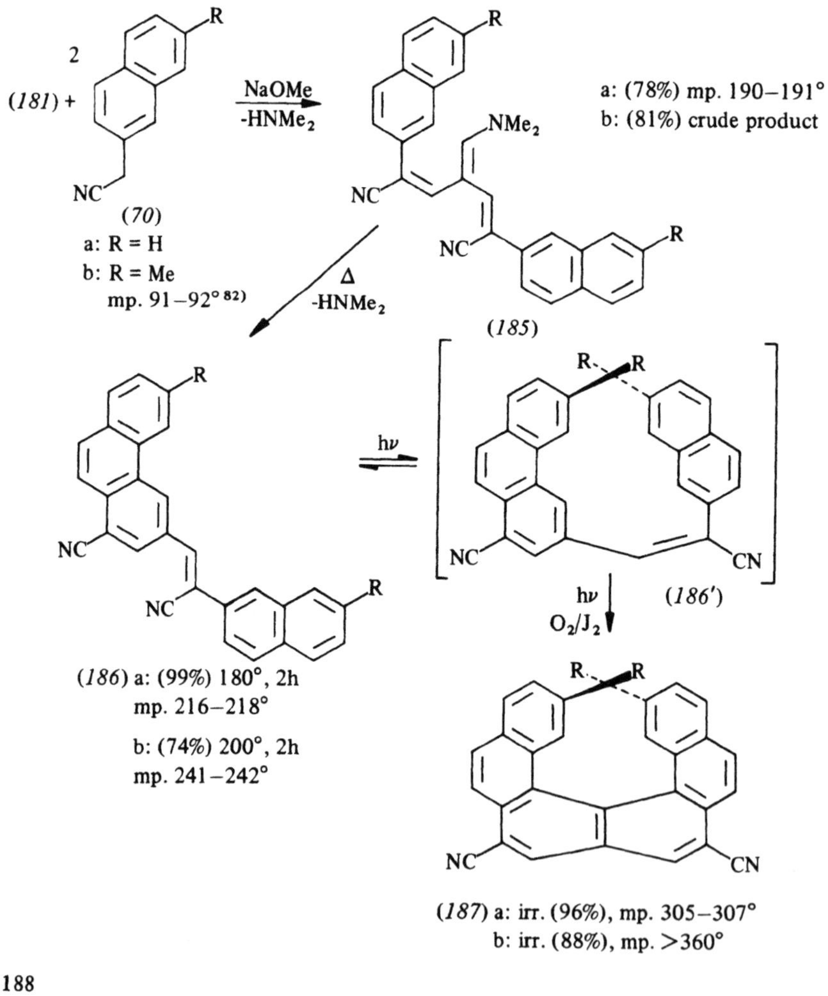

the condensation with (*181*). 3,6-Dimethylnaphthalene-1-carbonitrile (*56/6*) (see Section III., C. 1.; Table 12) served as starting material for the preparation of (*70 b*).

By hydrolysis and decarboxylation in the usual manner, the dinitriles (*187 a*) and (*187 b*) are transformed to the hydrocarbons, to hexahelicene (74%), mp. 227–228°, pale-yellow plates[83], and to 1,15-dimethylhexahelicene [numbering by Newman[83]] (81%), mp. 215–216°, pale-yellow needles.

In the latter hydrocarbon, the two methyl groups are partially lying above the plane of an aromatic ring, similarly, to (*184 e*), and hence they experience an shielding effect by the diamagnetic anisotropy of the system. The methyl signal in the ^{1}H NMR spectrum, as with (*184 e*), is found at $\tau = 8.28$ ppm.

3. Octahelicene [Naphtho(2,1-c)phenanthro(4,3-g)phenanthrene]

Phenanthrene-3-acetonitrile (*88*) condenses smóothly with (*181*) in 81% yield to the aminonitrile (*188*), mp. 240–242° (dec.). But, as we have just discussed for the related aminonitriles (*89*) and (*94*) (see Section III. C. 6.) the thermolysis of (*188*) at 200° (2 h) leads on sterical reasons to a mixture (93%) of (*189*) and (*190*), a yellow powder, mp. 264–275°.

Without further attempts to separate this poorly soluble product, the mixture of (*189*) and (*190*) has been subjected to dehydrophotocyclization in benzene (5 h, 600 W), yielding presumably a mixture of (*191*) and (*192*) (58%). After hydrolysis (KOH in triglycol, 210°, 6 h) and decarboxylation (boiling quinoline, Cu powder, 5 h), pure octahelicene in pale-yellow plates, mp. 325–327°[84] could be separated in a 20% yield, like (*188*) by chromatography in benzone on alumina[32].

No effort has been made to isolate the hydrocarbon which must be derived from (*192*).

For preparative work in the helicene field in the manner mentioned, a handicap may arise in the irradiation step, owing to the often very small solubility of some diarylacrylonitriles of types (*183*), (*186*), (*189*), and (*190*) in the customary solvents suitable for the photodehydrocyclization.

The helicenoid hydrocarbons: benz(c)phenanthrene, dibenzo(c,g)phenanthrene, hexahelicene, 1,15-dimethylhexahelicene, and octahelicene, show a characteristic pattern of signals of their aromatic protons in the ^{1}H NMR spectra, owing to their symmetry[85], qualifying the spectra for use in determination.

IV. Syntheses of Heterocyclic Systems

1. Heterocycles by Benzannelation

Provided a suitable acid methylene group, *e.g.*, the cyanomethyl group, may be introduced conveniently in the heterocyclic nucleus, the benzannelation by condensation with a vinamidinium salt (*7*), followed by thermolytic cyclization, offers

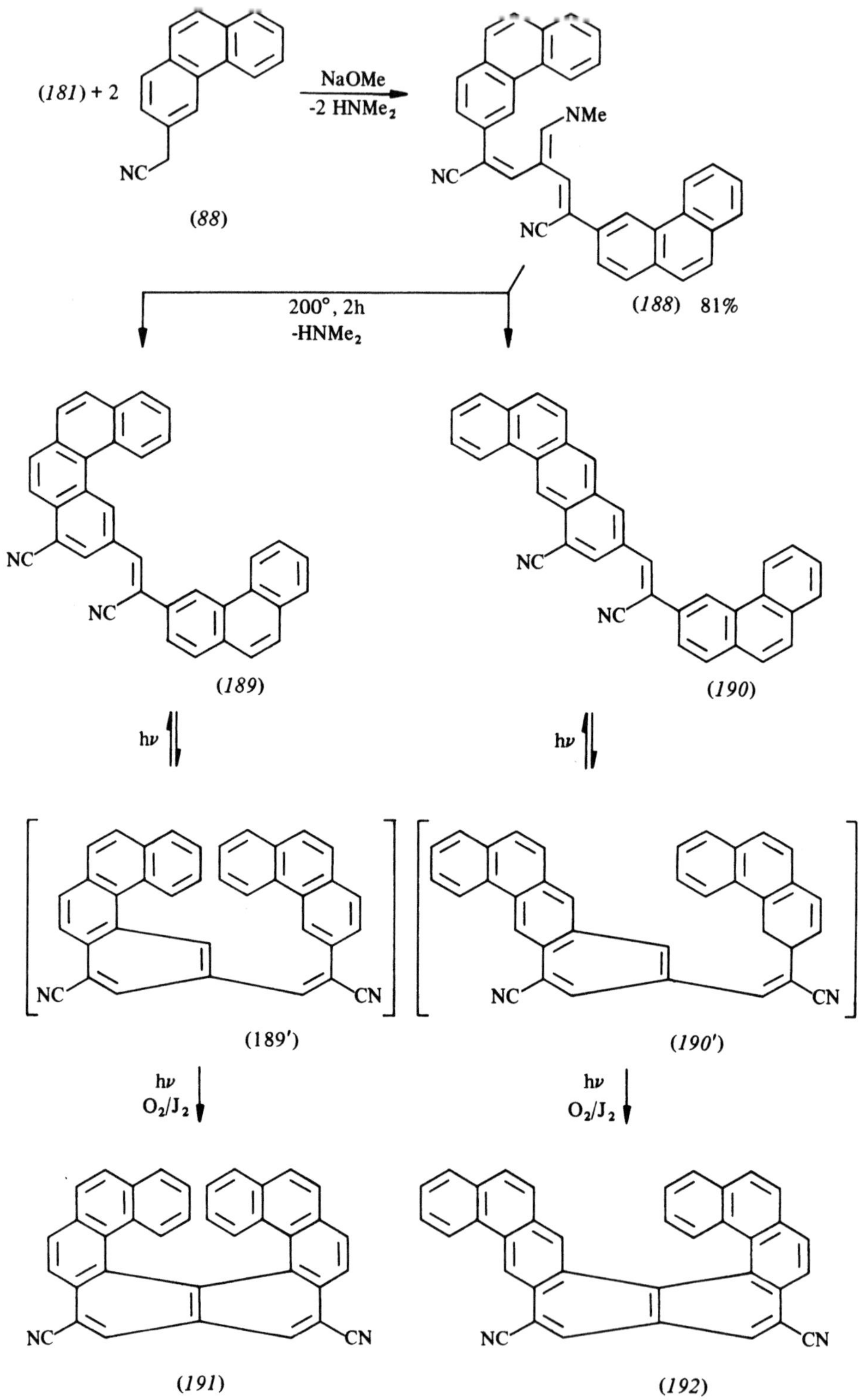
(181) + 2
NC
(88)
NaOMe
-2 HNMe2
NMe
NC
NC
(188) 81%
200°, 2h
-HNMe2
NC
NC
(189)
NC
NC
(190)
hν
hν
NC
CN
(189′)
NC
CN
(190′)
hν
O2/J2
hν
O2/J2
NC
CN
(191)
NC
CN
(192)

no special problem. The formation of 1,2,3,4-tetrahydroisoquinolines (*39*) from 1,2,3,6-tetrahydropyridine-4-acetonitriles (*38*) or xanthones (*43*) from 2-methylchromone (*19*) or 2,6-dimethyl-4-pyrone, are examples of such preparations (Section III. B., 2.).

As shown in the following reaction scheme, the 2-acetonitriles of furan (*193a*)[87], thiophen (*193b*), and N-methylpyrrole[88] are readily transformed to their corresponding benzoderivatives, the benzo(b)furans (*196a*), benzo(b)thiophens (*196b*), and 1-methylindoles (*196c*), with or without intermediate isolation of the condensation products (*194a–c*):

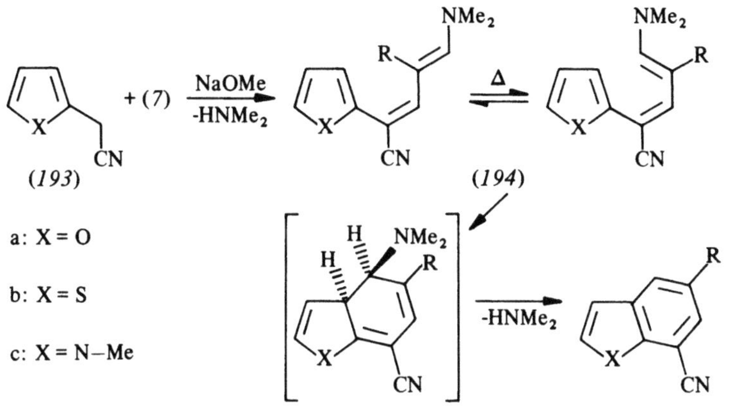

(*193*) (*194*) (*195*) (*196*)

a: X = O

b: X = S

c: X = N–Me

To achieve a further benzannelation, benzo(b)thiophen-7-carbonitrile (*196b*) has been converted by a four-step sequence – hydrolysis by KOH in cellosolve (150°, 1 h), reduction of the acid by sodium bis(methoxyethoxy)aluminiumdihydride, action of thionyl chloride on the carbinol, and reaction of the chloromethyl compound with sodium cyanide in DMSO – in a 75% overall yield to the oily benzo(b)thiophen-7-acetonitrile (*197*), which finally afforded the naphtho(1,2-b)-thiophen-9-carbonitriles (*198a, b*):

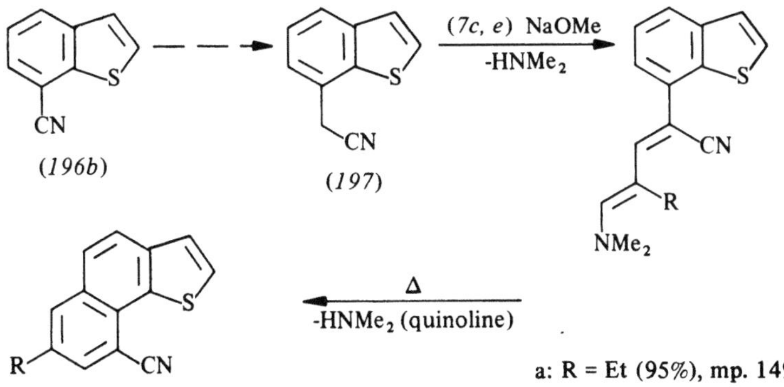

(*196b*) (*197*) (*198*)

a: R = Et (95%), mp. 149° (180°, 3 h)
b: R = OMe (72%), mp. 156° (170–180°, 5 h)

Table 19. Benzannelation to 5-ring heterocycles

						Reaction conditions	
(196 a)	*Benzo(b)furan-7-carbonitriles*	Yield	R	Components	Condensation product	(°C)	Time
1)	--, mp. 59° (subl.) colorless needles	81	H	*(193 a)* + *(7 a)*	*(194 a)* 96%, mp. 99° orange needles	200–220	4 h
2)	5-Methoxy--, mp. 103° colorless needles	93	OMe	*(193 a)* + *(7 e)*		175–180	2 h
3)	5-Benzyl--, mp. 73° colorless needles	82	CH_2Ph	*(193 a)* + *(7*)*		160–170	1.5 h
(196 b)	*Benzo(b)thiophen-7-carbonitriles*						
1)	--, mp. 67° (subl.) colorless needles	98	H	*(193 b)* + *(7 a)*	*(194 b)* 90%, mp. 74–76° yellow needles	210–220	6 h
2)	5-methoxy--, mp. 84–85° colorless needles	83	OMe	*(193 b)* + *(7 e)*		180–190	3 h
3)	5-benzyl--, mp. 76–77° colorless needles	88	CH_2Ph	*(193 b)* + *(7*)*		170	2 h
(196 c)	*1-Methylindole-7-carbonitriles*						
1)	--, mp. 68–69 colorless needles	94	H	*(193 c)* + *(7 a)*	*(194 c)* 98%, mp. 108–109°, yellow needles	200–220	6 h
2)	5-methyl--, mp. 87° colorless needles	82	Me	*(193 c)* + *(7 b)*		165–170	1.5 h
3)	5-methoxy--, mp. 72–73° colorless needles	86	OMe	*(193 c)* + *(7 e)*		165–170	1.5 h
4)	5-phenyl--, mp. 126° colorless needles	78	Ph	*(193 c)* + *(7 i)*		160–170	1 h

(7*) See footnote Table 16 (Section III. D. 1.).

Analogously, the benzthiazole (*199*) and the indazole (*200*) have been obtained, starting with the corresponding thiazole- and pyrazole-4-acetonitriles:

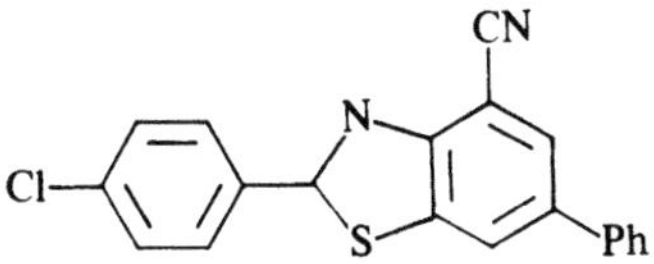

(*199*)

mp. 171–172.5° (98%)
2-(4-chlorophenyl)thiazole-4-acetonitrile + (*7i*) 120°, 4 h

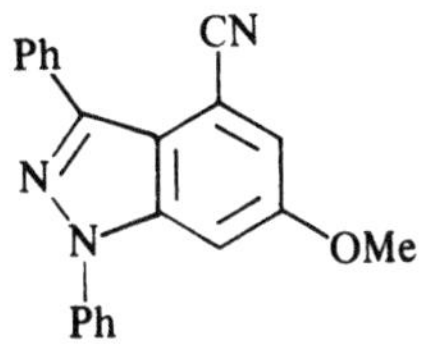

(*200*)

mp. 157–158° (84%)
1,3-diphenylpyrazole-4-acetonitrile + (*7e*) 140°, 4 h

For the synthesis of dibenzofuran (*201*), dibenzothiophenes (*202*), and carbazoles (*203*) by benzannelation, in the same manner, the 3-acetonitrile of benzo(b)furan, benzo(b)thiophen, and of indole are used:

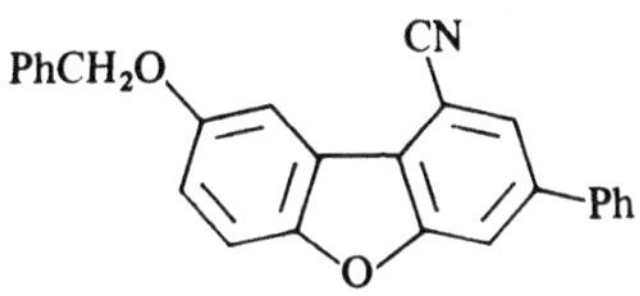

(*201*)

mp. 135–136° (86%)
5-benzylbenzofuran-3-acetonitrile + (*7i*) 170°, 5 h

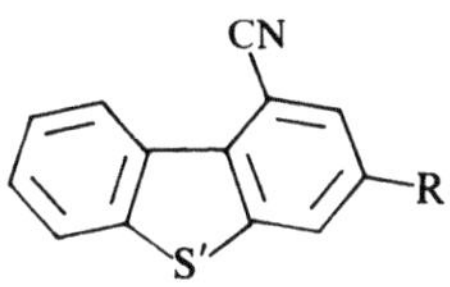

(*202*)

Dibenzothiophen-1-carbonitriles (*201*)

R	mp.	(Yield)	Benzo(b)thiophen-3-acetonitrile +	(°C)	Time
Me	139°	(93%)	(*7b*)	180	1.5 h
Et	104°	(78%)	(*7c*)	175	1.5 h
OMe	145°	(95%)	(*7e*)	170	1.5 h

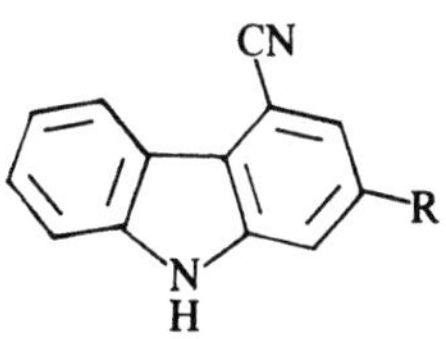

(*203*)

Carbazole-4-carbonitriles (*202*)

R	mp.	(Yield)	Indole-3-acetonitrile +	(°C)	Time
H	156–158°	(68%)	(*7a*)	190	10 h
OMe	142–143°	(71%)	(*7e*)	180	4 h
Ph	225–226°	(78%)	(*7i*)	160	3 h

Whereas pyridine-4-acetonitrile (*204*) leads by the condensation-cyclization sequence to derivatives of isoquinoline (*205*), pyridine-3-acetonitrile (*206*) gives rise to quinoline (*207*) as the main product, beside some isomeric isoquinoline (*208*):

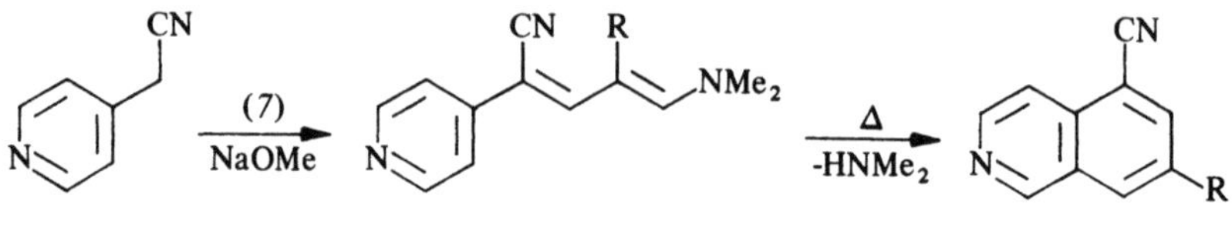

(*204*) (*205*)

7-methylisoquinoline-5-carbonitrile mp. 150–151° (91%)
(*204*) + (*7b*), 170°, 2 h

7-benzylisoquinoline-5-carbonitrile mp. 128–129° (90%)
(*204*) + (*7**), 160°, 3.5 h

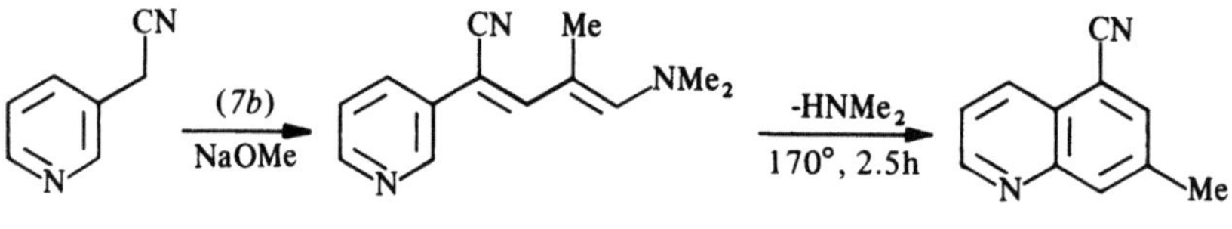

(*206*) (*207*)

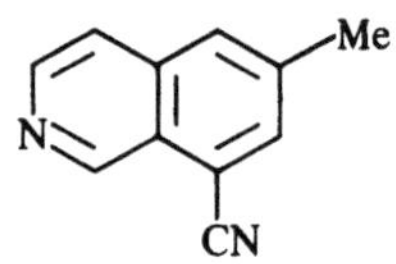

(*208*)

mixture of 7-methylquinoline-5-carbonitrile and 6- methylisoquinoline-8-carbonitrile mp. 103–125° (78%)

2. Heterocycles by Ring-Closure of Azahexatrienes

It has been suggested by kinetic measurements that the ring-closure reaction of derivatives of glutacondialdehyde (*209*) to pyridines (*211*) (*e.g.*, of the glutacondialdehyde dianil to the 1-phenylpyridinium ion), a long-known reaction[89], is an electrocyclic process[90]. The same should be valid also for the reverse process, the ring-opening of pyridinium salts by primary and secondary amines[89]. Indeed, the disrotatory mode of the ring-opening step of the intermediate dihydropridine (*210*) was evidenced recently by the stereoconfiguration *trans-cis-trans* of a derivative obtained (*209*)[91]:

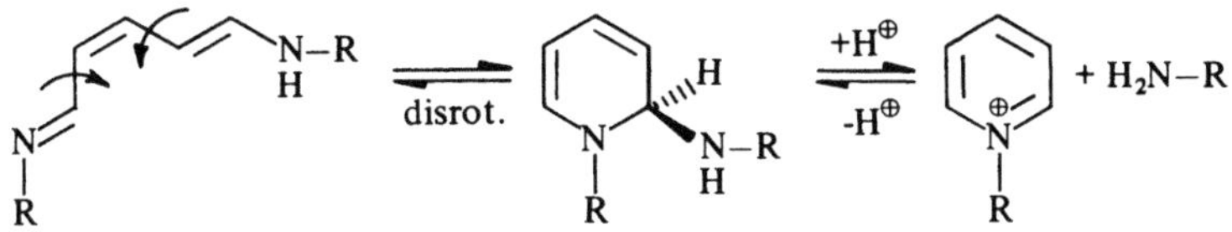

(*209*) (*210*) (*211*)

2.1. Pyridines. The ring-closure reaction of pentamethinium salts (*8*) by buffered ammonia (method A) or, better, fused ammonium acetate[92] (method B) to pyridines occurs through glutacondialdehyde derivatives related to (*209*) (R = H) and falls in the category of electrocyclic ring-closure with elimination dealt with, if we interpret (*209*) as types of amino-azahexatriene. The synthesis of pyridines should be of considerable synthetic value using pentamethinium salts (*8*) not accessible by ring-opening reactions of pyridines themselves. New salts (*8*) have been obtained by condensation reactions of vinamidinium salts and sodium methoxide (*7*) or their equivalents (*35*) with reactive methylene ketones[6], followed by transformation of the 5-dimethylaminopentadienones (*212*) to the salts (*8*)[6] (see also Section III. A.):

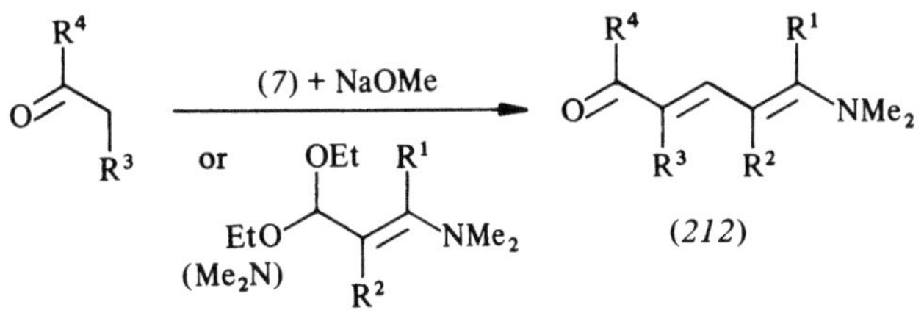

(*212*)

(*35*)

Some more complex pyridines can be obtained in the same manner, *e.g.*, acenaphtho(1,2-b)pyridine (7-azafluoranthene) (*214*)[93] mp. 96–97°, 5H-indeno(1,2-b)-pyridine (*215*), mp. 96–97°, and 5,6-dihydrobenz(h)quinoline (*216*), oil, picrate mp. 186–188°, starting from acenaphthenone (*25*) or 1-indanone and 1-tetralone.

(*212*)

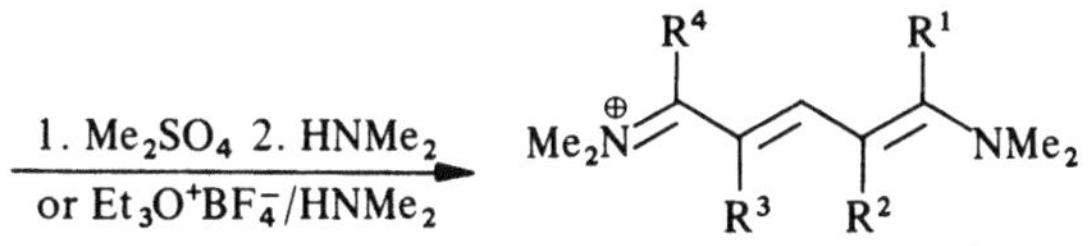

(*8*)

X: ClO_4, BF_4

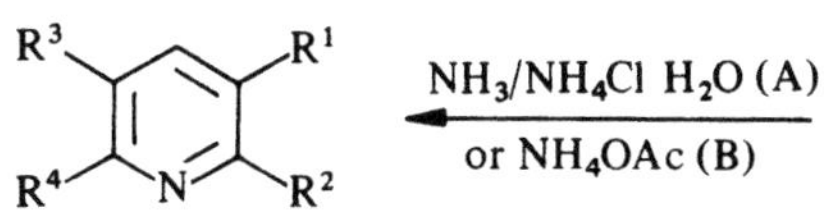

(*213*)

Some pyridines (*213*) have been prepared by this route[6]:

Table 20. Pyridines (*213*)

R^1	R^2	R^3	R^4		Yield	Method	Components for preparation of (*212*)
H	H	H	Ph	Oil picrate: mp. 174–175°	81%	A	Acetophenone + (*35 a*) or (*7 a*)
H	Ph	H	Ph	mp. 168–171°	98%	B	Acetophenone + (*7 i*)
Ph	H	H	Ph	mp. 80–81°	98%	B	Acetophenone + (*7 j*)
H	OMe	H	Ph	mp. 54–56°	85%	B	Acetophenone + (*7 e*)
H	H	H	Ph–CH=CH	mp. 90–91°	56%	A	Benzalacetone + (*35 a*) or (*7 a*)
H	H	Me	Ph	Oil picrate: mp. 166°	73%	A	Propiophenone + (*35 a*) or (*7 a*)
H	H	H	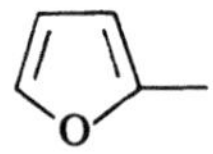	mp. 20° picrate: F. 174°	95%	B	2-Acetylfuran + (*35 a*) or (*7 a*)
H	H	H	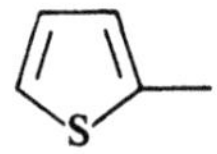	mp. 62°	60%	A	2-Acetylthiophene + (*35 a*) or (*7 a*)

The pentamethinium salt needed for the preparation of the 9H-indeno(2,1-b)pyridine (*217*), mp. 79–81°, could not be obtained by condensation of 2-indanone, but was obtained by substitution of 2-dimethylaminoindene with dimethyl-3-methoxy-allylideneammonium methylsulfate (the adduct of dimethylsulfate and 3-dimethyl-aminoacroleine) (see also Section III. D. 2.)[6].

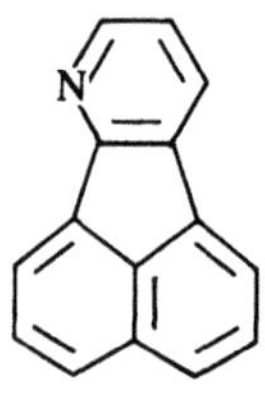

(*214*) mp. 96–97°
Acenaphto(1,2–b)pyridine
(7–azafluoranthene)

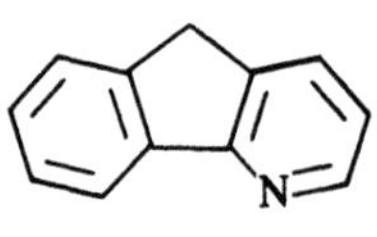

(*215*) mp. 96–97°
5H–Indeno(1,2–b)pyridine
(4–azafluorene)

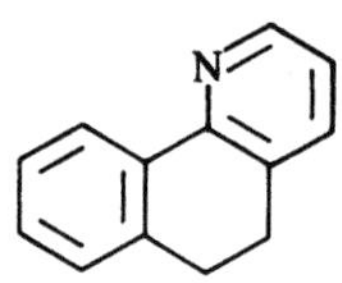

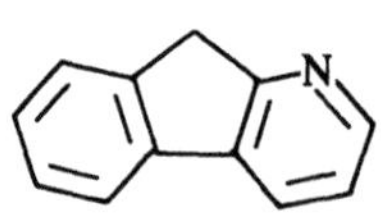

(*217*) mp. 79–81°
9H–Indeno(2,1–b)pyridine
(1–azafluorene)

2.2. Quinolines and Benzoquinolines. N.N′-Diphenylvinamidinium salts (*219*) cyclize on heating in boiling acetic acid to quinolines (*221*)[94, 95], by elimination of one aniline molecule.

The salts (*219*) have been prepared advantageously by action of anilines on 3-chloroacraldehydes (*218*) obtained in a Vilsmeier-Haack-Arnold formylation[25] of methyleneketones without intermediate isolation[94]. The ring-closure has been accounted for by an intramolecular electrophilic aromatic substitution, as shown in the following scheme:

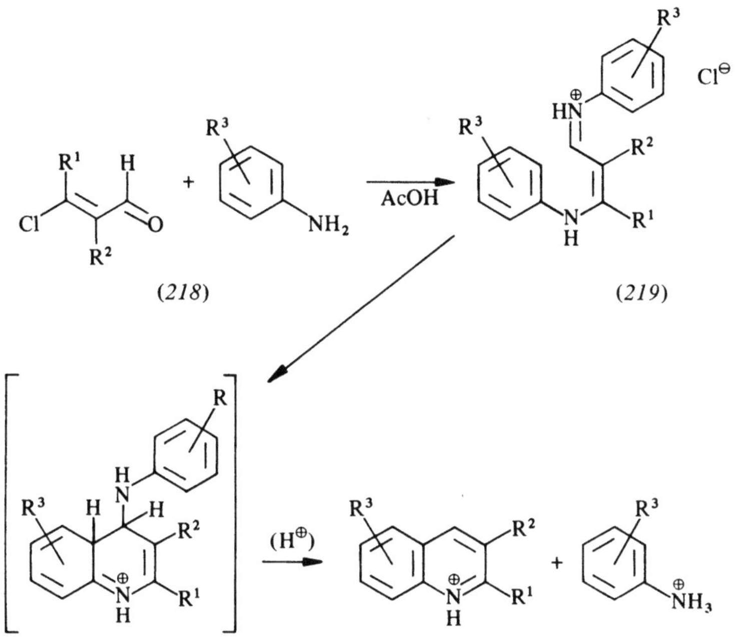

(*218*) (*219*)

(*220*) (*221*)

The electrophilic mechanism is suggested by the fact that salts (*219*) which bear a nitro group *para* or *meta* in the phenyl residues do not cyclize, whereas the methoxy group in this position as a donor substituent strongly enhances the cyclization of (*219*), stabilizing the intermediate cation (*220*)[94]. However, an electrocyclic course of the ring-closure is also not inconsistent with these findings. Thus the formation of the benzo(a)phenazine (*223*) from the protonated 1-benzeneazo-2-anilinonaphthalene (*22*) has been shown by kinetic studies to be an electrocyclic process[96]. See page 74.

The formation of (*221*) from (*219*) seems to be a borderline case. Inspection of the formulas used suggests that no rigorous discrimination between the two mechanisms, electrophilic substitution versus electrocyclic ring-closure, may be relevant.

Writing the mesomeric formula for (*219*) we will see a protonated azahexatriene system in (*219′*) borrowing two π-electrons from the benzene nucleus. We may rewrite this, with the corresponding stereochemical consequences. (See page 74).

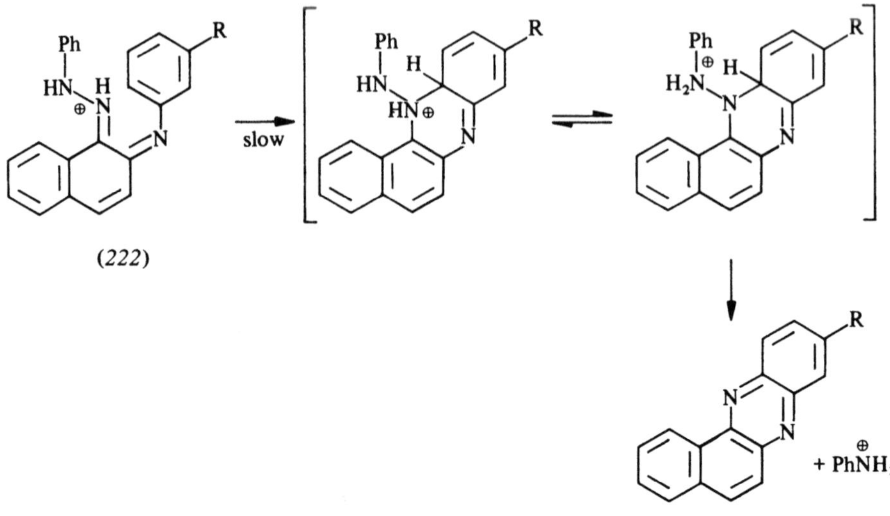

(222) (223)

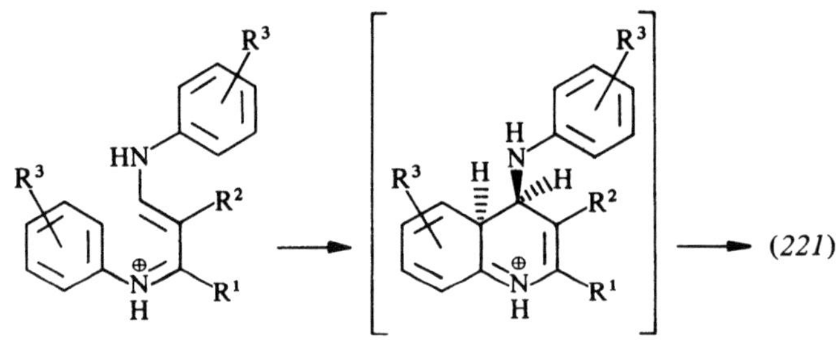

(219') (220')

Indeed, uncharged vinamidines, *e.g.* 3-dimethylaminoallylidenearylamines (*225*), split off dimethylamine smoothly on heating, either neat or in a high-boiling solvent, such as diglycol or quinoline, forming derivatives of quinolines (*227*), (*229*), and (*231*).

The vinamidines (*225*) and their hydroperchlorate salts are easy obtained by reaction of N-dimethyl-3-methoxyallylideneammonium methylsulfates (*224*), the adducts of the 3-dimethylaminoacraldehydes and dimethylsulfate with arylamines in dichloromethane. Purified in the form of the hydroperchlorates, the yellow-orange, slightly basic vinamidines (*225*) may be liberated from their salts by sodium methoxide.

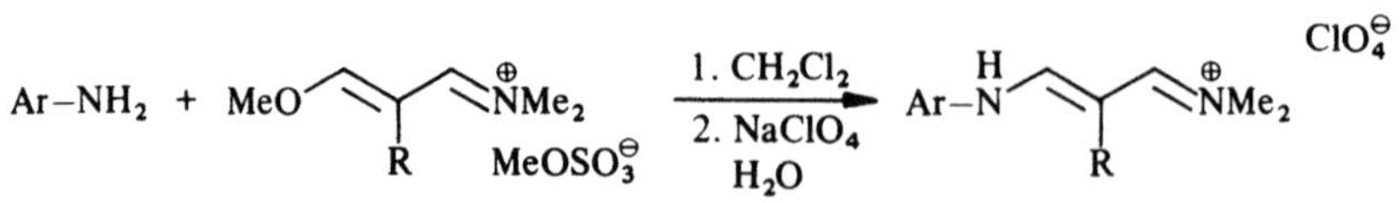

(*224*) (*225*, $HClO_4$)

The 3-dimethylamino-2-phenyl-allylideneanilines (*225 a–e*) cyclize to 3-phenylquinolines (*227*) and as expected (*225 c–e*) derived from *meta*-substituted anilines, yield mixtures of the 5- and 7-substituted quinolines (*227 c–e*):

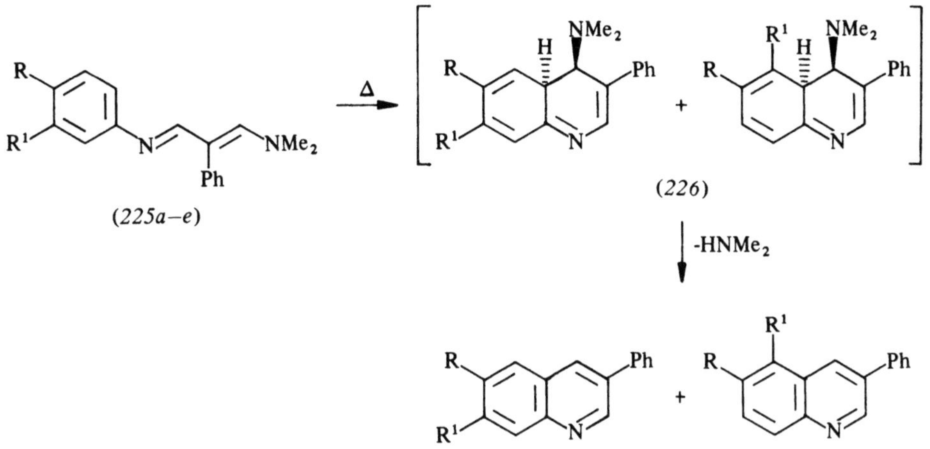

(*225a–e*) (*226*) (*227*)

Table 21. 3-Phenylquinolines (*227*)

	(*227*)	R	R′	Yield	Reaction conditions (°C)	Time	[(*22*), $HClO_4$]
a	-, mp. 52°, picrate F. mp. 203°	H	H	68%	200	2 h	mp. 189–192° (80%)
b	6-Methoxy-, mp. 122–3°	OMe	H	65%	200	2 h	mp. 180–1° (85%)
c	Mixture, mp. 90° 5-methoxy- 7-methoxy-	H	OMe	97%	180	1 h	mp. 142–3° (75%)
d	Mixture, mp. 79–85° 5-chloro-/7-chloro-	H	Cl	97%	200	1 h	mp. 218–20° (85%)
e	Mixture, mp. 127–155° 5-nitro-/7-nitro-	H	NO_2	86%	200	2 h	mp. 184–5° (86%)

In an analogous manner, benzo(f)quinolines (*229*) and benzo(h)quinolines (*231*), starting from β-naphthylamine and α-naphthylamine *via* the corresponding vinamidines (*225 f–h*) and (*225 k–l*), have been synthesized (Method A) or, much more simply, in a one-pot process, by boiling a mixture of equivalent amounts of arylamine, vinamidinium salt (*7*), and sodium methoxide in pyridine (Method B).

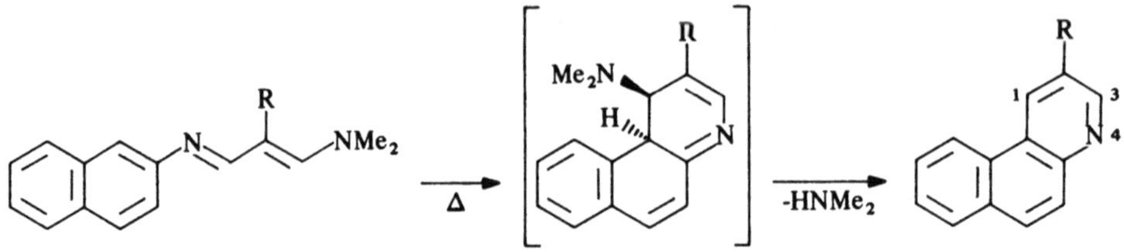

(*225f–j*) (*228*) (*229*)

Table 22. Benzo(f)quinolines (*229*)

	(*229*)	R	Yield	Reaction conditions (°C)	Time	[(*225*), $HClO_4$]	Method
f	-, mp. 92° picrate mp. 258°	H	72%	230	1 h	mp. 174–6° (95%)	A
g	2-Methoxy-, mp. 88–9°	OMe	94%	180	1 h	mp. 158–159° (75%)	A
h	2-Phenyl-, mp. 115–6°	Ph	81%	150	1 h	mp. 175–177° (83%)	A
			84%	115	5 h	β-Naphthylamine + (7*i*)/NaOMe in pyridine	B
i	2-Cyano-, mp. 157–8°	CN	90%	115	4 h	β-Naphthylamine + (7*f*)/NaOMe in pyridine	B
j	2-Nitro-, mp. 154°	NO_2	89%	115	1 h	β-Naphthylamine + (7*n*)/NaOMe in pyridine	B

The readiness with which the vinamidines (*225*) cyclize thermally follows the rules we discussed above (Section II. B. C.) for the carbocycles. Thus, the 3-dimethylaminoallylidene aniline, the parent compound of (*225 a–e*) without any bulky substituent in the methin chain, fails to cyclize even at higher temperatures. All factors

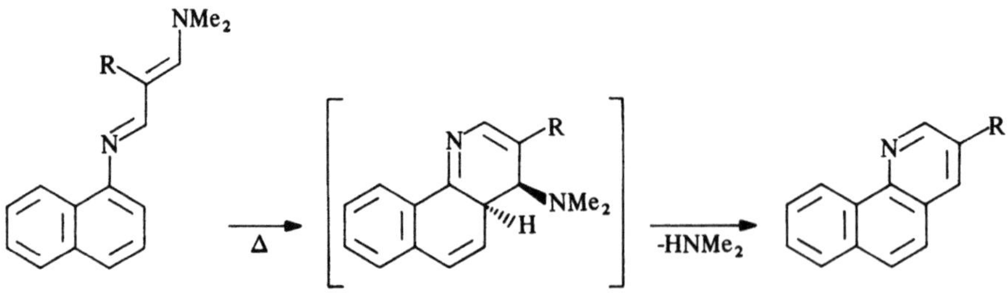

(*225k–m*) (*230*) (*231*)

Table 23. Benzo(h)quinolines (*231*)

				Reaction conditions			
	(*231*)	R	Yield	(°C)	Time	[(*225*), $HClO_4$]	Method
k	-, mp. 52° picrate mp. 191°	H	83%	230	1 h	mp. 156–158° (94%)	A
l	3-Phenyl-, mp. 113–4°	Ph	80%	150	1 h	mp. 167–169° (91%)	A
			76%	115	5 h	α-Naphthylamine + (*7i*)/NaOMe in pyridine	B
m	3-Nitro-, mp. 166–7°	NO_2	71%	115	3 h	α-Naphthylamine + (*7n*)/NaOMe in pyridine	B

which decrease the barrier of activation energy, *e.g.* decrease of the localization energy by benzanelation in (*225 f–j*), facilitate cyclization and are indicated by the temperature of thermolysis. The nature of the substituents R and R′ in (*225 a–e*) exerts only a little influence on the formation of the quinolines (*227*), in contrast to the observations in the acid ring-closure of (*219*) to (*221*).

The thermolysis of (*225*) and related compounds, superior in yields to the acid ring-closure *via* (*219*) to (*221*), opens a general way for pyridoannelation of structurally appropriate, also acid-sensible amino derivatives of aromatic and heterocyclic compounds.

The highly reactive vinamidinium salt (*7s*) has been prepared starting from N.N-dimethylcanoacetamide by reaction with dimethylformamide diethylacetal, followed by treatment of the acrylamide derivative obtained with phosphoryl chloride and conversion to the stable perchlorate[34].

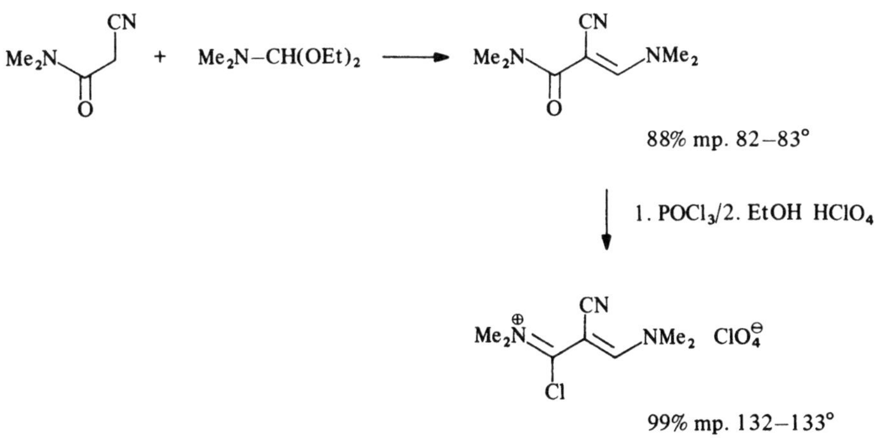

(*7s*)

The chlorine in (*7s*) is quite labile and was very readily replaced by nucleophiles. Treatment of (*7s*) with an excess of arylamine leads to hydroperchlorates of the strongly basic 3-dimethylaminoacrylamidines (*232*), vinylogues of guanidines. The neat free bases (*232*) cyclize smoothly on heating, yielding quinolines (*234*) and benzoquinolines (*236*), (*238*) having the cyano and the dimethylamino groups in adjacent positions of the hetero ring.

Similar to the vinylogous amidines (*225 a–e*), variation of the substitutent *para* in the parent aniline has only a little effect upon the cyclization tendency of the vinylogous guanidines (*232 a–e*), giving the 6-substituted quinolines (*234 a–e*) quantitatively[34].

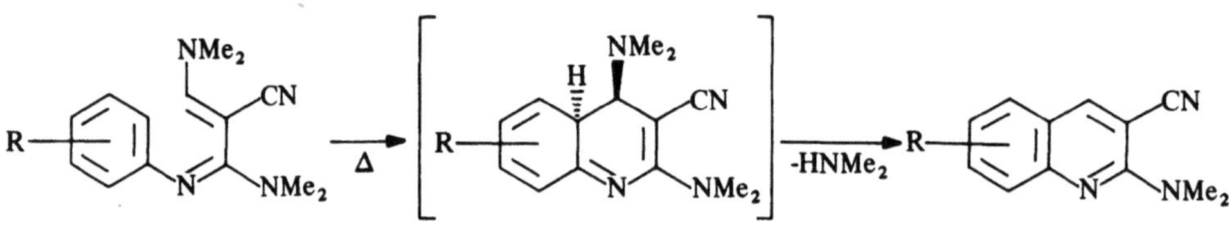

(*232*) (*233*) (*234*)

The acrylamidines (*232 g–i*), related to *meta*-substituted anilines, potentially yielding 5- and 7-substituted quinolines, give rise to the latter only. The 7-substituted quinolines (*234 g–i*) are isolated in high yield, however the substituents exert a pronounced influence on the ring-closure conditions. Electron-withdrawing residues impede, electron-releasing groups strongly enhance the cyclization. The isolation of (*332 f*) upon treatment of (*7s*) with m-dimethylaminoaniline failed. Instead, the corresponding quinoline (*234 f*) has been obtained, but in low yield. On the other hand, (*232 j*) with the *meta*-nitro group could not be cyclized successfully to (*234 j*). On heating (*232 j*) at 210° for a long time charring occurred.

The quinolines (*234*), (*236*), and (*238*) are pale-yellow or yellow crystalline substances, owing to their substitution pattern in the pyridine nucleus.

The exceptional behavior of the acryamidines (*232 f–j*) may be explained by the energy content of the cyclic intermediate (*233*), which determines the demand for the activation energy of the ring-closure reaction. It is immediately apparent in the corresponding (*233′*) that the substituent R is found in conjugation with the cyano group as an acceptor. Donor substituents stabilizing the intermediate by resonance should therefore lower the activation energy, and the reverse may be valid for electron-withdrawing substituents R.

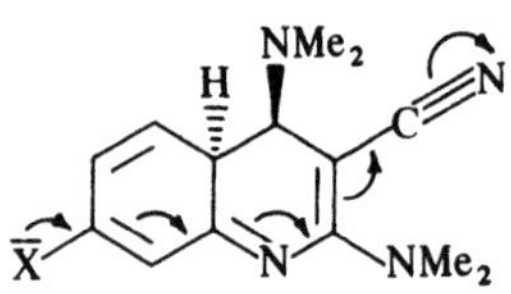

(*233′*)

Table 24. 3-Cyano-2-dimethylaminoquinolines (*234*)

	(234) R-,	Parent aniline	Yield	Reaction conditions (°C)	Time	(*232*)
a	-, mp. 81–82°	-H	quant.	180	1 h	– +$HClO_4$ (94%) mp. 177–178
b	6-Methyl-, mp. 111–112°	p-Me	quant.	180	1 h	-, (79%) mp. 109–111° (dec.)
c	6-Methoxy-, mp. 119–120°	p-MeO	quant.	180	1 h	– +$HClO_4$ (78%) mp. 179–181°
d	6-Chloro-, mp. 115–116°	p-Cl	quant.	180	1 h	-, (79%) mp. 115° (dec.)
e	6-Nitro-, mp. 188–189°	p-NO_2	90%	240	2 h	-, (79%) mp. 130–132°
f	7-Dimethylamino-, mp. 115°	m-NMe_2	27%	–	–	– – – – – –
g	7-Methoxy-, mp. 119°	m-MeO	quant.	160	7 h	– +$HClO_4$ (82%) mp. 155–156°
h	7-Chloro-, mp. 100–108°	m-Cl	91%	160	9 h	– +$HClO_4$ (82%) mp. 159–160°
i	7-Trifluoromethyl-, mp. 98–99°	m-CF_3	79%	190	53 h	– +$HClO_4$ (84%) mp. 153°
j	7-Nitro-,	m-NO_2	0%	210	dec.	– +$HClO_4$ (78%) mp. 197–200°

Acrylamidines (*235*) and (*237*) formed from *β*- or *α*-naphthylamine upon treatment of (*7s*) cyclize very smoothly and in quantitative yield to 2-cyano-3-dimethylaminobenzo(f)quinoline (*236*) or 3-cyano-2-dimethylaminobenzo(h)quinoline (*238*), respectively:

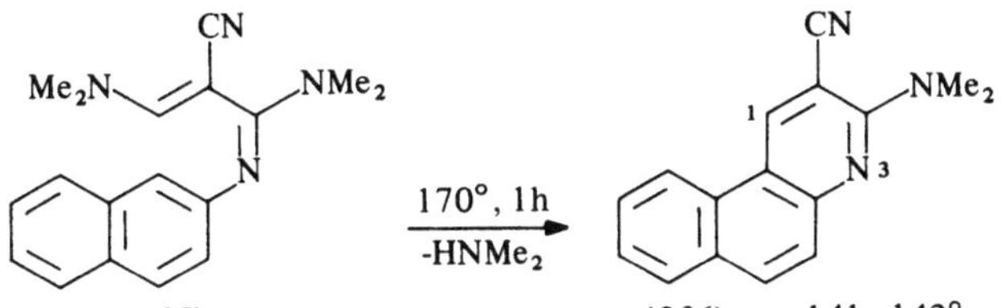

(*235*) (*236*) mp. 141–142°

– +$HClO_4$ (78%), mp. 198–199°

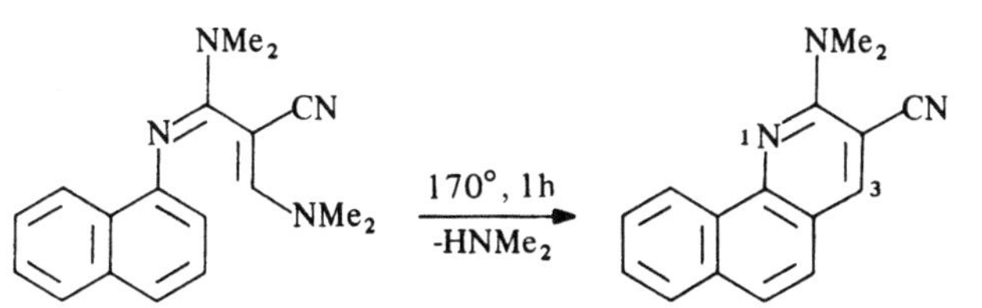

(*237*) (*238*) mp. 148–150°

2.3. Pyridoannelation. The cyclizations shown in the preceding section could accurately be called "pyridoannelations" and this reaction sequence has been applied to other amines also, especially to those not feasible by Skraup or related syntheses performed under acid conditions. 1-Amino-4,6,8-trimethylazulene (*239*)[97] has been transformed, thermolyzing the vinamidine (*240*) to 6,8,10-trimethyl-3-phenylazuleno-(1,2-b)pyridine (*241*)[18]:

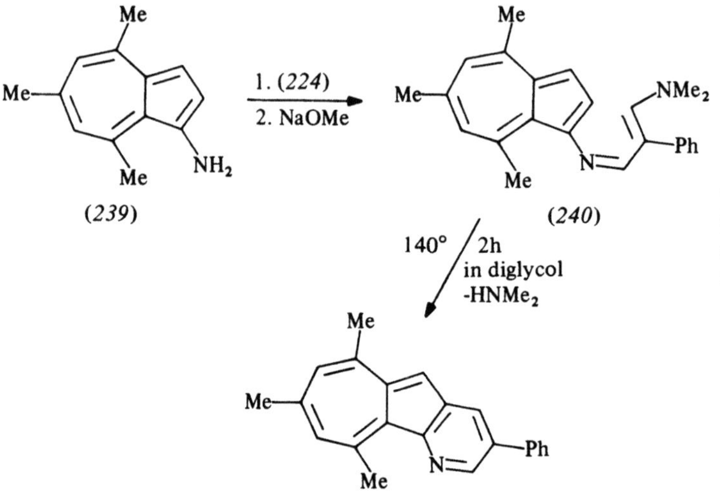

(*239*) (*240*)

—+$HClO_4$ (81%) Dark-green needles, mp. 179–181° (dec.)

(*241*) (61%) Black, glistening plates, mp. 134–135°
λ_{max} (log ϵ) in CH_2Cl_2 = 546 (2.79), 395 (3.89), 374 (3.97), 327 nm (4.88)

2-Nitrobenzo(f)(1,7)naphthyridine (*242*) was obtained in moderate yield by heating 3-aminoquinoline and (*7n*), after addition of sodium methoxide, first to 50°, then to 120°, in quinoline, for 2.5 h[34]:

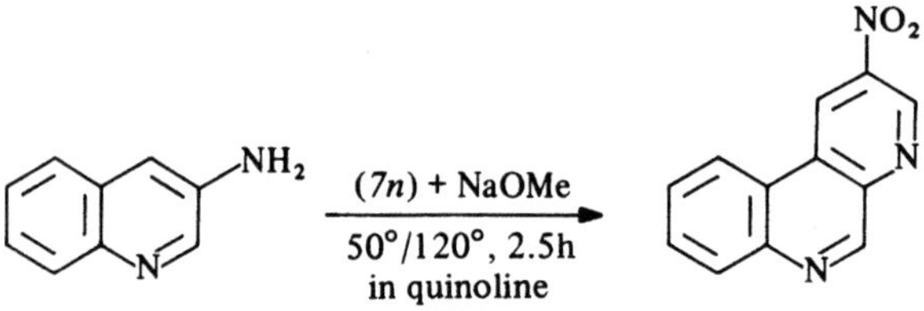

(*242*)

(36%) Pale-yellow needles mp. 199–200°

The pyrido(2,3-d)pyrimidinediones (*244a, b*)[34] are formed from 6-amino-1,3-dimethyluracil (*243*)[98], which contains a stable enamine moiety with a free amino group. If in (*243*) the condensation with vinamidinium salts (*7*) takes place exclusively at the amino group or at the β-enamine carbon also may be questionable.

2.4. Pyridines by Enamines. Like 6-aminouracil (*243*) 3-aminocrotonitrile (*245a*) and 3-aminocrotonic acid esters (*245b, c*), vinylogues of cyanamide or urethane represent stable but acyclic members of enamines with a free amino group. Vinamidinium salts (*7*) in the presence of sodium methoxide or the related 3,3-diethoxy-

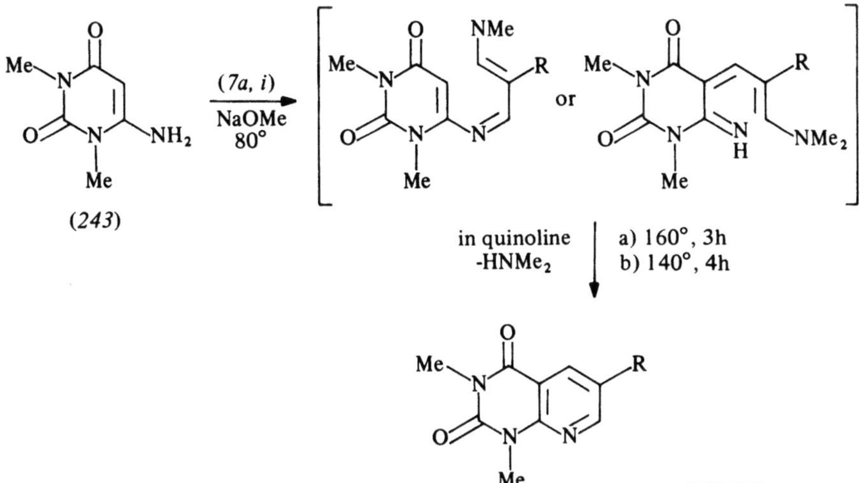

(243)

(244)

a: R = H (79%) mp. 158–159°
b: R = Ph (84%) mp. 157–158°

1-dimethylamino-propenes (*35a, b*) readily condense with (*245a*) at room temperature, and with (*245b, c*) by slightly heating, to yellow solutions of the suspected azahexatrienes, which cyclize by loss of dimethylamine to the nicotonitriles (*246a–e*) at 80°, or to the nicotinic acid esters (*246f–j*) at 120–130°:[99]

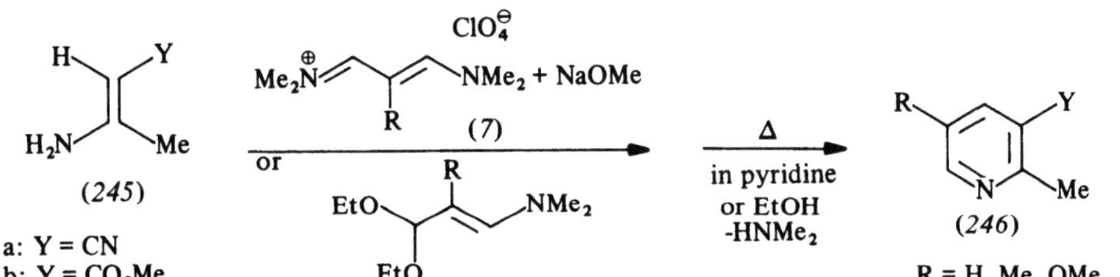

(245)

a: Y = CN
b: Y = CO_2Me
c: Y = CO_2Et

(246)

R = H, Me, OMe, CH_2Ph, Ph

Table 25. Pyridines (*246a–j*)[99]

2-Methylnicotonitriles (Y = CN)

	(*246*)	R	Yield	Components	Reaction conditions (°C)	Time	Solvent
a	-, mp. 56–57° picrate: mp. 170–171°	H	86%	(*7a*) + (*245a*) NaOMe	80	8 h	EtOH
b	5-Methyl-, mp. 55–56° picrate: mp. 146–147°	Me	91%	(*7b*) + (*245a*) NaOMe	80	8 h	EtOH
c	5-Methoxy-, mp. 95–6° picrate: mp. 118°	OMe	74%	(*7e*) + (*245a*) NaOMe	80	8 h	EtOH

Table 25. (continued)

	2-Methylnicotonitriles (Y = CN)				Reaction conditions		
	(246)	R	Yield	Components	(°C)	Time	Solvent
d	5-Benzyl-, oil bp. 121–4°/0.4 Torr. picrate: mp. 127–128°	CH_2Ph	82%	*(7*)* + *(245 a)* NaOMe	80	8 h	EtOH
e	5-Phenyl-, mp. 98–99°	Ph	95%	*(7 i)* + *(245 a)* NaOMe	80	8 h	EtOH
	Methyl-2-methylnicotinates (Y = CO_2Me)						
f	5-Methyl-, oil bp. 60–61°/0.3 Torr.	Me	52%	*(7 b)* + *(245 b)* NaOMe	115	5 h	Pyridine
g	5-Methoxy-, mp. 71–72°	OMe	81%	*(7 e)* + *(245 b)* NaOMe	115	5 h	Pyridine
h	5-Phenyl-, mp. 83–84° bp. 134–6°/0.2 Torr. Acid: mp. 275–280°	Ph	83%	*(7 i)* + *(245 b)* NaOMe	115	5 h	Pyridine
	Ethyl-2-methylnicotinates (Y = CO_2Et)						
i	-, bp. 52–56°/0.3 Torr. picrate: mp. 145–146°	H	71%	*(35 a)* + *(245 c)*	130	3 h	No solvent
j	5-Methyl-, oil bp. 70–71°/0.4 Torr. picrate: mp. 148–149°	Me	80%	*(35 b)* + *(245 c)*	130	4 h	No solvent

No statement for the site of the electrophilic attack of *(7)* on *(245 a–c)* or for the structure of the intermediate azahexatrienes can be made in the reaction to *(246 a–j)* above, owing to the symmetry of the vinamidinium salts *(7)* used.

Performing the pyridine synthesis with the cyclic vinamidinium salts *(7 o)*, *(7 p)* and the related 2-dimethylamino-dimethyliminiomethylcycloheptene perchlorate *(7 t)* should allow some insight into the course of the reaction.

The enaminonitrile *(245 a)* reacts smoothly with all three salts *(7 p)*, *(7 o)*, and *(7 t)* and leads finally to a 1 : 1 mixture of the corresponding cycloalkeno(c)pyridines *(248 a–c)* and cycloalkeno(b)pyridines *(250 a–c)*. Owing to the sensitivity of *(7 p)*, *(7 t)* to strong bases, the more inert enaminoester *(245 b)* could converted successfully only with *(7 o)*, yielding the isoquinoline derivative *(248 b')* as the single product. This indicates an attack on the enaminonitrile *(245 a)* at the amino group to *(247)* as well as at the β-carbon atom to *(249)*, whereas the enaminoester *(245 b)* is substituted at the amino group exclusively, possibly by the shielding effect of the bulky methoxycarbonyl group.

2.5. Heterocycles by Azavinamidinium Salts. Treatment of cyanuric chloride with dimethylformamide produces N-dimethyl(2-aza-3-dimethylamino)allylidene ammonium chloride[100], used as perchlorate *(7 u)*, which can serve as a component

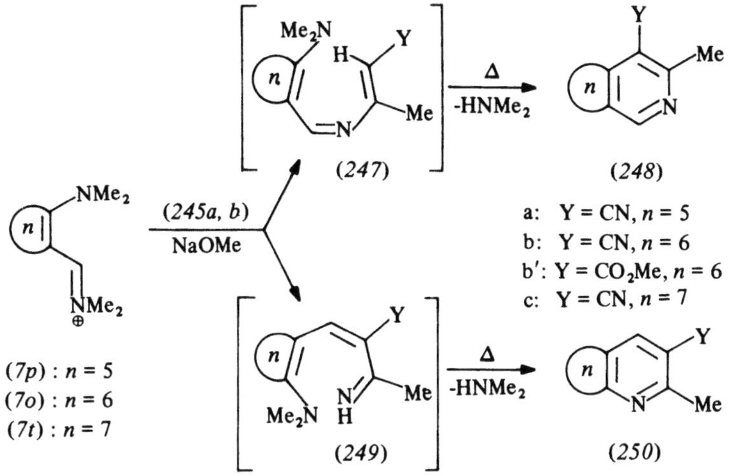

(247) (248) (249) (250)

(7p) : $n = 5$
(7o) : $n = 6$
(7t) : $n = 7$

(248) a: Y = CN, $n = 5$
b: Y = CN, $n = 6$
b′: Y = CO_2Me, $n = 6$
c: Y = CN, $n = 7$

(250) a: Y = CN, $n = 5$
b: Y = CN, $n = 6$
c: Y = CN, $n = 7$

Table 26. Cycloalkenopyridines (*248*, *250*)[99]

	Y	*n*	Yield	Components	Reaction conditions (°C)	Time	Solvent
Mixture: (*248a*) + (*250a*) mp. 95–99°, picrate: mp. 150°	CN	5	70%	(*245a*) + (*7p*) NaOMe	80	12 h	EtOH
Mixture: (*248b*) + (*250b*) mp. 72–8°, picrate: mp. 185–7°	CN	6	82%	(*245a*) + (*7o*) NaOMe	80	12 h	EtOH
Mixture: (*248c*) + (*250c*) mp. 112–6°, picrate: mp. 191–5°	CN	7	72%	(*245a*) + (*7t*) NaOMe	80	12 h	EtOH
(*248b′*): bp 100–101°/0.2 Torr. acid: mp. 205–206°	CO_2Me	6	61%	(*245b*) + (*7o*)	115	5 h	Pyridine

to build cyclizable "aza"-hexatrienes. However, (*7u*) possesses a highly electrophilic potential and is attacked by nucleophiles, *e.g.*, carbanions, splittings off one of the two carbon-nitrogen bonds, forming either a) a dimethylaminomethylene compound. or b) a 2-aza-3-dimethylaminoallylidene derivative:

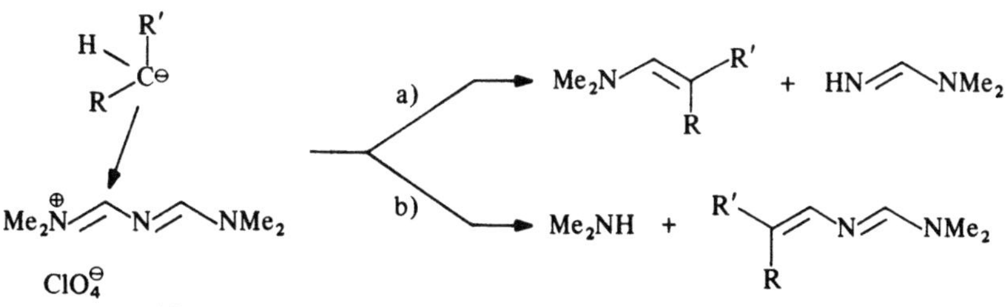

(*7u*)

Provided that a carbanion reacts with (*7u*) according to b) – and a limited number of compounds do so – the synthesis of heterocyclic rings is possible. Benzo(h)isoquinoline-4-carbonitrile (*251*) has been prepared in moderate yield from naphthalene-2-acetonitrile (*70*), and 2-azapyrene (*252*) from phenalene (*103*)[101].

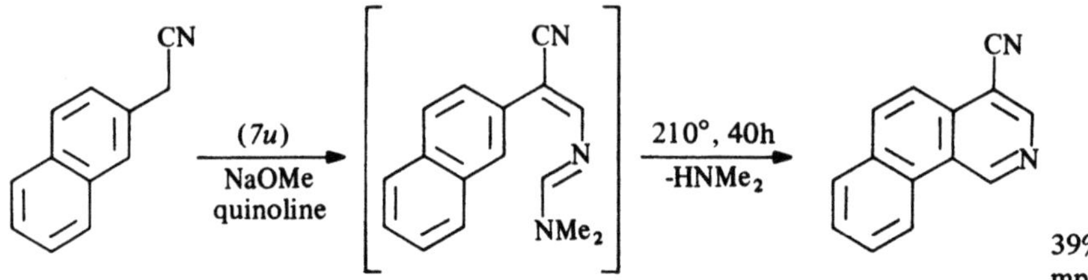

(*70*) (*251*) 39% mp. 184–185°

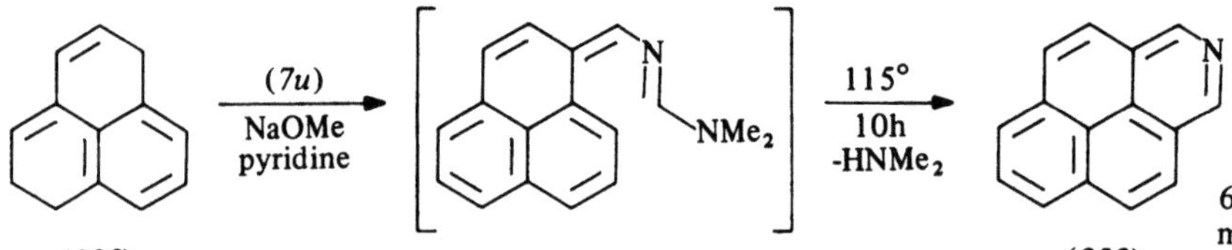

(*103*) (*252*) 61% mp. 162–163°

Dimethylchloromethylene ammonium chloride (dimethylformamide chloride) or the adduct from dimethylformamide and phorphoryl chloride adds to N-dimethylcyanamide, yielding dimethyl(2-aza-1-chloro-3-dimethylamino)allylidene ammonium chloride, isolated as perchlorate (*7v*)[102]. The highly reactive chlorine in (*7v*) is very readily substituted by bases and affords, with β-naphthylamine, the guanidine derivative (*253*), which on heating cyclizes smoothly to 2-dimethylaminobenzo(f)quinazoline (*254*):

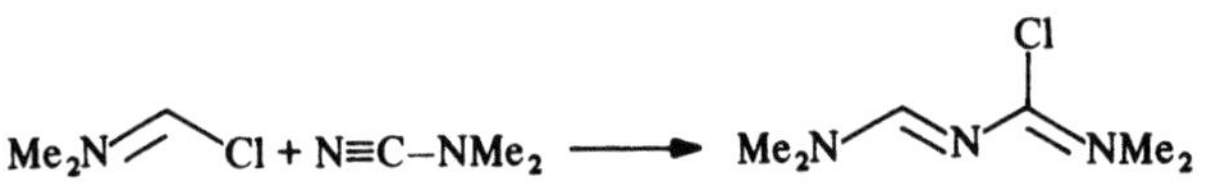

$Cl^{\ominus}$ (*7v*) $(Cl^{\ominus})\ ClO_4^{\ominus}$

mp. 136–137° (90%)

(*253*) (*254*)

2.6. Related Cyclizations to Heteroaromatics. Some examples, best explained by an electrocyclic process with subsequent elimination, have been found in the literature, but we do not claim to cover them completely.

2-Trifluoromethyl-3,1-benzoxazine-4-ones (*255*) react rapidly with enamines (*256*) at low temperatures to quinoline-8-carboxylic acids (*258*) and a secondary amine. Nucleophilic attack of (*256*) on (*255*) has been assumed, forming an intermediate vinamidine (*257*) which immediately undergoes cyclization[103]:

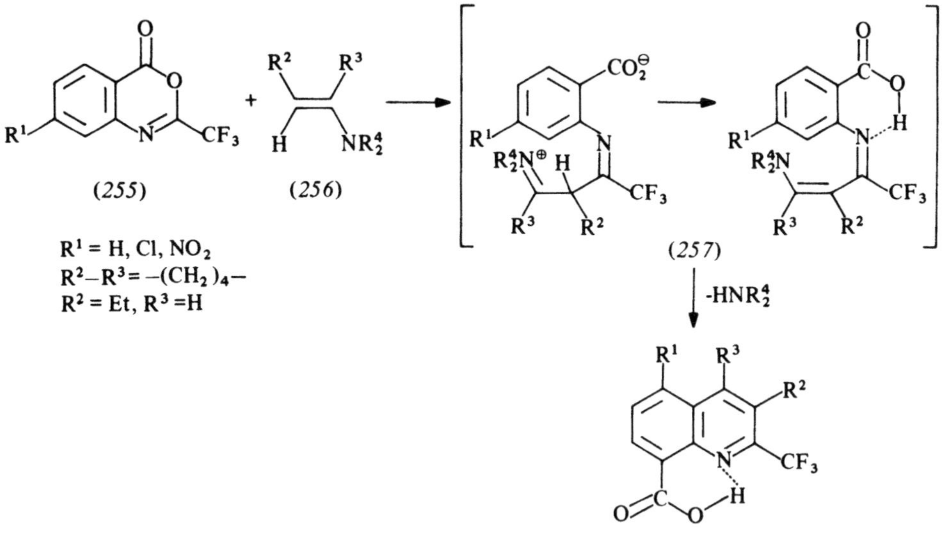

(255) (256) (257) (258)

$R^1 = H, Cl, NO_2$
$R^2–R^3 = –(CH_2)_4–$
$R^2 = Et, R^3 = H$

The aminouracil derivative (*259*) condenses with dimethylformamide diethylacetal to the dimethylaminomethyleneamino compound (*260*) the thermolysis of which leads to the pyrimido(4,5-b)pyrazine (*261*)[104]:

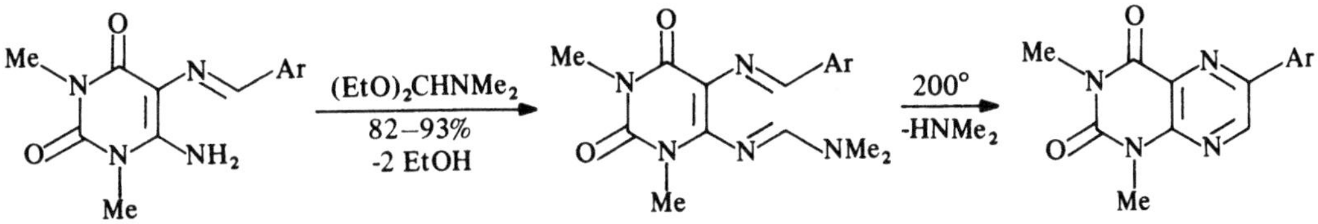

(259) (260) (261)

In a similar mode, dimethylformamide diethylacetal condenses with the enamidoamides (*262*) to the corresponding dimethylamino-methyleneamides (*263*). On heating the latter, cyclization to the 2-pyridones (*265*) takes place[105]. The protomeric (*264*) should be an intermediate if the ring-closure follows an electrocyclic course:

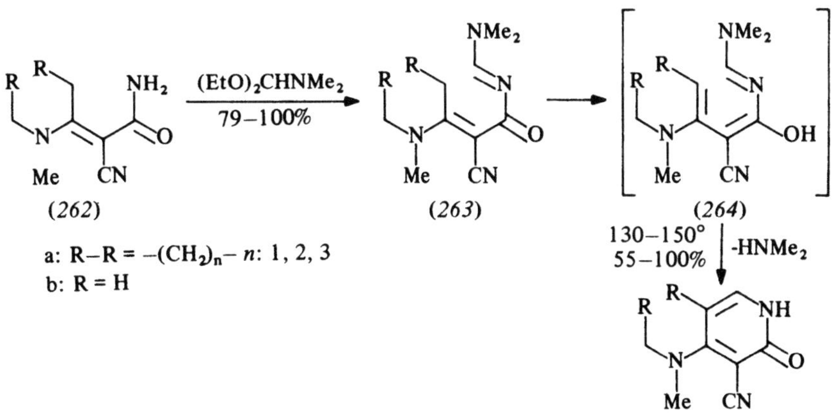

(262) (263) (264) (265)

a: $R–R = –(CH_2)_n–$ *n*: 1, 2, 3
b: R = H

The transformations of the product from 4-aza-3-chloro-2-cyano 5 dimethyl-aminopentadien-2,4-al with methyl-hydrazinecarboxylate may be of a more complex nature, involving electrocyclic ring-closures and ring-openings attended by protonations and deprotonations.

The chlorohydrazone (*266*) cyclizes in aqueous methanol, losing dimethylamine hydrochloride to the 4-pyrimidone (*267*). The latter is reopened by the action of a secondary amine in methanol to (*268*) and recyclized on boiling in acetic acid to (*267*)[106]:

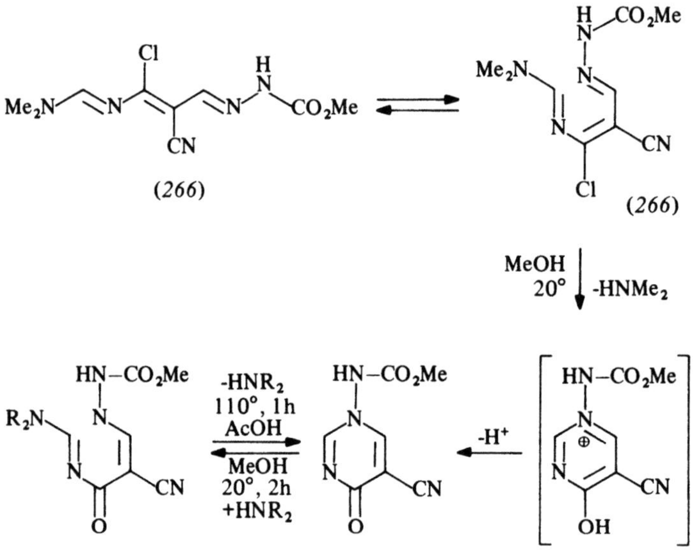

(*266*) (*266*)

(*268*) (*267*)

Diene-one oximes (*269*) of right configuration are capable to form pyridines by loss of water. The intermediate (*270*) has been assumed[107]:

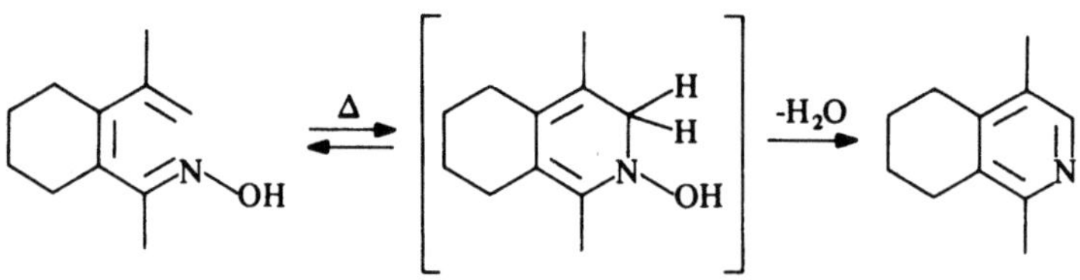

(*269*) (*270*) (*271*)

Diene isocyanates (*273*) generated by Curtius rearrangement of several dienoic acid azides (*272*) represent another type of "aza"-hexatrienes. They are of elusive existence and cyclize smoothly to 2-pyridones (*274*)[108]. One may suggest an electrocyclic ring-closure to (*275*) followed by a 1,5 sigmatropic hydrogen shift to (*274*):

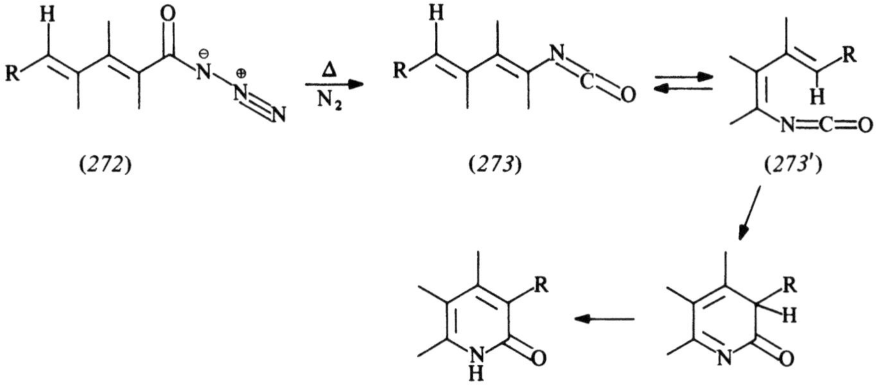

(272) (273) (273') (274) (275)

In a very similar manner, the imidoylisothiocyanate (*276*), containing formally a diazahexatriene system, cyclizes to the quinazolinethion (*278*) on heating[109]. The ring-closure step involves the localization of an aromatic double bond to the intermediate (*277*).

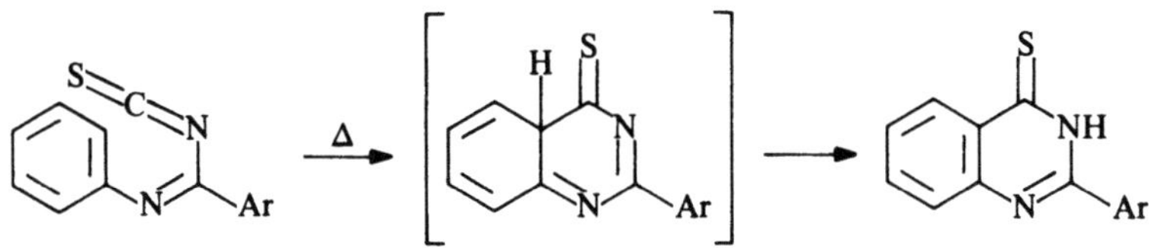

(276) (277) (278)

The authors could show also that the thermolysis of the imidoylurethane (*279*) to the corresponding quinazolineone (*282*) is connected in a preceding equilibrium with intermediate formation of the imidoylisocyanate (*280*). Introduction of an additional methyl group at the urethane nitrogen in (*279*), blocking the elimination of ethanol to an isocyanate, prevents the ring-closure[109].

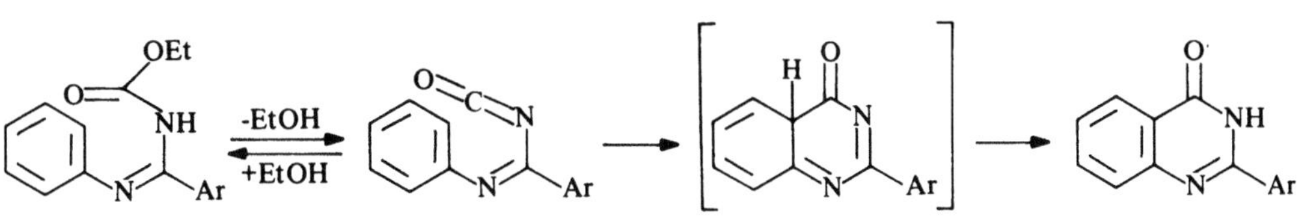

(279) (280) (281) (282)

This sequence calls to mind an old, long-known reaction: the thermal version of the Conrad-Limpach synthesis of quinoline-4-one derivatives by thermolysis of arylaminocrotonates[110a)], arylaminomethylenemalonates[110b, c)], arylaminomethylenecyanoacetates[110d)], and related compounds[111–114)].

The chances are that the ester (*283*), analogously to (*279*), may be transformed by elimination of alcohol to the intermediate ketene (*284*) which cyclizes *via* (*285*) attended by 1,5-sigmatropic hydrogen shift to (*286*):

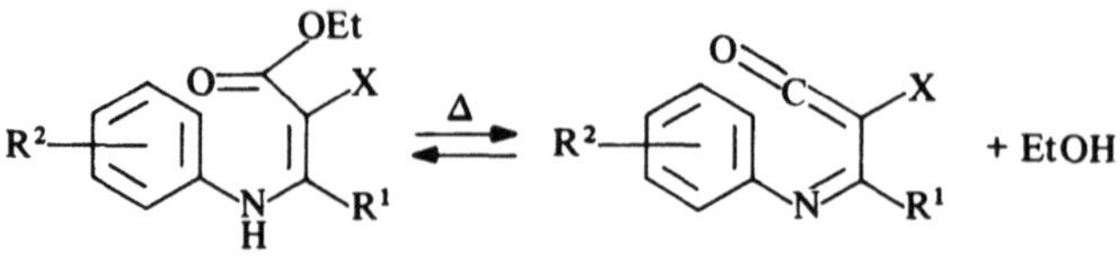

(283) (284)

X : H, CO_2Et, CN
R^1: H, Me, Ar, CO_2Et
R^2: H, NO_2, CN ...

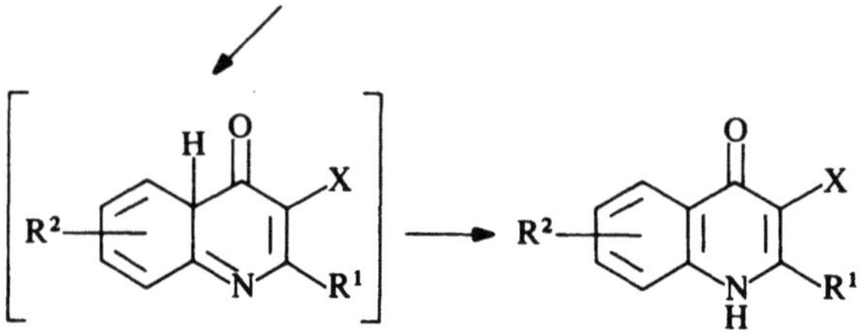

(285) (286)

Mixtures of 5- and 7-substituted quinolineones (*286*) can be formed, and the relation of the one to the other depends on steric requirements and the nature of the residue R^2 if a *meta*-substituted aniline has been used in starting to (*283*)[110c, d].

A twofold ring-closure to the dibenzo(b,g)-1,8-naphthyridinedione (*288*) has been observed thermolyzing the stable keteneaminal (*287*) in diphenyloxide, obtained by base-catalyzed addition of diethylmalonate to diphenylcarbodiimide[111]:

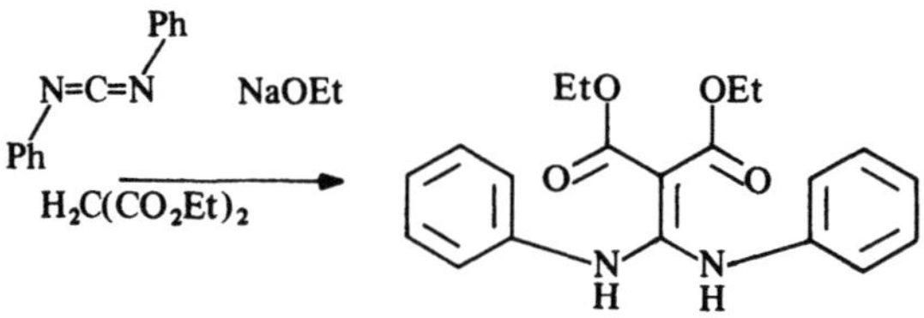

(287)

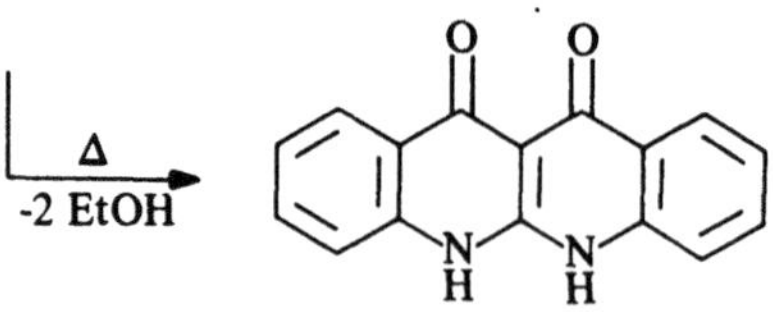

(288)

The Conrad-Limpach synthesis has been extended also to corresponding derivatives of 3-aminopyridines. With position 2 blocked in the starting aminocrotonate (*289*), one obtains 1,7-naphthyridineones (*290*)[112], but if this position is free, as in the aminomethylenemalonate derivative (*291*), only 1,5-naphthyridineones (*292*) are formed[113]:

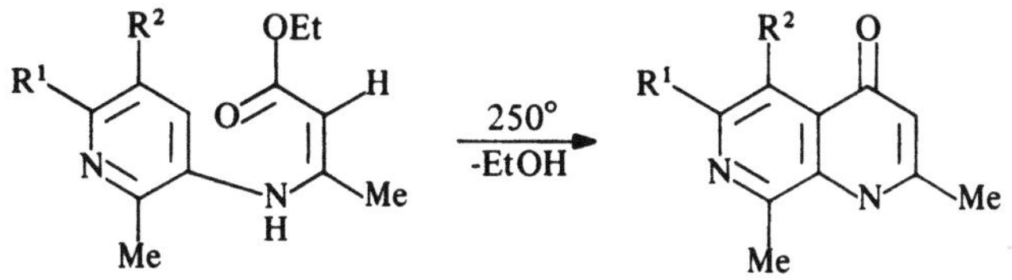

(289) (290)

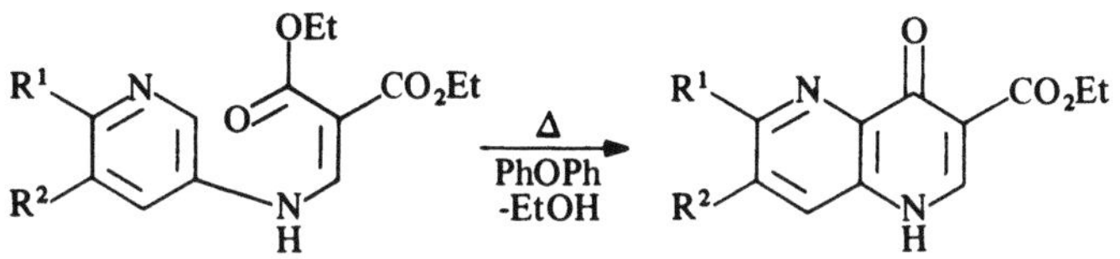

(291) (292)

The vinylaminomethylenemalonate derivatives (*295*) obtained by the action of α-methyleneketones (*293*) on aminomethylenemalonate (*294*) or of ethoxymethylenemalonates (*297*) on the corresponding enamines (*296*) of the ketones (*293*) lead finally, on heating, to the 4-pyridones (*298*)[114]:

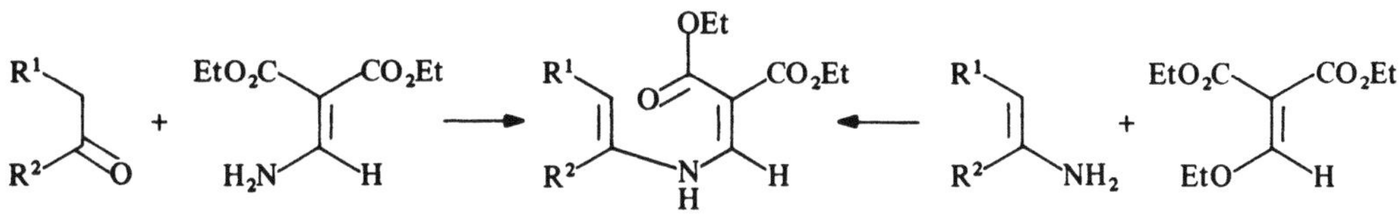

(293) (294) (295) (296) (297)

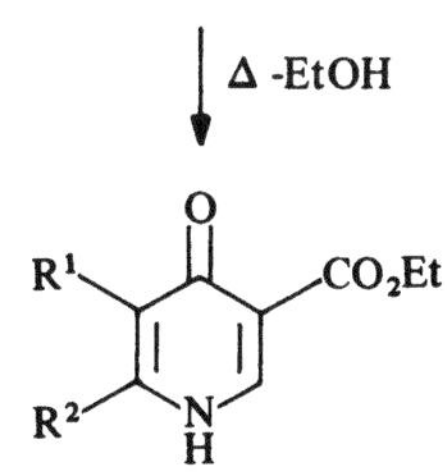

R^1: COMe, CO_2Et
R^2: Me, Ph
R^1–R^2: $-(CH_2)_n-$
$n = 3, 4, 5$

(298)

V. N-Bridged Heterocycles

1. Synthesis by N-Heteroaromatic Acetonitriles

In an attempt to prepare N-unsubstituted indoles by benzannelation according to the sequence (*193c*) → (*194c*) shown for the synthesis of 1-methylindole (*196c*) (Section IV., 1.), the yellow 5-dimethylamino-2-(2-pyrrolyl)-2,4-pentadienenitrile

(*300*) was made, by condensation of pyrrole-2-acetonitrile (*299*)[115] with (*7a*). Compound (*300*) splits off dimethylamine on heating in neat form above 200° and a cyclization takes place. Instead of the expected indole, however, yellow needles of indolizine-8-carbonitrile (*302a*) are formed in a 80% yield. To fulfil the structural preliminaries for an electrocyclic ring-closure we postulate a preceding 1,5-sigmatropic hydrogen shift to (*301 a*). The formation of (*301*) in an equilibrium may possibly represent the rate-determining step of the whole reaction sequence:

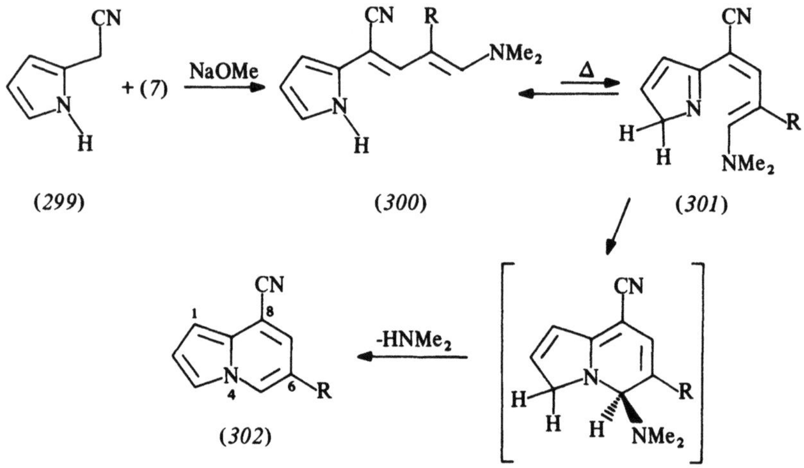

Table 27. Indolizine-8-carbonitriles [pyrrolo(1,2-a)pyridines] (*302*)[32, 86]

(*302*)		R	Yield	Components	Reaction conditions (°C)	Time
a	--, mp. 74–75° (subl.) yellow needles	H	80%	(*299*) + (*7a*)	200–220 In quinoline	3 h[1])
b	6-Methoxy-, mp. 96–97° yellow needles	OMe	83%	(*299*) + (*7e*)	180–185 In quinoline	3 h
c	6-Phenyl-, mp. 102–103° yellow needles	Ph	79%	(*299*) + (*7i*)	170–180 In quinoline	2 h

[1]) By thermolysis of isolated (*300a*): mp. 147–148° (91%).

In the same manner, the azavinamidinium salt (*7u*) und (*299*) (160–170°, 5 h in quinoline) afford pyrrolo(1,2-c)pyrimidine-4-carbonitrile (*303*): mp. 124°, yellow needles (55%).

With the cyclic heptamethinium salt (*44*) and (*299*) one can obtain the peculiar system of pyrrolo(1,2-a)cyclopenta(e)azepine-10-carbonitrile (*304*): mp. 149–151°, red-brown crystals in a small yield (14%).

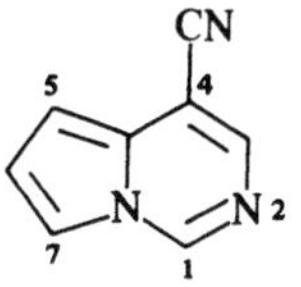

(303)

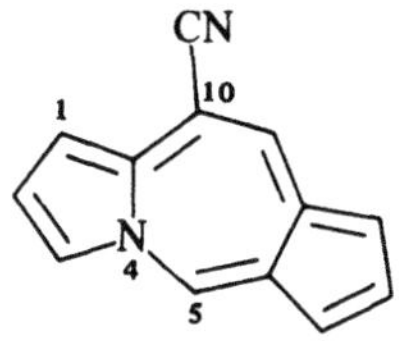

(304)

UV (CH_2Cl_2): λ_{max} (log ε)
473 (3.04), 425 (3.36),
402 (3.32), 380 (3.21),
294 (4.61), 288 nm (4.63)

The benzo derivative of pyrrole-2-acetonitrile (*299*), the indole-2-acetonitrile (*305*)[116)] leads in high yield to the orange-red pyrido(1,2-a)indole-9-carbonitriles (*308*). The requirement of energy for the conversion of (*306*) to (*307*) may be smaller than the corresponding gap from (*300*) to (*301*). Consequently the thermolysis temperature is found distinctly lowered.

With (*305*) the azavinamidinium salt (*7u*) yields (150–160°, 2 h in quinoline) the pyrimido(3,4-c)indole-4-carbonitrile (*309*): mp. 160–161°, orange needles (47%), besides a not very soluble, yellow substance, not identified.

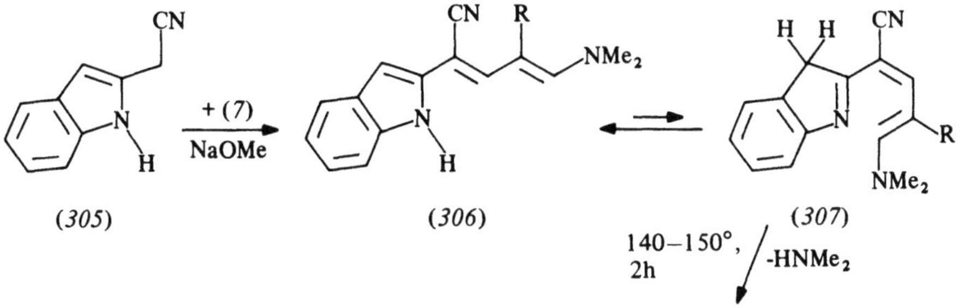

(numbering according IUPAC)

Pyrido(1,2-a)indole-9-carbonitriles (*308*)[32)]

a) R = H : mp. 85–86° (95%) from the isolated (*306a*) in quinoline (*306a*): yellow needles, mp. 150–155° (dec.) (85%)

b) R = Me : mp. 165° (87%) without isolation of (*306b*) in quinoline

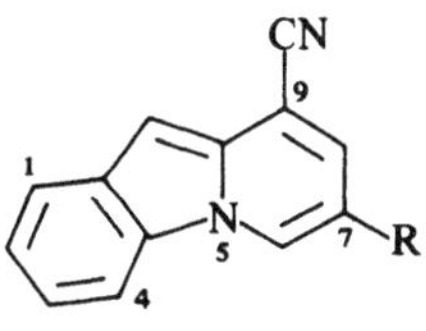

(308)

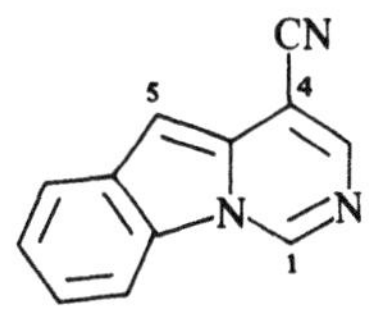

(309)

Pyrazole-3-acetonitrile (*310*)[117)] and imidazole-4(5)-acetonitrile (*315*)[118)] react by just this means to pyrazolo(1,5-a)pyridine-4-carbonitriles (*312*) and imidazo(1,5-a)-pyridine-8-carbonitriles (*318*) respectively. Only (*310*) can be reacted with (*7u*), giving pyrazolo(1,5-c)pyrimidine-4-carbonitrile (*313*): mp. 164°, colorless plates (150°, 3 h in quinoline, 56%) and yielding with (*44*) red needles of the pyrazolo(1,5-a)-cyclopent(e)azepine-4-carbonitrile (*314*): mp. 160–161° (150°, 4 h in quinoline 15%).

The different colors of the various "azaindolizines" synthesized have attracted our attention: (*302*), (*303*), and (*318*) form yellow solutions, (*312*) and (*313*) colorless ones. Some comments about the spectra of these compounds may be useful[123)].

The band of longest wavelength absorption in the spectrum of indolizine which shows some relation in shape to that of azulene corresponds to the $\pi^* \leftarrow \pi$ transition

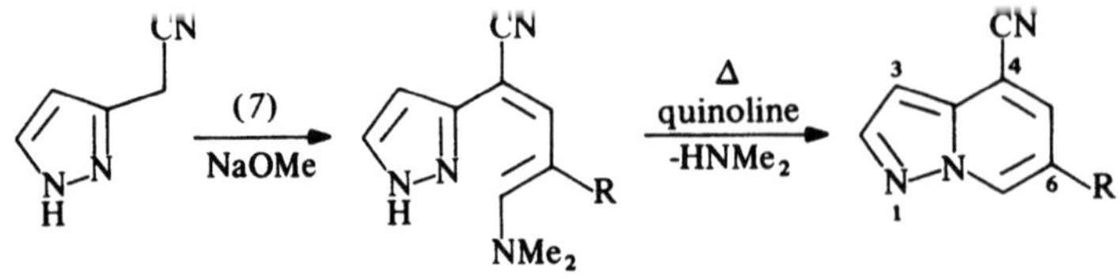

(310) (311) (312)

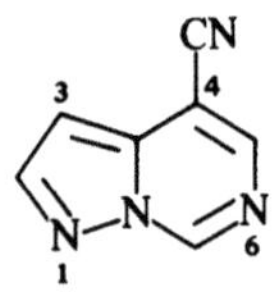

(313)

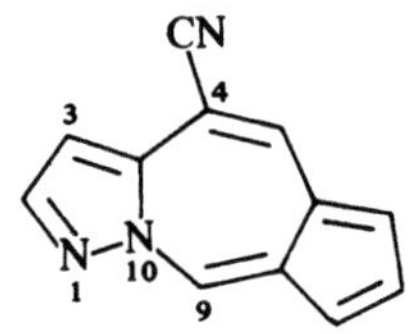

(314)

(*314*) UV (CH_2Cl_2): λ_{max} (log ϵ)
500 (2.65), 425 (3.48), 403 (3.50), 322 (3.69), 295 (4.77), 236 nm (3.90)

Table 28. Pyrazolo(1,5-a)pyridine-4-carbonitriles (*312*)[32]

(*312*)		R	Yield	Components	Reaction conditions
a	--, mp. 102–103° (subl.) colorless needles	H	79%	(*7a*) + (*310*)	170–190°, 2 h from (*311 a*), oil
b	6-Methyl-, mp. 140° (subl.), colorless needles	Me	89%	(*7b*) + (*310*)	150°, 2 h
c	6-Phenyl-, mp. 180–181° colorless needles	Ph	78%	(*7i*) + (*310*)	150°, 1 h

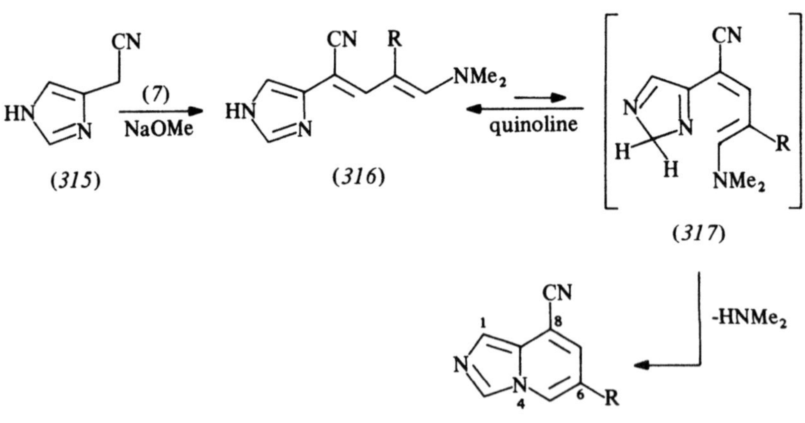

(315) (316) (317) (318)

Table 29. Imidazo(1,5-a)pyridine-8-carbonitriles (*318*)[32]

(*318*)		R	Yield	Reaction conditions
a	--, mp. 133–134° (subl.) yellow needles	H	84%	210°, 3 h in quinoline from isolated (*316 a*): mp. 182–4°
b	6-Methyl-, mp. 151–152° yellow needles (subl.)	Me	66%	180°, 2 h in quinoline
c	6-Benzyl-, mp. 154–155° yellow needles	$PhCH_2$	65%	180°, 2 h in quinoline

according to the transfer of a π-electron from HOMO (highest occupied molecular orbital) to LUMO (lowest unoccupied molecular orbital).

The diagram (Fig. 2) shows the calculated alteration of the one-π-electron densities at the single positions in indolizine-8-carbonitrile (*302 a*) going from HOMO to LUMO.

Excitation of (*302 a*) removes electrons from atoms 1, 3, and 8 a of the five-membered ring in different amounts to atoms of the six-membered ring. Electron-withdrawing substitutents or replacement of the methine group by nitrogen at positions 1 and 3 stabilizing the ground state makes the transition state more difficult and therefore effects a hypsochromic shift of the long wavelength band. For position 3 the effect should be most pronounced. The opposite should be true, facilitating electron abstraction in the transition for positions 5 and 8 to cause a bathochromic shift. For electron-releasing substitutents this may be valid in the converse sense.

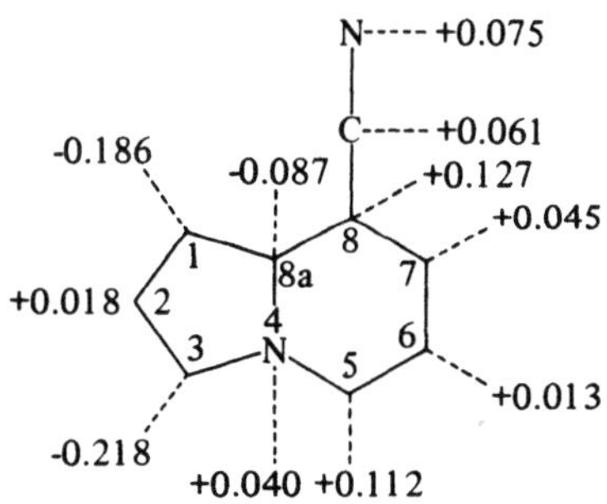

Fig. 2. Calculated (SCF) alteration of the one-π-electron densities ($c^2_{LUMO}-c^2_{HOMO}$) in indolizine-8-carbonitrile (*302a*) for the π*←π transition HOMO to LUMO

The predicted shift of the absorption band correlates well with the variation of the experimentally measured longest wavelength absorption:

Indolizine		Predicted shift	Measured λ_{max} (log ε)
Indolizine-8-carbonitrile	(*302 a*)		390 nm (3.32)
2-Aza--	(*318 a*)	(+)	378 nm (3.36)
3-Aza--	(*312 a*)	---	345 nm (3.44)
6-Aza--	(*303*)	(+)	387 nm (3.21)
3,6-Diaza--	(*313*)	--	355 nm (3.22)

2. Synthesis by N-Heteroaromatic Amines

A further extension synthesizing N-bridged hereocycles can be achieved by use of suited N-5-ring heteroaromatic amines. In general the ring-closures occur under much milder conditions, compared with the N-5-ring heteroaromatic acetonitriles, and can be successfully performed as a one-pot reaction even in boiling ethanol or pyridine. Although the isolation of the intermediate amidine, *e.g.* (*320*), failed, its formation during the reaction is nonetheless indicated by the structure of the isolated products if unsymmetrical amidinium salts, such as (*7o*) and (*7p*), are put in for the synthesis. The electrophilic attack of (*7*) on the amino nitrogen takes place exclusively at the unsubstituted methine iminium group of (*7*).

A series of easily accessible aminoheteroaromatic compounds has been used: 3-aminopyrazole (*319*), obtained by new mode in a 70% yield, reacting 1 mol 3-dimethylaminocrylnitrile[119] with 2 mol hydrazine hydrate in boiling ethanol[44], and 3-amino-5-methylpyrazole (*319 b*)[120] afford the pyrazolo(1,5-a)pyrimidines (*321*):

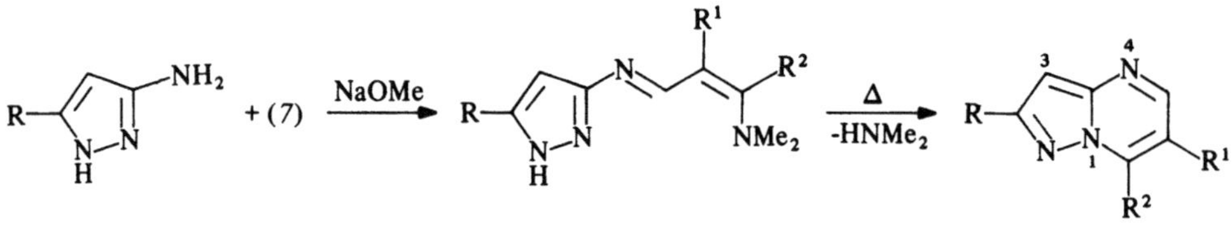

(*319*) (*320*) (*321*)

a: R = H
b: R = Me

Table 30. Pyrazolo(1,5-a)pyrimidines (*321*)[32,44]

(*321*)		R	R^1	R^2	Yield	Components	Reaction conditions
a	mp. 103° (subl.) colorless needles	H	H	H	84%	(*319 a*) + (*7 a*)	80°, 5 h ethanol
b	mp. 58–9° (subl.) colorless needles	Me	H	H	83%	(*319 b*) + (*7 a*)	115°, 4 h pyridine
c	mp. 111–2° (subl.) colorless plates	Me	Me	H	82%	(*319 b*) + (*7 b*)	115°, 4 h pyridine
d	mp. 94° (subl.) colorless needles	H	OMe	H	91%	(*319 a*) + (*7 e*)	80°, 1.5 h ethanol
e	mp. 86–7° colorless plates	Me	OMe	H	78%	(*319 b*) + (*7 e*)	115°, 4 h pyridine
f	mp. 163–4° colorless plates	Me	Ph	H	77%	(*319 b*) + (*7 i*)	130°, 3 h quinoline
g	mp. 75–6° (subl.) colorless needles	Me	$-(CH_2)_3-$		58%	(*319 b*) + (*7 p*)	130°, 1 h quinoline

Table 30. (continued)

(*321*)		R	R^1	R^2	Yield	Components	Reaction conditions
h	mp. 89–90° (subl.) colorless plates	H	$-(CH_2)_4-$		93%	(*319a*) + (*7o*)	80°, 2 h ethanol
i	mp. 84–85° colorless plates	Me	N=CH–NMe_2	H	81%	(*319b*) + (*7h*)	115°, 4 h pyridine
j	mp. 134° yellowish needles	Me	NH_2	H	69%	By alkaline of (*321 i*)	hydrolysis KOH/EtOH

Application of pyridoannelation to the amino derivative (*321 j*) with (*7e*) (150–170°, 1 h in quinoline) gives the 8-methoxy-2-methylpyrido(1′,2′–6′.7′)-pyrazolo(1,5-a)pyrimidine (*322*): mp. 199–202°, colorless needles (87%):

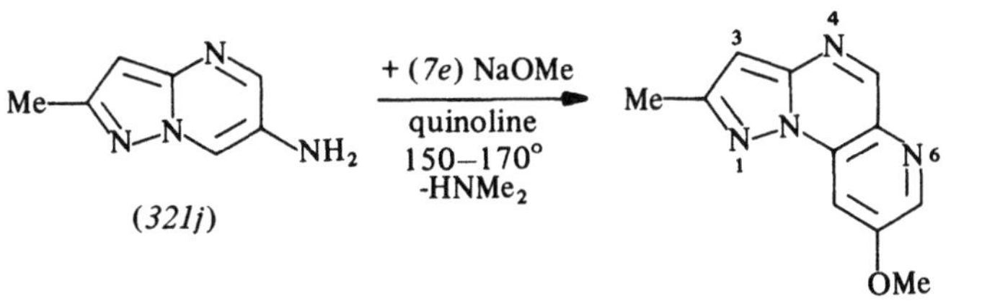

(*321j*) (*322*)

The 2-aminobenzimidazole (*323*) serves as starting compound for the preparation of pyrimido(1,2-a)benzimidazoles (*325*)[32]:

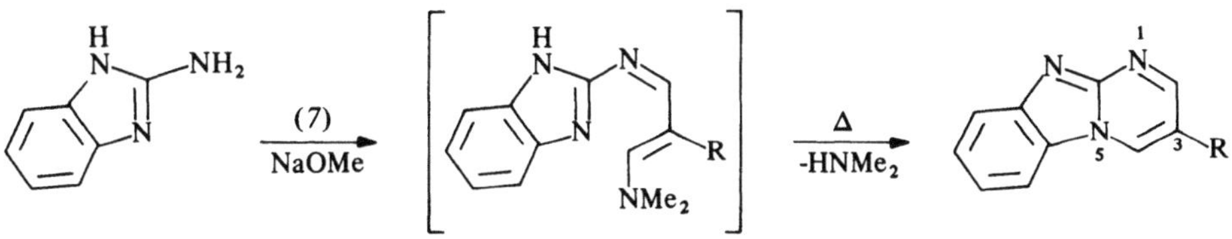

(*323*) (*324*) (*325*)

Table 31. Pyrimido(1,2-a)benzimidazoles (*325*)

(*325*)		R	Yield	Components	Reaction conditions
a	-, mp. 204–206° yellow needles	H	74%	(*323*) + (*7a*)	60–90°, 5 h in quinoline
b	3-Methyl-, mp. 243–6° pale-yellow plates	Me	76%	(*323*) + (*7b*)	60–90°, 5 h in quinoline
c	3-Ethyl-, mp. 169–171° yellow needles	Et	71%	(*323*) + (*7c*)	60–90°, 5 h in quinoline
d	3-Methoxy-, mp. 234–7° yellow needles	OMe	78%	(*323*) + (*7e*)	60–90°, 5 h in quinoline

A series of s-triazolo(1,5-a)pyrimidines (*327*) has been obtained with 3-amino-1,2,4-triazole (*326*)[121] and vinamidinium salts (*7*) in the presence of sodium methoxide:

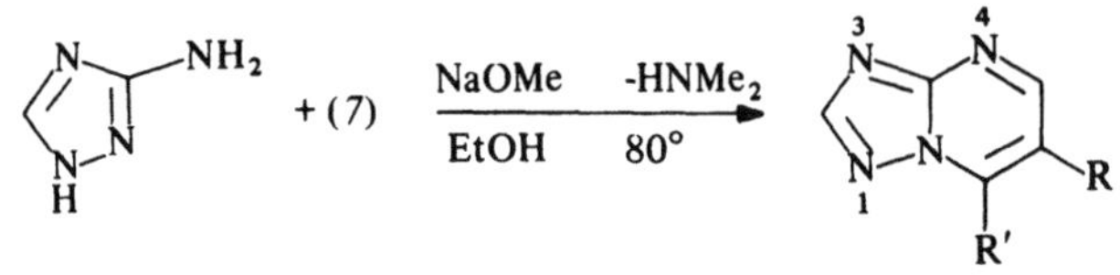

(*326*) (*327*)

Table 32. s-Triazolo(1,5-a)pyrimidines (*327*)

(*327*)		R	R'	Yield	Components	Reaction conditions
a	-, mp. 143–5° colorless needles	H	H	50%	(*326*) + (*7a*)	80°, 7 h ethanol
b	6-Methyl-, mp. 157° colorless plates	Me	H	76%	(*326*) + (*7b*)	80°, 4 h ethanol
c	6-Methoxy-, mp. 205° colorless needles	OMe	H	70%	(*326*) + (*7e*)	80°, 4 h ethanol
d	6-Phenyl-, mp. 167° colorless plates	Ph	H	87%	(*326*) + (*7i*)	80°, 7 h ethanol
e	6.7-Trimethylene-, mp. 125–7° colorless plates	$-(CH_2)_3-$		47%	(*326*) + (*7p*)	80°, 10 h ethanol
f	6.7-Tetramethylene-, mp. 117–118° colorless plates	$-(CH_2)_4-$		86%	(*326*) + (*7o*)	80°, 3 h ethanol

Attempts to prepare tetrazolo(1,5-a)pyrimidines in the same manner, from 5-aminotetrazole (*329*), failed. As a weak acid (*329*) forms a scarcely soluble sodium salt which in addition shows only poor nucleophilic properties. A conversion could be achieved with 3,3-diethoxy-1-dimethylamino-2-methylpropene (*35 b*) and (*329*) (whereas the unsubstituted (*35 a*) does not give reproducible results) leading to colorless needles of sharp melting point, mp. 116° (82%) which shows by ^1H NMR examination to be in deuteriochloroform an equilibrium mixture of 6-methyltetra-

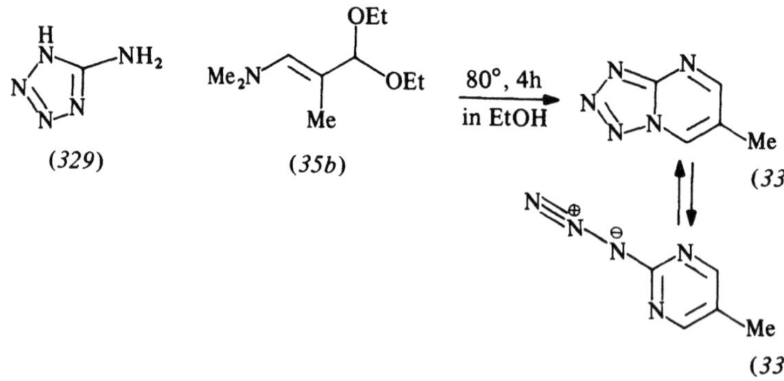

(*329*) (*35b*) (*330*) (*331*)

zolo(1,5-a)pyrimidine (*330*) and the isomeric 2-azido-5-methylpyrimidine (*331*). The valence tautomerism between tetrazolo(1,5-a)pyrimidine and 2-azidopyrimidine, also an electrocyclic ring-opening–ring-closure process, is a well-described phenomenon[122].

VI. Epilogue

The transformations given in our essay might have been designated essentially as a part of the chemistry of vinamidinium salts (*7*) and their vinylogues (*8*) or more simply as the chemistry of malondialdehyde and of its vinylogue glutaconic aldehyde.

In spite of their versatility and practical significance, all the preparations start with a condensation reaction of (*7*) or (*8*), and end with the thermolysis of a 1,3,5-hexatriene-like intermediate. In our treatise we have therefore given preference to the electrocyclic aspect of the reaction course, to use them as the heuristic principle. We have been quick to go in search if there already exist in the literature examples of such reactions, of which the mechanistic course might be best explained in terms involving a thermal electrocyclic ring-closure attended by elimination, as in our instances. The thermal Conrad-Limpach synthesis of quinolineones (Section IV., 2.6.) represents such an example. But some reactions involving carbocyclic ring-closures have also been found, which we feel forced to interpret as electrocyclic processes.

Attempting to hydrolyze 2-dimethylaminovinyl-4,6-diphenylpyryliumperchlorate (*332*) by aqueous sodium hydroxide results in the formation of the expected 2-formylmethylene-4,6-diphenyl-2H-pyran, besides 4-dimethylamino-2-phenylbenzophenone (*335*). The latter emerges by the action of dimethylamine on (*332*), which is also formed during the hydrolysis as the main product[124]. This conversion may involve, after nucleophilic addition of dimethylamine, first an electrocyclic ring-opening of the intermediate (*333*), and then a recyclization of the aminoheptatrienone (*334*), followed by elimination to (*335*):

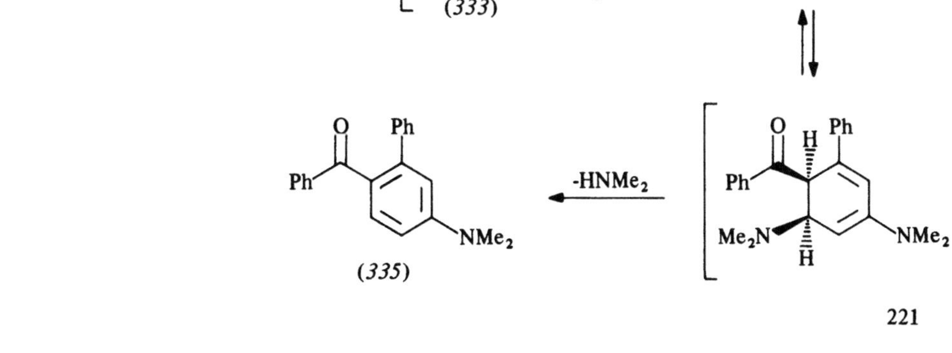

A series of cyclizations to benzene derivatives during formylation reactions with dimethylchloromethyleneammonium chloride or the related adduct of dimethylformamide-phosphorylchloride has been observed[125].

Treatment of the heptamethinium perchlorate (*336*) with dimethylchloromethyleneammonium chloride and dimethylformamide in chloroform (80°, 2 h) leads after work-up, to 1,3,5-triformylbenzene (*339*) in 40% yield. The acyclic intermediate (*337*), initially formed by double formylation, undergoes a cyclization that might well be seen as an intramolecular, electrophilic attack on the enamine β-position of (*337*). The alternative interpretation of this as an electrocyclic process in the mesomeric formula (*337'*) to (*338*), however, also seems to be more convincing:

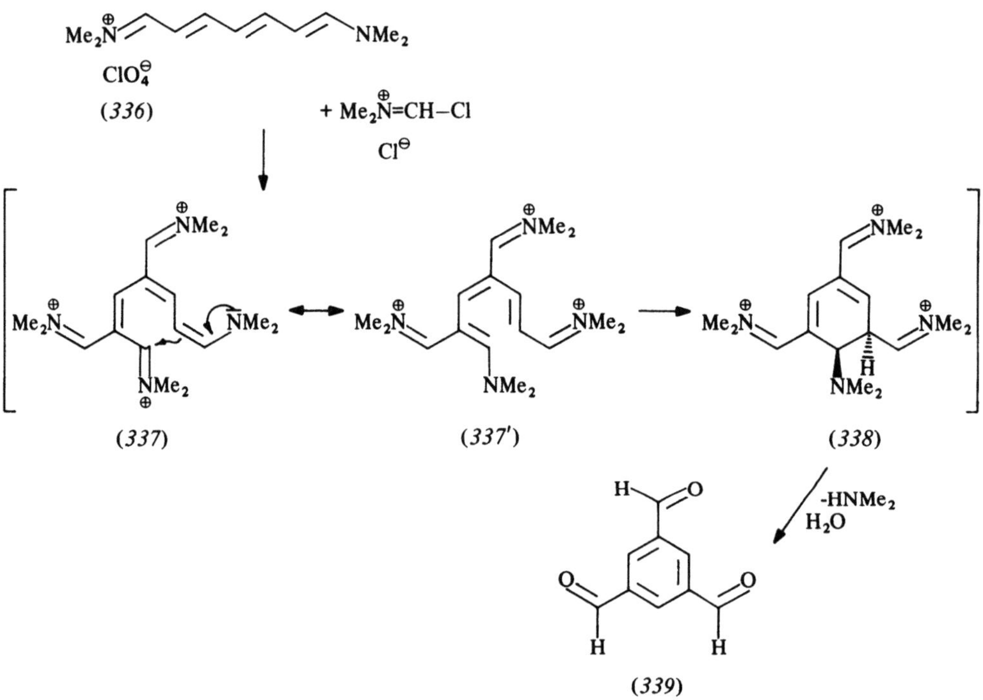

In a similar manner, acetylacetone (*340*) reacts with dimethylformamide-phosphorylchloride *via* the heptamethinium derivative (*341*) by double formylation and leads to 2,4-dichlorobenzaldehyde (*342*) (84%). 3-Penten-2-one (*343 a*), mesityloxide (*343 b*) and 4-dimethylamino-3-penten-2-one (*343 c*) undergo triformylation and yield, by subsequent ring-closure, 4-chloroisophthalic dialdehydes (*345 a–c*):

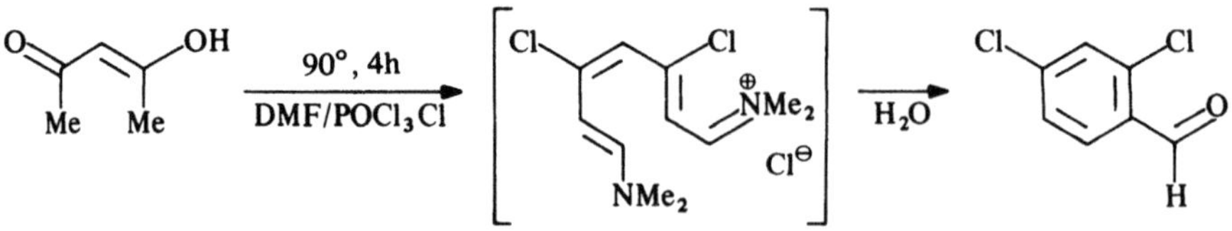

(*340*) (*341*) (*342*)

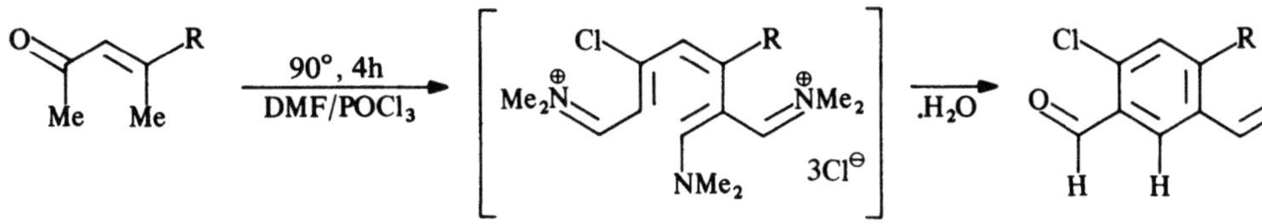

(343) (344) (345)

a: R = H
b: R = Me
c: R = NMe_2

a: 44%
b: 33%
c: 48%

Diformylation also takes place with the 1-methylpentamethinium salt (*8g*) and leads finally to the 4-dimethylaminoisophthalicdialdehyde (*347*) (64%):

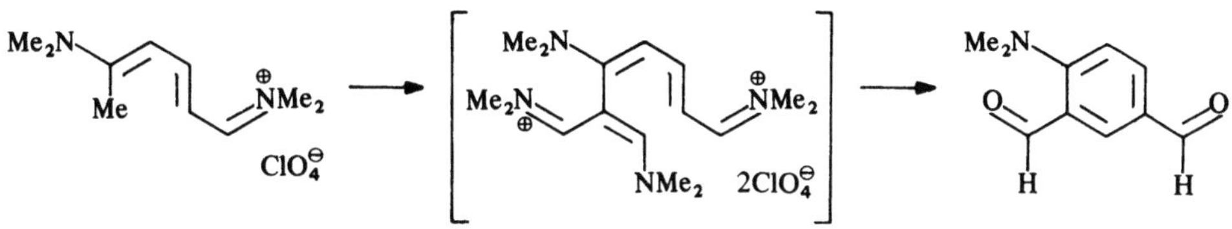

(8g) (346) (347)

In contrast to the foregoing examples, the formation of the 3-chloroanisole (*350*) (42%) in addition to (*342*) (23%) during the formylation of the 3-methoxy-3-penten-2-one (*343d*) (methyl ether of acetylacetone) can be satisfactorily explained only by an electrocyclic ring-closure of the assumed intermediate (*349*):

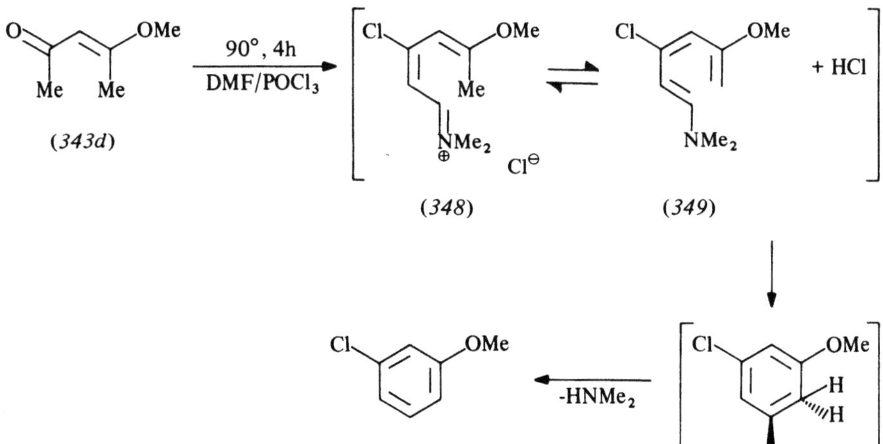

(343d) (348) (349) (350)

3,5-Diphenyl-4-pyrone (*356*) has been obtained (35%), along with recovered starting ketone (60%), in the formylation of dibenzylketone (*351*) with dimethylformamide-phosphorylchloride[126]. As known for other methylene ketones, a double formylation to the chloropentamethinium salt (*352*) should take place at first, but the sterically overcrowded structure of (*352*) lacks resonance stabilization and hydrolyzes readily to the protonated 5-dimethylaminopentadienal (*353*). The most important step in the reaction course may be the electrocyclic ring-closure of (*353*) to the 2H-pyran (*354*), followed by elimination to the 4-chloropyrylium salt (*355*), which then hydrolyzes to (*356*):

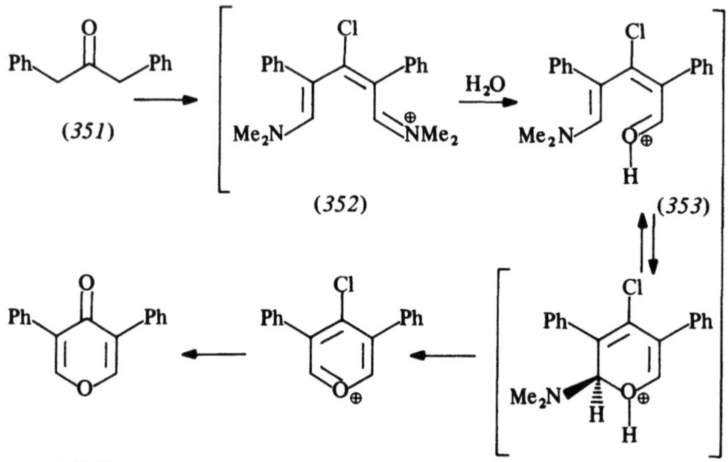

(*351*) (*352*) (*353*) (*356*) (*355*) (*354*)

A real counterpart to this imagined reaction course is the formation of 2-methoxy-4,6-diphenyl-2H-pyran (*358*) (99%) by treatment of 5-dimethylamino-1,4-diphenylpentadienone (*212*) with a small amount of hydrochloric acid in methanol. Addition of perchloric acid to (*358*) immediately precipitates 2,5-diphenylpyrylium perchlorate (*357*) and conversely, base and methanol reform (*358*) from (*357*):[6)]

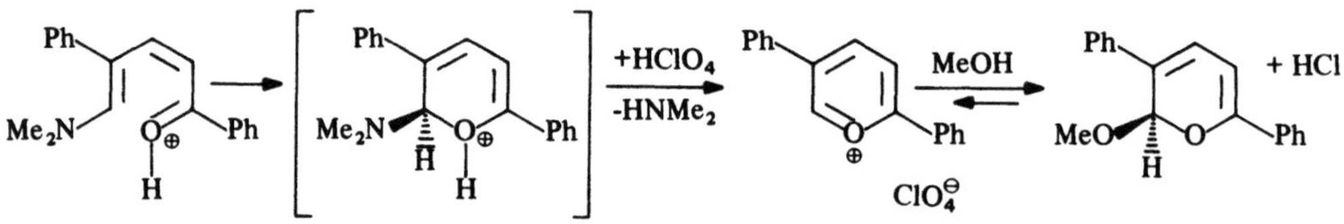

(*212*) (*357*) (*358*)

The α-2′-chloro (or bromo) phenylcinnamic acid produces 9-phenanthroic acid by refluxing with potassium hydroxide in boiling quinoline[127)]. For the preparation of polycyclic hydrocarbons this valuable method has been extended to a number of more complex systems of related, halogenated diarylacrylic acids – *e.g.*, α-2-(1-bromo-naphthyl)-β-1′-naphthylacrylic acid (*359*) gives dibenzo(a,g)phenanthroic acid (*360*) (46%)[128)] – and is generally named the "Hewett method" in the literature[128)]:

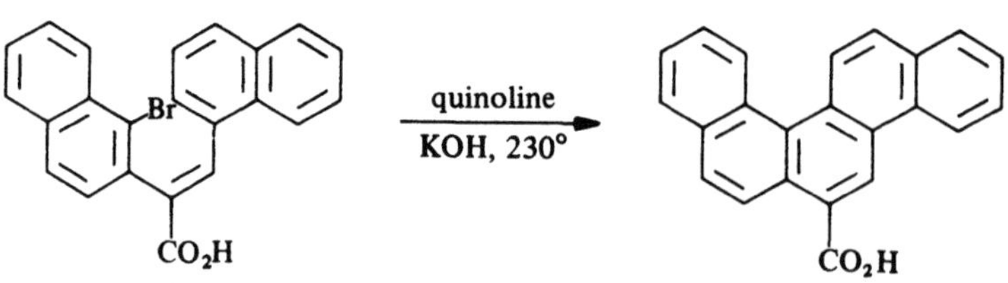

(*359*) (*360*)

The same cyclization also succeeds in the absence of a carboxylic group. Heating the fulvene (*361*), prepared by condensation of o-chlorobenzaldehyde with fluorene, with potassium hydroxide in boiling quinoline leads to benzo(e)acephenanthrylene (*362*)[127]:

(*361*) (*362*)

A cyclization of this type can be effected even under much milder conditions, using an excess of potassium amide in liquid ammonia[129]. The halogenated α-phenylcinnamic acids (*363a, b*) and the 2-chloro-β-phenylstilbene (*363c*) produce the corresponding phenanthrene derivatives (*364*) in good yields:

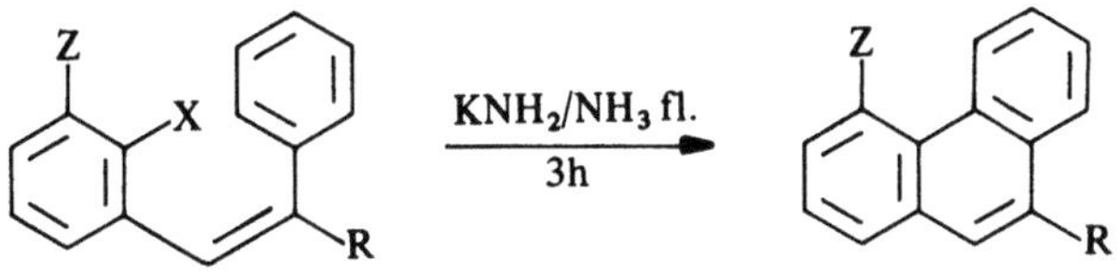

(*363*) (*364*)

a: R = CO_2H, X = Cl
b: R = CO_2H, X = Br Z = H
c: R = Ph, X = Cl

A suggested benzyne as intermediate during the reaction could be ruled out by the authors. If the position of the halogen was changed in (*363a, b*) using the *meta*-halogenated compounds, X = H, Y = Cl (Br), no detectable amounts of (*364*) could be obtained. In the successful cyclization of (*359*), an intermediate formation of a benzyne is also excluded. A cyclization mechanism, based on addition of amide ion across the carbon double bond, followed by nucleophilic displacement of the halogen by the highly activated, negatively charged *ortho* carbon of the other ring, formula (*363'*), has also been envisaged. But this mechanism cannot work in compounds such as (*359*) and has been discounted in view of the failure of the fluoro compound (*363*): R = CO_2H, X = F to react.

What remains after all is the indication that this reaction also follows an electrocyclic mode, passing through the energy-rich intermediate of formula (*363''*) with subsequent *cis*-elimination of hydrogen halide.

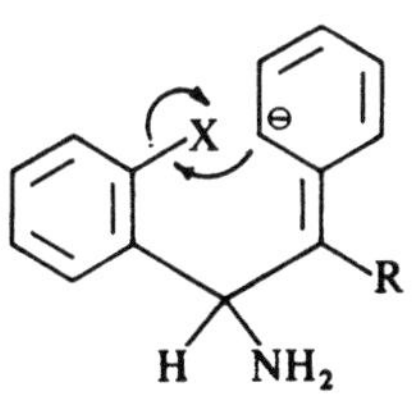

(*363'*)

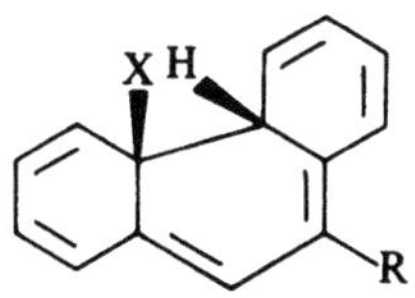

(*363''*)

To conclude this chapter we will be occupied with an old, long-known reaction: The Elbs reaction[130)], suited for a simple preparation of polycyclic aromatic hydrocarbons containing the anthracene moiety, by heating diarylketones with an ortho-placed methyl group at 400–450°. Our mechanistic proposal, applied to the cyclization of the 2-naphthyl-(2′-methylnaphthyl-1′)ketone (*365*) which yields the dibenzo-(a,h)anthracene (*369*) may involve an electrocyclic step followed by an elimination:

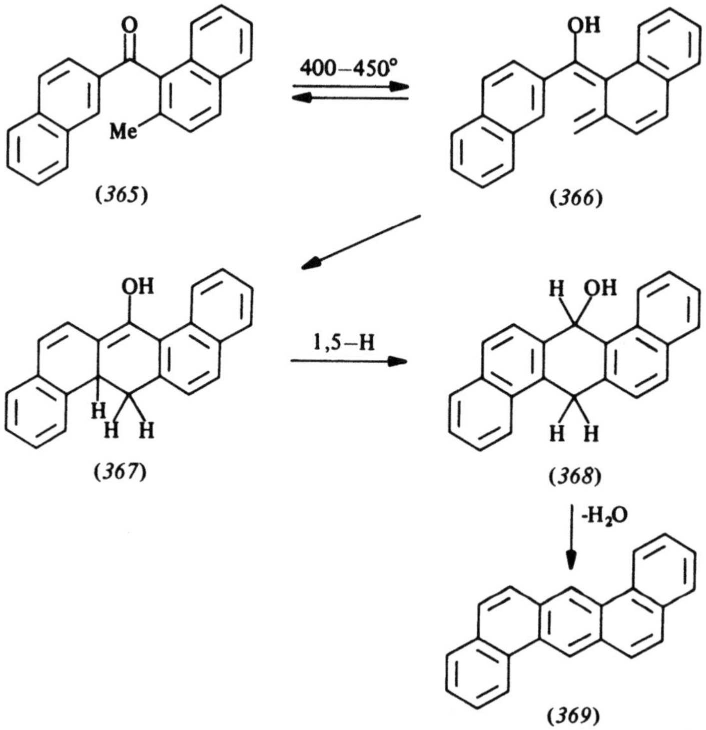

Accordingly the reaction starts with a thermal enolization to (*366*), generating the needed hexatriene system. The cyclization to (*367*) is followed by the energetically strongly favored 1,5-sigmatropic hydrogen shift to (*368*). Finally the elimination of water to the hydrocarbon (*369*) takes place.

VII. References

1) Clar, E.: Polycyclic hydrocarbons. New York: Academic Press 1964

2a) Woodward, R. B., Hoffmann, R.: Angew. Chem. *81*, 797 (1969); Angew. Chem. Int. Ed. Engl. *8*, 781 (1969)

b) Dewar, M. J. S.: Angew. Chem. *83*, 859 (1971); Angew. Chem. Int. Ed. Engl. *10*, 761 (1971)

3a) Wood, C. S., Mallory, F. B.: J. Org. Chem. *29*, 3373 (1964)
Stermitz, F. R.: In: Organic Photochemistry. Chapman, O. L. (ed.). New York: Dekker 1967, Vol. 1, p. 247

b) Martin, R. H.: Angew. Chem. *86*, 727 (1974); Angew. Chem. Int. Ed. Engl. *13*, 649 (1974)
c) Hayward, R. J., Leznoff, C. C.: Tetrahedron *27*, 2085 (1971); *27*, 5115 (1971); Carruthers, W., Evans, N., Pooranamoorthy, R.: J. C. S. Perkin Trans., *I*, 44(1973); Leznoff, C. C., Lilie, W., Manning, C.: Can. J. Chem. *52*, 132 (1974); Carruthers, W., Evans, N., Withmarsh, D.: J. C. S. Chem. Commun. *1974*, 526; Carruthers, W., Evans, N., Pooranamoorthy, R.: J. C. S. Perkin Trans. I, 76 (1975)
4) Ziegler, K., Hafner, K.: Angew. Chem. *67*, 301 (1955); Hafner, K.: Justus Liebigs Ann. Chem. *606*, 79 (1957); Hafner, K., Asmus, K.-D.: Justus Liebigs Ann. Chem. *671*, 31 (1964)
5) Downing, A. P., Ollis, W. D., Sutherland, I. O.: J. C. S. Chem. Commun. *1967*, 143; *1968*, 1053; J. Chem. Soc. *1969* B, 111
6) Jutz, C., Wagner, R.-M., Kraatz, A., Löbering, H.-G.: Justus Liebigs Ann. Chem. *1975*, 874
7) Spangler, C. W., Jondahl, T. P., Spangler, B.: J. Org. Chem. *38*, 2478 (1973)
8) Streitwieser, A., in: Molecular orbital theory for organic chemists. New York-London: Wiley 1961, p. 335
Heilbronner, E., Bock, H., in: Vol. 1, Das HMO-Modell und seine Anwendung. Weinheim: Verlag Chemie 1968, p. 355
9) Brown, R. D.: Trans. Faraday Soc. *45*, 1013 (1950)
10) Lloyd, D., McNab, H.: Angew. Chem. *88*, 496 (1976); Angew. Chem. Int. Ed. Engl. *15*, 459 (1967)
11) Malhotra, S. S., Whiting, M. C.: J. Chem. Soc. *1960*, 3812
12) Arnold, Z., Sorm, F.: Coll. Czech. Chem. Commun. *23*, 452 (1958)
13) Martin, G. J., Martin, M.: Bull. Soc. Chim. Fr. *1963*, 1637; Markin, S. M., Shavrygin, O., Bereznaya, M. T., Kolobova, T. P.: Zh. Org. Khim. *8*, 1394 (1972)
14) Arnold, Z., Holy, A.: Coll. Czech. Chem. Commun. *28*, 2040 (1963); Bredereck, H., Effenberger, F., Zeyfang, D.: Angew. Chem. *77*, 219 (1965); Angew. Chem. Int. Ed. Engl. *4*, 242 (1965); Bredereck, H., Effenberger, D., Zeyfang, D., Hirsch, K. A.: Chem. Ber. *101*, 4036 (1968)
15) Jutz, C., Kirchlechner, R., Seidel, H.-J.: Chem. Ber. *102*, 2301 (1969)
16) Wagner, R. M., Jutz, C.: Chem. Ber. *104*, 2975 (1971)
17) Kucera, J., Arnold, Z.: Coll. Czech. Chem. Commun. *32*, 3792 (1967); *32*, 1704 (1967)
18) Jutz, C., Schweiger, E.: Chem. Ber. *107*, 2383 (1974)
19) Arnold, Z.: Coll. Czech. Chem. Commun. *26*, 3051 (1961)
20) Arnold, Z.: Coll. Czech. Chem. Commun. *30*, 2125 (1965)
21) Zemlicka, J., Arnold, Z.: Coll. Czech. Chem. Commun. *26*, 2852 (1961)
22) Arnold, Z.: Coll. Czech. Chem. Commun. *38*, 1168 (1973); Arnold, Z., Sauliova, J., Krchnak, V.: Coll. Czech. Chem. Commun. *38*, 2633 (1973)
23) Brechbühler, H., Büchi, H., Hatz, E., Schneider, J., Eschenmoser, A.: Helv. Chim. Acta *48*, 1747 (1965) footnote
24) Jutz, C., Amschler, H.: Chem. Ber. *97*, 3331 (1964)
25) Jutz, C.: C–C-Bond forming reactions of chloromethylene iminium ions. In: Advances in Organic Chemistry, Vol. 9, Iminium Salts in Organic Chemistry. Viehe, H. G., Böhm, H. (eds.) New York: Wiley 1976, pp. 225–343
26) Arnold, Z., Holy, A.: Coll. Czech. Chem. Commun. *28*, 869 (1963)
27) Arnold, Z., Zemlicka, J.: Proc. Chem. Soc. *1958*, 227; Coll. Czech. Chem. Commun. *24*, 2385 (1959); Gagan, J. M. F., Lloyd, D.: J. C. S. Chem. Commun. *1967*, 1043
28) Benary, E.: Ber. deutsch. chem. Ges. *63*, 1573 (1930)
29) Zincke, T.: Liebigs Ann. Chem. *333*, 296 (1904); *339*, 139 (1905)
Zincke, T., Würker, W.: Liebigs Ann. Chem. *341*, 365 (1905); König, W.: J. prakt. Chem. (2) *69*, 105 (1904)
30) Köbrich, G.: Liebigs Ann. Chem. *648*, 114 (1961)
31) Arnold, Z.: Coll. Czech. Chem. Commun. *25*, 1308 (1960)
32) Löbering, H.-G.: Thesis, Technische Universität München 1976
33) Märkl, G., Baier, H.: Tetrahedron Lett. *1968*, 4379

34) Jutz, C., Wagner, R.-M.: Angew. Chem. *84,* 299 (1972); Angew. Chem. Int. Ed. Engl. *11,* 315 (1972)
Wagner, R.-M.: Thesis, Techn. Univers. München 1972
35) Kraatz, A.: Diplomarbeit, Techn. Univers. München 1974
36) Unpublished results
37) Kurz, P.: Liebigs Ann. Chem. *631,* 46 (1960)
38) Org. Synth. Coll. Vol. *IV,* 234 (1963); Ming-Chien-Chiang: J. Chinese Chem. Soc. *18,* 65 (1951); C. A. *46,* 4471 (1952)
39) Barnes, R. A.: J. Am. Chem. Soc. *70,* 145 (1948)
40) Hewett, C. L.: J. Chem. Soc. *1936,* 596
41) Weidlich, H. A.: Ber. dtsch. chem. Ges. *71,* 1203 (1938)
42) Hewett, C. L.: J. Chem. Soc. *1938,* 1286; Bergmann, E.: J. Chem. Soc. *1938,* 1291
43) Nightingale, D. V., Erickson, B., Shickelford, J. M.: J. Org. Chem. *17,* 1006 (1950); In: Vol. 10/I of Methoden der organischen Chemie. Houben-Weyl-Müller. Stuttgart: Thieme Verlag 1971, p. 5338
44) Kraatz, A.: Thesis, Techn. Univers. München 1976
45) Schroeter, G., Kindermann, E., Dietrich, C.: Liebigs Ann. Chem. *426,* 17 (1922)
46) Hafner, K., Vöpel, K. A., Ploss, G., König, C.: Liebigs Ann. Chem. *661,* 67 (1963)
47) Jutz, C., Schweiger, E., Löbering, H.-G., Kraatz, A., Kosbahn, W.: Chem. Ber. *107,* 2956 (1974)
48) Bausteine der Bayer-Forschung, Substance No. 7058: a gift from Bayer A. G.
49) Fieser, L. F., Jones, J. E.: J. Am. Chem. Soc. *64,* 1666 (1942); Saint-Ruf, G., Buu-Hoi, N. P., Jaquignon, P.: J. Chem. Soc. *1958,* 48
50) Gabard, J., Dvoloitzky, M., Jacques, J.: Bull. Soc. Chim. Fr. *31,* 27 (1964)
51) Haworth, R. D., Mavin, C. R., Sheldrick, G.: J. Chem. Soc. *1934,* 459
52) Storms, P. W., Taussig, R. R.: J. Chem. Eng. Data *11,* 272 (1966); C. A. *64,* 19509b (1966)
53) Rahmann, A. U., Khan, A. A.: Chem. Ber. *95,* 1786 (1962)
54) Org. Synth. Coll. Vol. *III,* 134 (1955)
55) Friedman, L., Shechter, H.: J. Org. Chem. *26,* 2522 (1961)
56) Haworth, R. D.: J. Chem. Soc. *1932,* 1132
57) Cook, J.: J. Chem. Soc. *1932,* 470
58) Cook, J.: J. Chem. Soc. *1932,* 456
59) Jutz, C., Peuker, H.-G.: Synthesis *1975,* 431
60) Jensen, F. R., Coleman, W. E.: Tetrahedron Lett. *1959,* 7
61) Fieser, L. F., Gates jr., M. D.: J. Am. Soc. *72,* 1245 (1950);
Boekelheide, V., Goldman, M.: J. Org. Chem. *19,* 575 (1954);
Reid, D. H.: The chemistry of phenalenes, Quart. Rev. Chem. Soc. *19,* 274 (1965)
62) Prinzbach, H., Freudenberger, V.: Angew. Chem. *77,* 346 (1965); Angew. Chem. Int. Ed. Engl. *4,* 243 (1965)
63) Norman, R. O. C., Thompson, G. A., Waters, W. A.: J. Chem. Soc. *1958,* 175
64) Lang, K. F., Buffleb, H.: Chem. Ber. *90,* 2894 (1957)
65) Buu-Hoi, N. P., Jacquignon, J.-P., Jutz, C.: Bull. Soc. Chim. Fr. *1972,* 2514
66) Adelfang, J. L., Daub, G. H.: J. Am. Chem. Soc. *79,* 1751 (1957)
67) Vollmann, H., Becker, H., Corell, M., Streek, H.: Justus Liebigs Ann. Chem. *531,* 1 (1937)
68) Hafner, K., Pelster, H., Patzelt, H.: Liebigs Ann. Chem. *650,* 80 (1961); McDonald, R. N., Wolfe, N. L., Petty, H. E.: J. Org. Chem. *38,* 1106 (1973)
69) Heinlein, K. H.: Thesis, Univers. Marburg 1958; Mayer, J., Sondheimer, F.: J. Am Chem. Soc. *88,* 602 (1966)
70) Anderson jr., A. G., Anderson, R. G., Fuijta, T. S.: J. Org. Chem. *27,* 4535 (1962)
71) Nozoe, T., Seto, S., Matsumura, S.: Bull. Chem. Soc. Jpn. *35,* 1990 (1962); McDonald, R. N., Richmond, J. M.: J. Org. Chem. *40,* 1689 (1975)
72) Hafner, K.: Pure Appl. Chem. *28,* 168 (1971)
73) Anderson jr., A. G., McDonald, A. A., Montana, A. F.: J. Am. Chem. Soc. *90,* 2993 (1968); Anderson jr., A. G., Masada, G. H., Montana, A. F.: J. Org. Chem. *38,* 1439 (1973); Anderson jr., A. G., Montana, A. F., McDonald, A. A., Masada, G. H.: J. Org. Chem. *38,* 1445 (1973)

74) Jutz, C., Schweiger, E.: Angew. Chem. *83,* 886 (1971); Angew. Chem. Int. Ed. Engl. *10,* 808 (1971); Jutz, C., Schweiger, E.: Synthesis *1974,* 193

75) Nikolajewski, H. E., Dähne, S., Leupold. D., Hirsch, B.: Tetrahedron *24,* 6685 (1968)

76) Schweiger, E.: Thesis, Techn. Univers. München 1973

77) Buu-Hoi, N. P., Giao, N. B., Jutz, C.: Naturwiss. *57,* 499 (1970); Buu-Hoi, N. P., Hien, D. P., Jutz, C.: Naturwiss. *54,* 470 (1967)

78) Jutz, C., Peuker, H.-G., Kosbahn, W.: Synthesis *1976,* 673

79) Hess, B. A., Schaad, L. J.: J. Org. Chem. *36,* 3418 (1971); J. Am. Chem. Soc. *93,* 2413 (1971)

80) Jutz, C., Löbering, H.-G.: Angew. Chem. *87,* 415 (1975); Angew. Chem. Int. Ed. Engl. *14,* 418 (1975)

81) Peuker, H.-G.: Thesis, Techn. Univers. München 1977

82) Buu-Hoi, N. P., Lecocq, J.: J. Chem. Soc. *1946,* 830

83) Newman, S. W., Lednicer, D.: J. Am. Chem. Soc. *78,* 4765 (1956)

84) Martin, R. H., Flammang-Barbieux, M., Cosyn, J. P., Gelbcke, M.: Tetrahedron Lett. *31,* 3507 (1968)

85) Martin, R. H., Defays, N., Figeys, H. P., Flammang-Barbieux, M., Cosyn, J. P., Gelbcke, M., Shurter, J. J.: Tetrahedron *25,* 4985 (1969)

86) Jutz, C., Wagner, R.-M., Löbering, H.-G.: Angew. Chem. *86,* 781 (1974); Angew. Chem. Int. Ed. Engl. *13,* 737 (1974)

87) Plucker, J., Amstutz, E. D.: J. Am. Chem. Soc.: *62,* 1512 (1940)

88) Herz, W., Rogers, J. L.: J. Am. Chem. Soc.: *73,* 4921 (1951)

89) Zincke, T.: Liebigs Ann. Chem. *330,* 361 (1903); *333,* 296 (1904); Ber. Deut. Chem. Ges. *38,* 3824 (1905); Borsche, W., Leditschke, H., Lange, K.: Ber. Deut. Chem. Ges. *71,* 951 (1938); König, W.: J. prakt. Chem. *69,* 105 (1904)

90) Marvell, E. N., Caple, G., Shahidi, I.: J. Am. Chem. Soc. *92,* 5641 (1970)

91) Sliwa, H., Tartar, A.: Tetrahedron Lett. *1976,* 4717

92) Kröhnke, F., Zecher, W.: Angew. Chem. *74,* 811 (1962); Angew. Chem. Int. Ed. Engl. *1,* 626 (1962)

93) Oberkobusch, R.: Chem. Ber. *86,* 976 (1953)

94) Gagan, J. M. F., Lloyd, D.: J. C. S. Chem. Commun. *1967,* 1043; Lloyd, D., Gagan, J. M. F.: J. Chem. Soc. (C) *1970,* 2488

95) Acheson, R. M., Bolton, R. G.: Tetrahedron Lett. *1973,* 2821

96) Schroeder, G., Lüttke, W.: Chem. Ber. *105,* 2175 (1972)

97) Hafner, K., Stephan, A., Bernhard, C.: Liebigs Ann. Chem. *650,* 42 (1961)

98) Speer, J. H., Raymond, A. L.: J. Am. Chem. Soc. *75,* 114 (1953)

99) Jutz, C., Löbering, H.-G., Trinkl, K.-H.: Synthesis *1977,* 326

100) Gold, H.: Angew. Chem. *72,* 956 (1960)

101) Kirchlechner, R., Jutz, C.: Angew. Chem. *80,* 367 (1968); Angew. Chem. Int. Ed. Engl. *7,* 367 (1968)

102) Müller, W.: Thesis, Techn. Univers. München 1967

103) Steglich, W., Hollitzer, O.: Angew. Chem. *85,* 505 (1973); Angew. Chem. Int. Ed. Engl. *12,* 495 (1973)

104) Yoneda, F., Higuchi, M., Kawamura, M.: Heterocycles (Jpn.) *4,* 1659 (1976)

105) Granik, V. G., Marchenko, N. B., Vlasova, I. F., Glushkov, R. G.: Khim. Geterotsikl. Soedin *1976,* 1509

106) DOS 2344757; 2440776, Byk-Gulden, 5.9.73

107) Schiess, P., Chia, H. L., Ringele, P.: Tetrahedron Lett. *1972,* 313

108) Eloy, F., Deryckere, A.: J. Heterocycl. Chem. *7,* 1191 (1970); McMillan, J. H., Washburn, S. S.: J. Org. Chem. *38,* 2982 (1973)

109) Blatter, H. M., Lukaszewski, H.: Tetrahedron Lett. *1964,* 855

110a) Conrad, M., Limpach, L.: Ber. Deut. Chem. Ges. *20,* 944, 948 (1887); *24,* 2990 (1891);

b) Gould, R. P., Jacobs, W. A.: J. Am. Chem. Soc. *61,* 2890 (1939); Harris, N. D.: Synthesis *1971,* 220; Price, C. C., Roberts, R. M.: J. Am. Chem. Soc. *68,* 1204 (1946);

c) Price, C. C., Snyder, H. R., Bullit jr., O. H., Kovacic, P.: J. Am. Chem. Soc. *69,* 374 (1947); Albrecht, R.: Liebigs Ann. Chem. *762,* 55 (1972);

d) Bredereck, H., Effenberger, F., Botsch, H., Rehm, H.: Chem. Ber. *98,* 1081 (1965); review: Weiss, M. J., Hauser, C. R., in: Elderfield 7, 204, 212, 217 (1961)
111) Gompper, R., Kunz, R.: Chem. Ber. *98,* 1391 (1965)
112) Achremowicz, L., Mlochowski, J.: Rocz. Chem. *47,* 1383 (1973)
113) Pomorski, J.: Rocz. Chem. *48,* 321 (1974)
114) Agui, H., Tobiki, H., Nakagome, T.: J. Heterocycl. Vhem. *12,* 1245 (1975)
115) Herz, W., Dittmer, K., Cristol, S. J.: J. Am. Chem. Soc. *69,* 1699 (1947); Herz, W.: J. Am. Chem. Soc. *75,* 483 (1953); Herz, W., Tocker, S.: J. Am. Chem. Soc. *77,* 6354 (1955)
116) Elks, J., Elliot, D. F., Herms, B. A.: J. Chem. Soc. *1944,* 629; Schindler, W.: Helv. Chim. Acta *40,* 2156 (1957)
117) Jones, R. G.: J. Am. Chem. Soc. *71,* 3994 (1949)
118) Totter, J. R., Darby, W. J.: Org. Synth. *24,* 64 (1949); Turner, R. A., Huebner, C. F., Scholz, C. R.: J. Am. Chem. Soc. *71,* 2801 (1949); Pyman, F. L.: J. Chem. Soc. *99,* 668 (1911)
119) Brechbühler, H., Büchi, H., Hatz, E., Schneider, J., Eschenmoser, A.: Helv. Chim. Acta *48,* 1747 footnote (1965)
120) Dewar, M. J. S., King, F. E.: J. Chem. Soc. *1945,* 114
121) Allen, C. F. H., Bell, A.: Org. Synth. *26,* 11 (1946)
122) Wentrup, C.: Tetrahedron *26,* 4969 (1970); Temple jr., C., Montgomery, J. A.: J. Org. Chem. *30,* 826 (1965); Temple jr., C., McKee, R. L., Montgomery, J. A.: J. Org. Chem. *30,* 829 (1965); Temple jr., C., Thorpe, M. C., Coburn jr., W. C., Montgomery, J. A.: J. Org. Chem. *31,* 935 (1966)
123) Mason, S. F.: Electronic absorption spectra. In: Vol. II, Physical Methods in Heterocyclic Chemistry. Katritzky, A. R. (ed.), New York and London: Academic Press 1963, pp. 61–66
124) Reynolds, G. A., Van Allen, J. A.: J. Org. Chem. *34,* 2736 (1969)
125) Holy, A., Arnold, Z.: Coll. Czech. Chem. Commun. *30,* 53 (1965)
126) Weissenfels, M., Pulst, M., Schneider, P.: Z. Chem. *13,* 175 (1973)
127) I. G. Farbenindustrie, Brit. Pat. *459,* 108 (1936); Z. *1937, II,* 2262; Franz. Pat. 807704; Z. *1937,* I, 5053
128) Hewett, C. L.: J. Chem. Soc. *1938,* 1286; Bogaert-Verhoogen, D., Martin, R. H.: Tetrahedron Lett. *1967,* 3045
129) Kessar, S. V., Sood, R. G., Nadir, U. K., Singh, M.: J. C. S. Chem. Commun. *1969,* 316; Kessar, S. V., Singh, M., Chander, R., Pal, D., Nadir, U. K.: Tetrahedron Lett. *1971,* 467; Kessar, S. V., Narula, S., Gandhi, S. S., Nadir, U. K.: Tetrahedron Lett. *1974,* 2905
130) Elbs, K.: Ber. Deut. Chem. Ges. *17,* 2847 (1884); J. prakt. Chem. *33,* 180 (1886); *35,* 465 (1887); *41,* 1, 121 (1890); Fieser, L. F.: Org. Reactions *1,* 129 (1942) (survey); Hurd, C. D., Azorlosa, J.: J. Am. Chem. Soc. *73,* 37 (1951); Badger, G. M.: J. Chem. Soc. *1953,* 2774; *1956,* 3435; Buu-Hoi, N. P., Lavit, D.: J. Chem. Soc. *1959,* 38; Mabille, P., Buu-Hoi, N. P.: J. Org. Chem. *25,* 1094 (1960); Bergmann, E. D., Blum, J.: J. Org. Chem. *25,* 474 (1960); *26,* 3214 (1961)

Received June 28, 1977

Some Newer Aspects of Mass Spectrometric *Ortho* Effects

Priv.-Doz. Dr. Helmut Schwarz*

Institut für Organische Chemie, Technische Universität Berlin, 1000 Berlin 12,
Straße des 17. Juni 135, W.-Germany

Table of Content

---

* Presented in part as a plenary lecture at the Annual Meeting of the "Arbeitsgemeinschaft Massenspektrometrie", Bad Kissingen, 1977.

I. Introduction

Beside the γ hydrogen transfer (*McLafferty* rearrangement) and the Retro-*Diels-Alder* reaction, the *ortho* effect belongs to the diagnostically most important processes occurring in the mass spectrometer[1–6]. An essential structural requirement for the rearrangement is the existence of a suitably 1.2-disubstituted double bond of *cis*-configuration. The double bond may be part of an aromatic system. In a stricter definition the *ortho* effect refers to the hydrogen transfer *via* six-membered transition states at vicinally substituted aromatic compounds with the result that the fragments *2* and *3* come into existence. More general definitions cover all aspects of interactions of 1.2-disubstituted compounds including higher-membered transition states as well as the transfer of atoms and molecules other than hydrogen onto suitable acceptors (1).

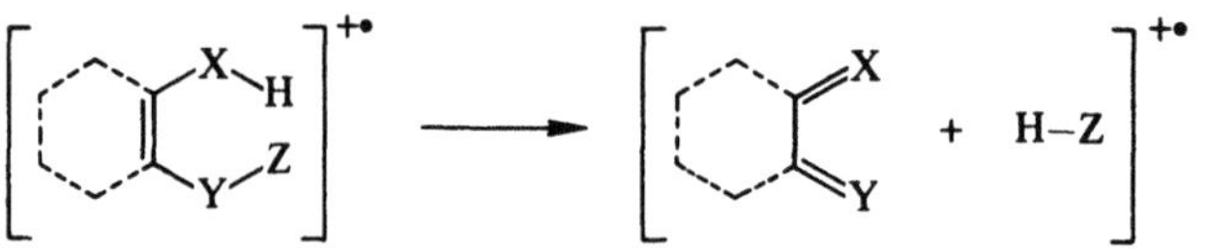

1 *2* *3* (1)

X = CH_2, NH, O, S
Y = C=O (Z=OH, O–Alkyl, O–Aryl, SH, NH_2, Aryl)
CH_2 (Z=OH, O–Alkyl, O–Aryl, NH_2, $N(Alkyl)_2$, S–Aryl

The present review intends neither to report on the analytical importance of electron impact induced *ortho* effects nor to summarize structurally different compounds, where the hydrogen transfer according to (1) has been observed. Instead, it is the purpose of this article to show which mechanisms are the basic principles for the interaction of vicinal substituents in compounds of the general structure, *1*. On the other hand, more general aspects will be discussed, which might lead to a better understanding of the gas phase chemistry of ionic species.

II. Classification of *Ortho* Effects

1. Influence of the Charge

As a consequence of the historical development of organic mass spectrometry, rearrangements according to (1) have been observed primarily only for *radical cations.* In the meantime it could be shown that this rearrangement is not restricted to radical cations. Bowie[7] demonstrated that the mass spectra of doubly charged molecular cations of salicylic and anthranilic acid eliminate H_2O with high intensity, while this effect does not operate for the *meta/para* isomers from which $OH^{\cdot}$ is exclusively eliminated. At suitably substituted *even* electron cations reactions according to (1) have been observed. Examples of this type of rearrangement are the $(M-CH_3)^+$ ions of *ortho* substituted isopropylbenzene derivatives. In the mass spectra of *ortho* isopropylbencoic acid, the $(M-CH_3-H_2O)^+$ ion represents the base peak while the same signal appears with only moderate intensity in the spectra of the other positional isomers[8]. Also, the secondary decompositions of the $(M-CH_3)^+$ ions from bis-trimethylsilylether of isomeric dihydroxybenzenes are fundamentally influenced by the substitution pattern[9]. An intense signal at *m/e* 73 (100%) is produced according to (2). In the case of the other isomers, this signal is observed with moderate intensity and moreover originates directly from the molecular ions and not from the $(M-CH_3)^+$ fragment.

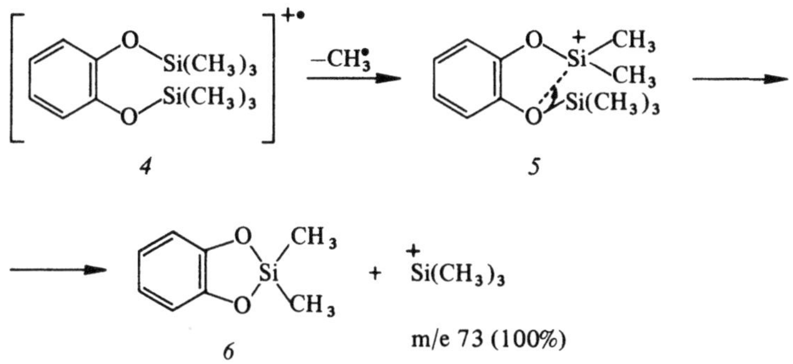

(2)

Detailed investigations of ^2H and ^{18}O labelled 2-ethoxybencoic acid methylester, 7, establish that oxonium ions (*e.g.*, *8*) rearrange *via ortho* effect as well[10, 11]. It is noteworthy that these reactions (3) take place unimolecularly as well as collision induced[12]. In both cases the CH_3OH elimination from *8* leads to a cyclic compound *9* [13].

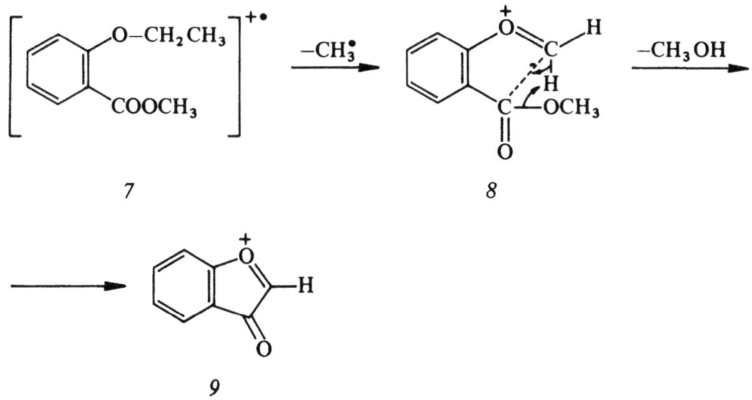

(3)

The mass spectra of *radical anions* of 1.2-disubstituted aromatic compounds sometimes contain signals which must be attributed to the operation of an *ortho* effect[14, 15]. Remarkably the spectra of the negative ions in some cases permit a better differentiation of the three positional isomers than do the mass spectra of the radical cations. A convincing explanation of these phenomena could not yet be found. As shown in Figs. 1 and 2, a differentiation of the isomeric 1.2.3-trisubstituted benzenes *10* and *11* is possible as a result of competing *ortho* effects (elimination of OH versus PhOH)[14].

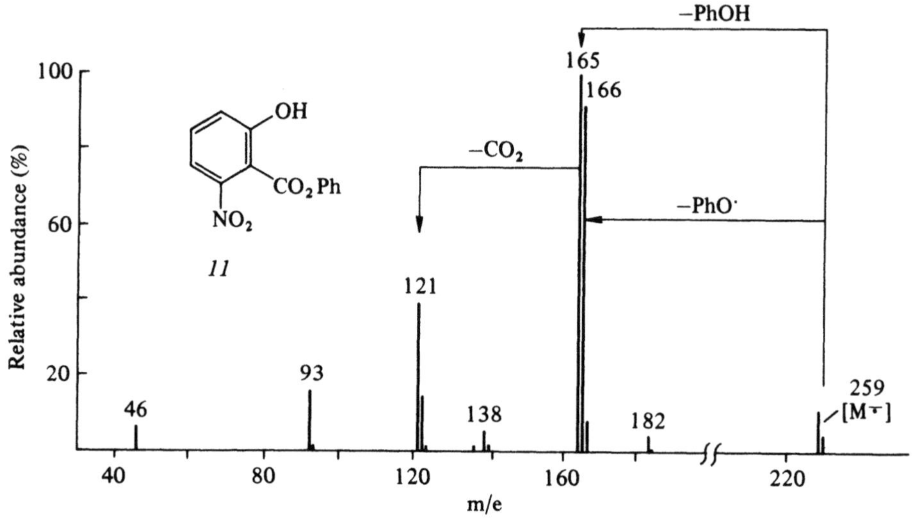

Fig. 1. Negative ion mass spectrum of phenyl-2-hydroxy 3-nitrobenzoate. [Reproduced from Org. Mass Spectrom. *9*, 1006 (1974).]

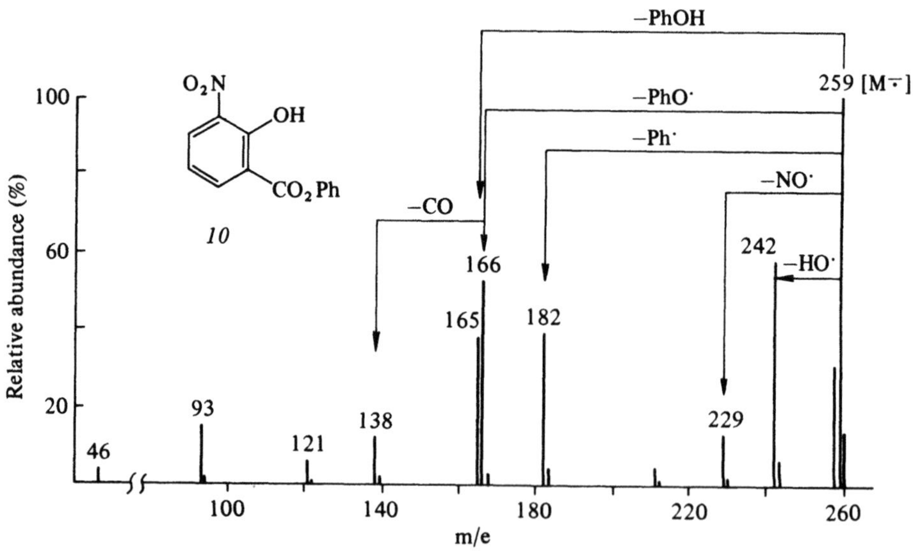

Fig. 2. Negative ion mass spectrum of phenyl 2-hydroxy-6-nitrobenzoate. [Reproduced from Org. Mass Spectrom. *9*, 1006 (1974).]

2. Mechanism of Hydrogen Transfer in Radical Cations

While the hydrogen transfer according to (1) in the original version[16] was formulated as a concerted process, Shannon[17] presented evidence that at least the water elimination from *12* must be described as a multistep reaction. Deuterium labelling establishes that water elimination is preceded by a reversible hydrogen exchange, the amount of which is determined by the lifetime of the molecular ions (4).

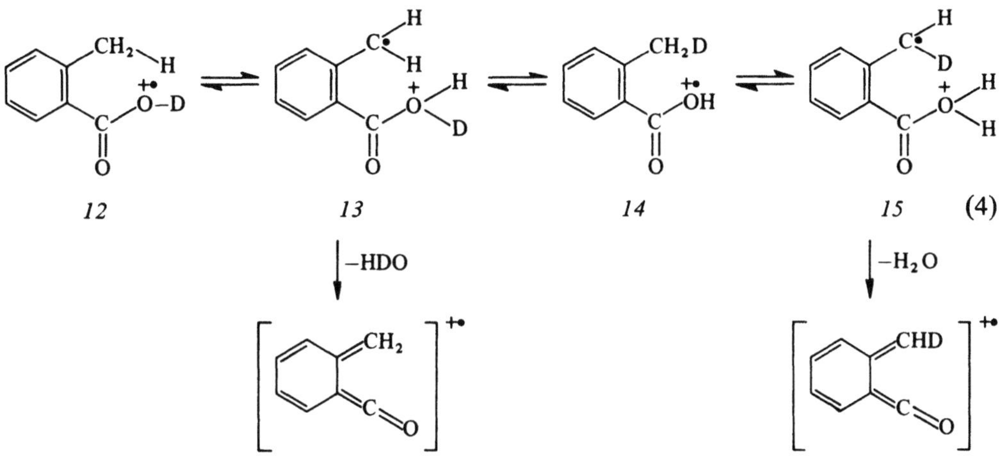

12 *13* *14* *15* (4)

16 *17*

In principle, rearrangement/elimination reactions *via ortho* effect may occur always as a multistep reaction if the molecular part *Y* (1) is able to act as a hydrogen acceptor (*e.g.*, a carbonyl function). In such a case, the hydrogen primarily can be transferred onto this function, whereas the elimination of *H–Z* would be the result of a four center cycloreversion. It cannot be stated with certainty that the arene elimination from *ortho* hydroxyphenylarylketones belongs to this class of stepwise reactions[6, 18, 19].

3. Energetic Aspects

Since the mass spectrum of a compound is the result of very fast consecutive and competitive unimolecular reactions[20], it seems reasonable that processes according to (1) will be suppressed if a much faster reaction, for instance a simple bond cleavage, is possible. Indeed it could be shown by an estimation of the energetic properties of *ortho* effect reactions versus α-cleavage (elimination of *Z*)[21] that the elimination of *H–Z* from substituted benzoic and thiobenzoic acid derivatives only takes place if extremely stable neutrals are formed (for instance, H_2O, CH_3OH). The formation of such stable species obviously reduces the activation energy of the rearrangement/elimination process (1). In the case of $Z = CH_3S$, PhS or PhO the elimination of $H-Z$ cannot compete with the α-cleavage reaction (loss of $Z^{\cdot}$). These results based on the investigation of model compounds[21] necessarily lead to the important conclusion that less intense or even missing signals for *2* do not exclude a 1.2-substitution pattern. This conclusion should be borne in mind in the structure elucidation of unknown compounds. The highly selective decomposition of *18* (*m/e* 104 represents the base peak whereas *m/e* 118 is absent) is probably a direct consequence of such energetic factors[22, 23]. Qualitative estimations show that the acid *20* is about 50 kcal · mol^{-1} more stable than *21*. This difference should influence decisively the activation energy of the competing reactions and should favour the sequence *18* → *19* (5).

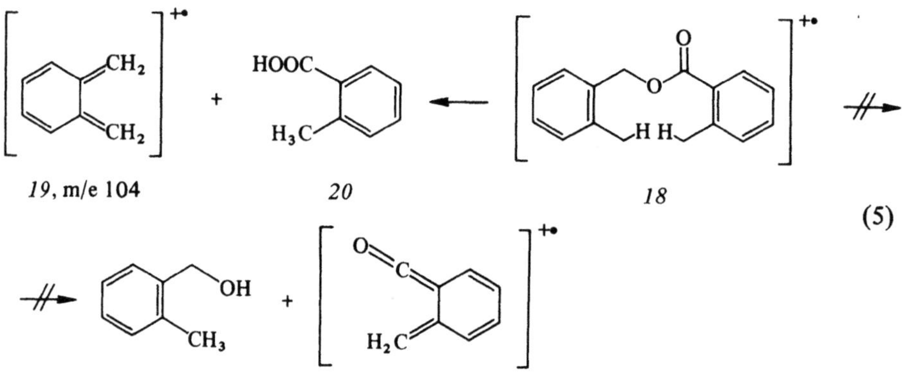

19, m/e 104 *20* *18* (5)

21 *16*, m/e 118

4. Hydrogen Transfer *via* Higher-membered Transition States

The first evidence that *ortho* effects can occur even if the hydrogen transfer proceeds *via* cyclic transition states with more than six members was presented by *Spiteller*[24]. He, and later Willhalm *et al.*[25] demonstrated that the mass spectra of O-ethylsalicyclic acid amides and those of the *meta/para* isomers are completely different. The *ortho* compound only eliminates NH_3 from the molecular ion, which cannot be explained assuming a 6-membered transition state for the hydrogen migration. Detailed investigations of ^2H labelled compounds established[26] that the elimination of ammonia from *22* does not take place by a simple hydrogen transfer/elimination reaction. The loss of the neutral is preceded by a partial exchange process of the NH_2 and $^{\alpha}CH_2$-hydrogens. The extent of this equilibration is determined by the lifetime of the molecular ion and is strongly catalyzed by the carbonyl group which acts as an intramolecular acceptor/donator function[26]. In the same way, the NH_3 resp. the R_2CO loss from the molecular ion of *23*[27] seems to be preceded by complex reactions. The decomposition of *24* leading to $C_6H_7N^{+\cdot}$ is interpreted by the authors as an example of a concerted hydrogen transfer/elimination process occurring *via* an 8-membered transition state[28]. Strictly speaking, however, there is no need or convincing evidence for such an interpretation.

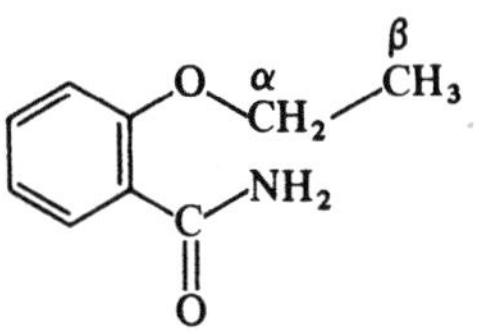

22

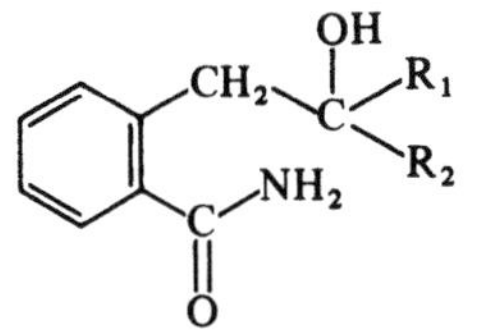

23 (R=H, Alkyl, Phenyl)

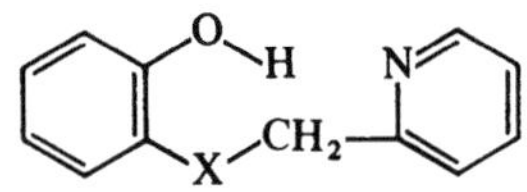

24 (X=CH_2, CO)

The mass spectra of the piperidine derivatives, *25–28*, deuterated in different positions demonstrate how strongly the selectivity of hydrogen transfers can be influenced by the nature of the substituents as well as by the ring size of transition states[29, 30, 31]. While *25* eliminates piperidine in a specific process according to (1), the analogous reaction from *26* is less selective. The hydrogen incorporated in the neutral originates from three different positions. Obviously the decreased selectivity for hydrogen transfer is the result of opposing energetic and conformational effects. The differences concerning the piperidine elimination from the amides *27* and *28* (*27*: hydrogen originates exclusively from the α'-position; *28*: participation of the α- as well as the α'-CH_2 groups) can be explained in a similar way. At the time of writing this article it seems to be impossible to correlate unequivocally the relevant parameters for the hydrogen transfer (for instance, activation energy, conformation and ring size of the transition state) with the gross structure of the compounds.

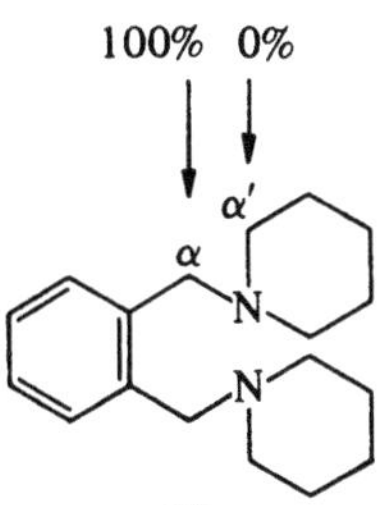

25

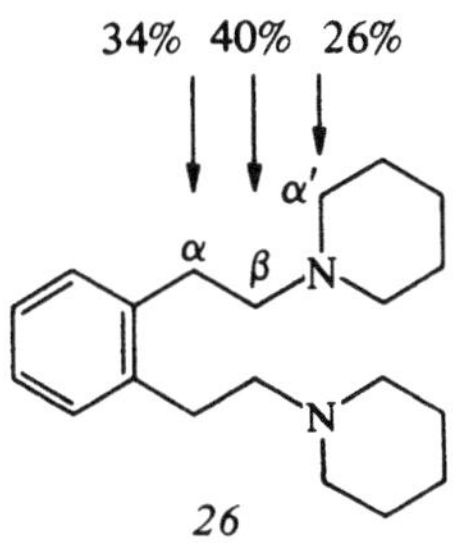

26

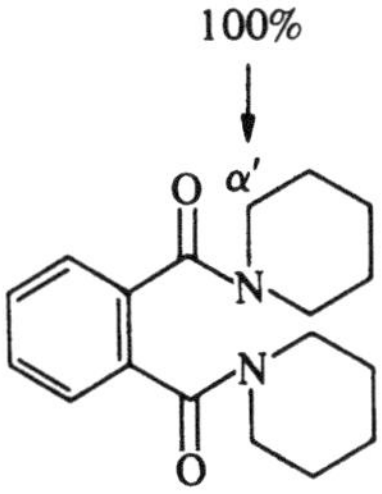

27

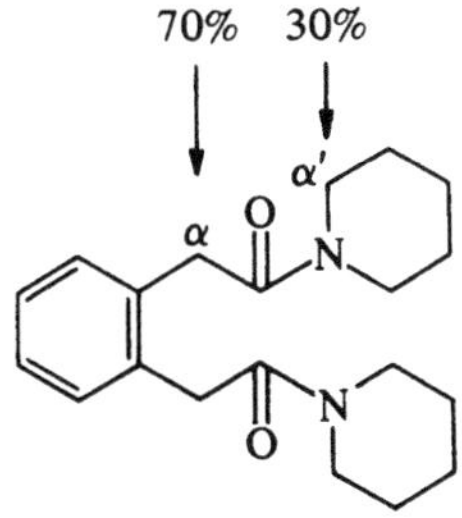

28

Investigations of Wünsche *et al.*[32] established that elimination of CH_3OH *via ortho* effect from *29* is possible only for $n = 0$. It is open to question whether the suppression of this reaction in the case of $n = 1$ is caused by the transconfiguration of the N=N double bond or whether in this system hydrogen transfer *via* an 8-membered transition state is in principle impossible (6). Seibl and co-workers[32], however, have presented evidence that such higher-membered transition states are possible: the molecular ion of *ortho*-hydroxycinnamic acid methylester shows intense CH_3OH-elimination irrespective of the configuration of the C=C double bond. ^2H labelling demonstrates that the phenolic hydrogen is exclusively transferred onto the ester function.

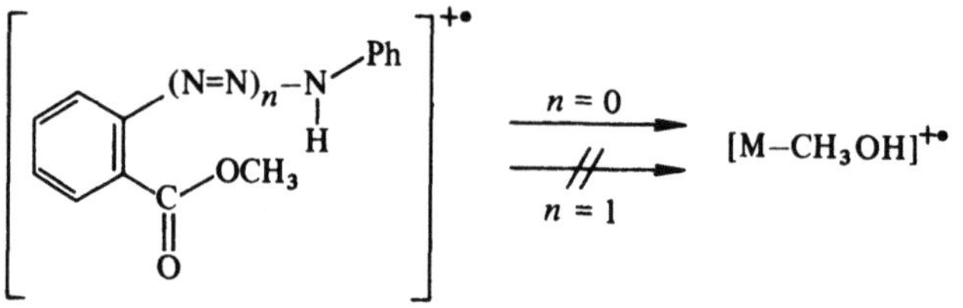

(6)

29

Roberts *et al.*[33] reported a remarkable hydrogen transfer in *ortho* substituted ferrocenyl derivatives. The specific elimination of CH_2O (a process highly improbable for normal esters) from *30* was announced by the authors as the first example of a new class of *ortho* effect reactions. Evidence for further reactions of this type is not available in the literature (7).

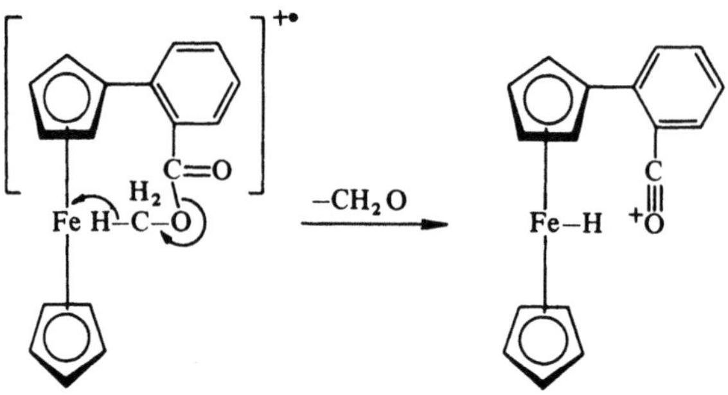

(7)

30 *31*

5. Apparent *Ortho* Effects

The fact that sometimes 1.3- or 1.4-disubstituted arenes exhibit a fragmentation behaviour which is characteristic for the *ortho* isomer may be caused by different effects. Among these the most familiar ones are the migration of substituents (*e.g.*, *via* valence isomerization) prior to elimination reactions. Alternatively the fragment $(M-HZ)^{+\cdot}$ might be generated *via* consecutive radical eliminations[14]. Both reactions have in common, that, in general, the spectra contain signals of apparent *ortho* products. The intensities of these signals, however, are moderate compared with those from the true 1.2-disubstituted derivatives. Much more difficult are those cases in which the wrong isomers produce intense fragments (as for instance *2* and *3*) and in which additionally the whole mass spectroscopic behaviour (including the metastable ion characteristic) seems to reflect an *ortho* effect. The OH elimination from *meta/para* nitrocyclopropylbenzene, *32*[34], or the elimination of HZ from substituted diphenylmethan derivatives, *33*[35], belongs to these trouble some cases. Whereas the OH loss could not be rationalized up to now, the behaviour of *33* can be easily explained, assuming a ring expansion *33* → *34*. In general, these eliminations as well as the fragmentation pattern of the positional isomers of *35* [M → $(M-CH_3)^+$ → $(M-CH_3-CD_3OH)^+$][36] can be elucidated only (if at all) through careful studies of labelled compounds combined with the application of sophisticated techniques (as for instance: collisional activation, peak shape analysis, determination of kinetic energy release, T_{kin} [36, 37]).

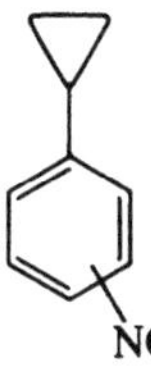

32

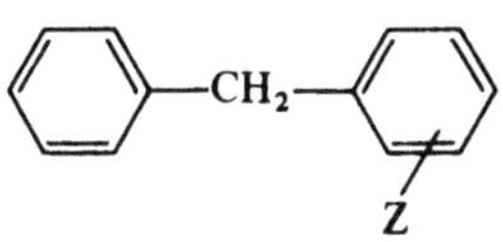

33

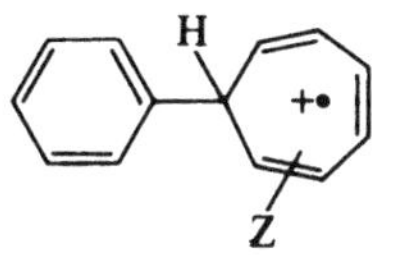

34

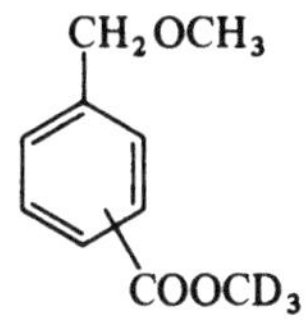

35

III. *Ortho* Effects as Internal Probes for Reaction Mechanisms

For the ketene elimination from *36* (X = O, NH; R = H, $COOCH_3$) as well as the loss of alkene (*e.g.*, ethylene) from *37* (R = H, $COOCH_3$), numerous mechanisms have been discussed in the literature[38–42]. Despite some differences concerning minor aspects, the central question is whether the side chain hydrogen is transferred directly onto the heteroatom *X* (4-membered transition state) or whether a migration onto the *ortho* position (6-membered transition state) takes place first, followed by a (1.3) hydrogen shift.

For the reactive (M-ketene)$^{+\cdot}$ ions from *36* (R = $COOCH_3$) and the (M-ethylene)$^{+\cdot}$ fragment from *37* (R = $COOCH_3$), this question could be unequivolcally answered in favour of the tighter 4-membered transition state[38, 40]. If a substituted cyclohexadienone radical cation, *40*, were produced as a reactive intermediate, the isomerization to *38* as well as *41* would occur with equal probability. This is a direct consequence of the constitutional and stereochemical equivalence of H/D in *40*. A kinetic isotope effect would favour the tautomerization *40* → *41* and consequently the elimination of CH_3OH. Because independent of the lifetime only CH_3OD is eliminated, one has to exclude *40* as a reactive intermediate in the primary fragmentation steps (8).

The observation that *39* is exclusively formed from a fragment and not from the molecular ions, *36* and *37*, excludes alternative decomposition pathways, for instance hydrogen transfer *via* 8-membered transition states and concerted elimination of CH_3OD and ketene resp. ethylene.

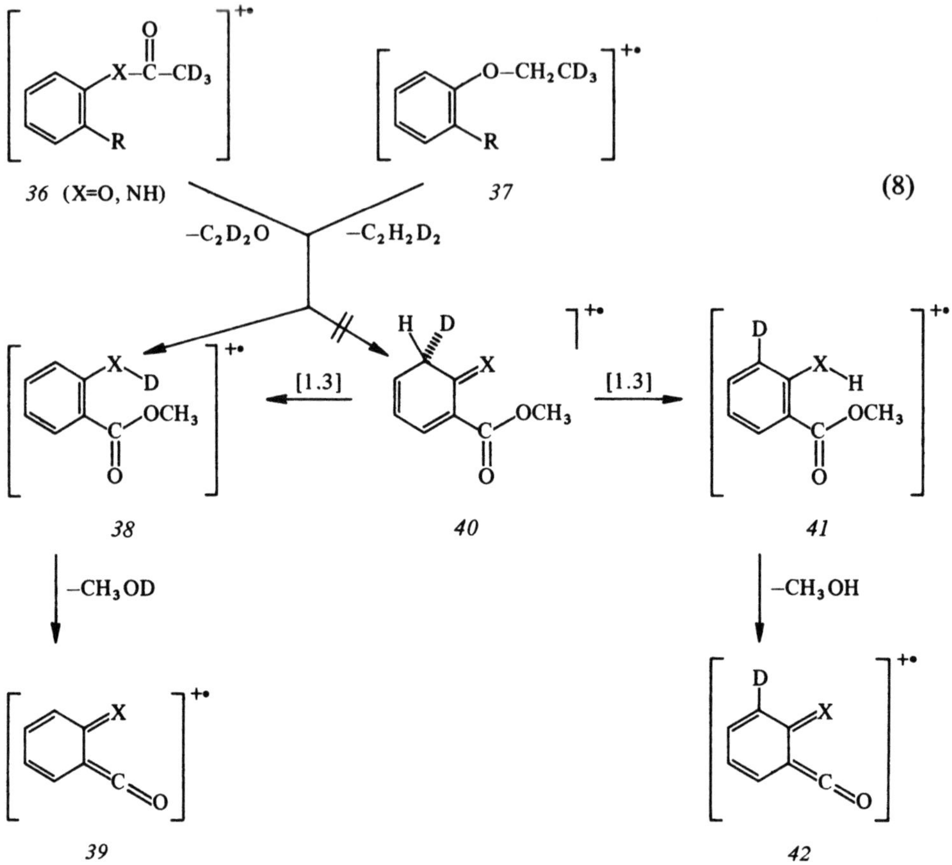

Analyzing the frequency factor ν, Bursey[43] was able to show that the details of the reaction mechanisms of the electron impact induced ketene elimination from *ortho* substituted phenyl acetates and from acetanilides, *36*, are dramatically influenced by the nature of the substituent R = F, Cl, Br, J. The results demonstrate that the *decrease* of ν going from the voluminous iodine to the small fluorine is not caused by steric effects (which should operate in an opposite direction) but is the result of an electronic interaction of both substituents, leading to a tighter transition state in the case of the more electronegative fluorine. Additional factors which are not yet completely understood play a decisive part in the fragmentation of the anilides.

IV. Anomalous Radical Eliminations *via Ortho* Effect

Electron impact induced cleavages of heteroatom-carbon bonds with elimination of radicals and charge localization on the heteroatom function require considerable energy and therefore provide intense fragment ions only in exceptional cases[44, 45]. At suitably substituted 5-membered heterocycles, for instance *43*[46–49], one always observes intense $(M-CH_3)^+$ ions, whereas 3.4-disubstituted compounds (*e.g.*, *44*) do not eliminate methyl radicals. Obviously the methyl elimination from the *ester* function of *43* is connected with hydrogen transfer from the activated and sterically accessible ring methyl group onto the carbomethoxy function[49]. According to the literature[50] such a transfer is possible only if the distance between the hydrogen donator and acceptor is about 1.8 Å. This condition is fulfilled for *43* and not for *44*, where the distance is about 2.8 Å according to *Dreiding* models.

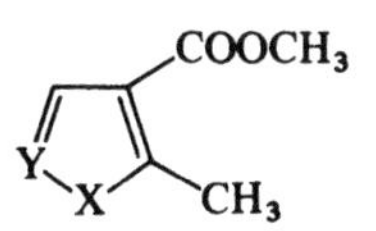

43 (X=S; Y=N; X=O, NH; Y=CH)

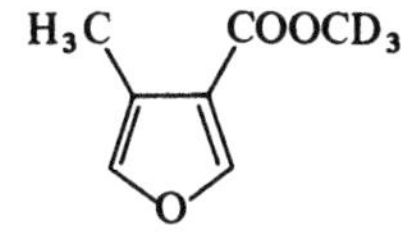

44

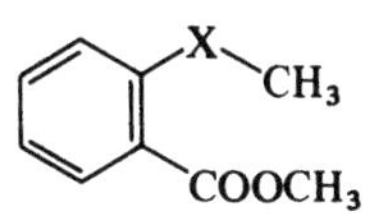

45 (X=NCH_3, S,O)

More precise information about the details of such anomalous radical eliminations resulted from the detailed investigation of labelled compounds of the general structure *45* [51]. Here the combined application of different methods (*e.g.*, appearance potential measurements, peak shape analysis, CA studies, kinetic energy release) established that the cleavage of a heteroatom-carbon bond can take place *via* two fundamentally different mechanisms:

(i) the methyl elimination is connected with a simultaneous hydrogen migration from an activated donator onto a suitable acceptor (as for the cases of CD_3 as well as CH_3 eliminations from *46*);

(ii) the cleavage is accompanied by a cyclization process (*e.g.*, loss of CD_3 from *49*).

Both reaction types differ considerably in the peak shape of the γ curves[52] as well as the amount of T_{kin} (9), (10). Independent collisional activation studies[11, 13] confirm the proposed concept to a high degree.

Compounds of the general structure *51* produce anomalous cleavage products under electron impact as well. While the methoxymethyl ester *51* (n = 1) shows an intense $(M-CH_3)^+$ ion, which fragments further by CD_3OH elimination, one does not observe any signal for methyl elimination from the homologues ester *51* (n = 2)[53]. Instead, the molecular ion eliminates CH_3OH, and the hydroxyl hydrogen of the neutral originates with nearly equal probabilities from both methylene groups. The complete equilibration of the α and β methylene hydrogens in the low voltage spectra can be the result of a fast (1.2) hydrogen shift, which should be classified as an example of electron impact induced *dyotropic* rearrangement[44, 45, 54]. Alternatively it cannot be excluded with certainty that this positional exchange is effectively catalyzed by the ester function (see Section II. 4). The fact that substituents other than the carbomethoxy function, for instance *52*, prevent methyl elimination from the molecular ions shows that this cleavage is the result of a specific intramolecular catalysis of the $COOCH_3$ function[36].

A further example of the outstanding activity of an ester group is represented by the DCl elimination from *53a*. The study demonstrates that from a mechanistic

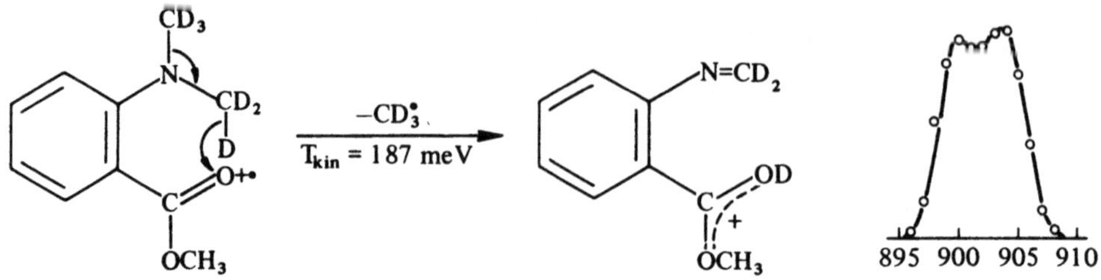

46 47

(9)

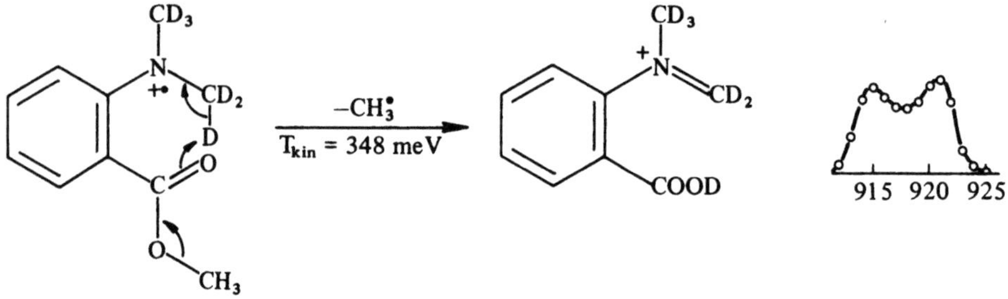

46 48

(10)

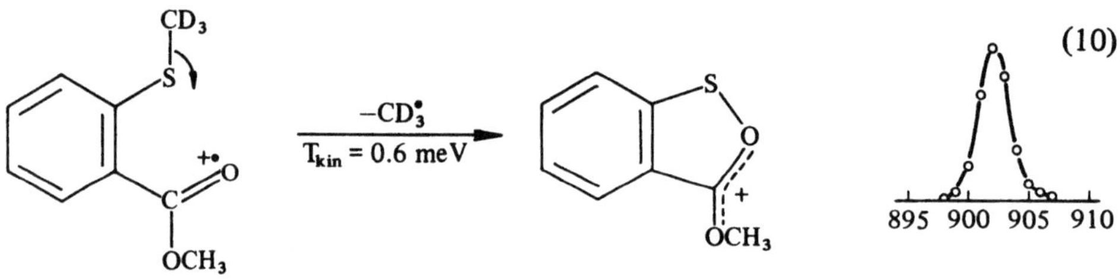

49 50

point of view this reaction cannot be described as a (1.2) elimination[55]. Other examples of radical induced multistep reactions are the OD resp. CDO eliminations from *53b*[13, 51].

These and many other reactions clearly establish that the interaction of vicinal substituents should be considered as radical initiated neighbouring group participation. It seems that the future study of such reactions will shed light on the complexity of gas phase chemistry of radical cations.

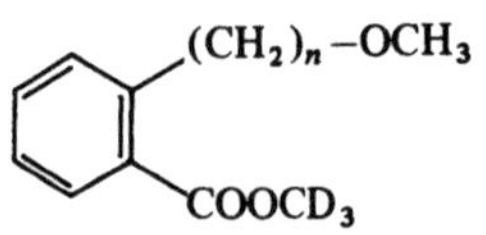

51 (n=1,2)

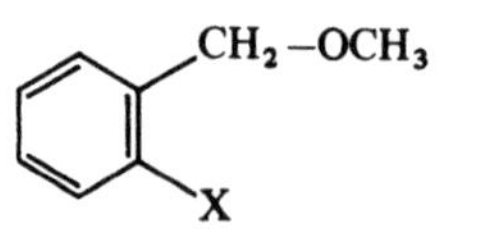

52 (X=OH, CH_2OH, NO_2)

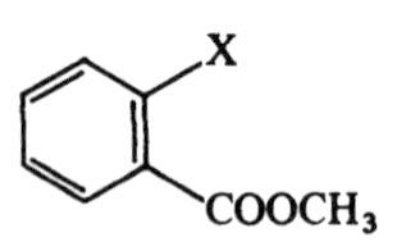

53a: X= CD_2CH_2Cl
53b: X= OCD_3

V. Redox Reactions of *Ortho* Substituted Nitrobenzene Derivatives

According to Holmes[56], the electron impact induced fragmentation of *ortho* substituted nitroarenes may be classified in three groups depending on the nature of X:

(i) Elimination of the NO_2 function and migration of atoms or molecular parts of the remaining substituent onto the vacant *ortho* position. These rearrangements can be completed by a set of secondary decompositions (11).

(ii) Hydrogen transfer from an activated position onto the nitro group which is followed by the loss of small neutral particles.

(iii) Oxygen transfer, which for the first time has been described by Seibl[34].

In the following chapters some newer results of the last two mentioned reaction types will be discussed preferentially.

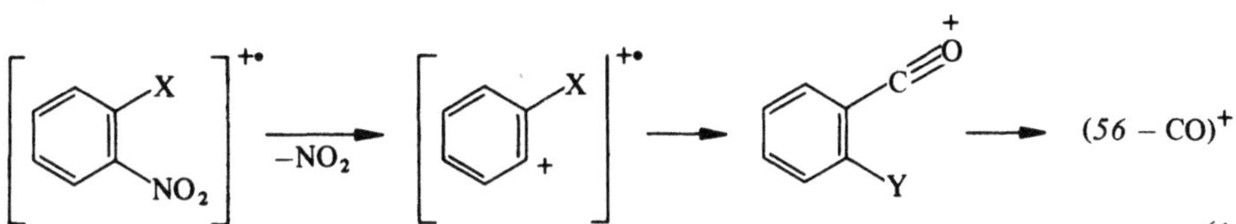

54 (X=COOH, $CONH_2$) *55* *56* (Y=OH,NH_2) (11)

1. Hydrogen Rearrangements

If the substituent *X*, *54*, contains activated hydrogen atoms, for instance CH_3 [57,58], CH_2Ph [59], NH_2 [60], HC=NAr[61] or $CH_2N(alkyl)_2$ [62], the mass spectra always show abundant signals for $(M-OH)^+$ ions. The genesis of this fragment depends strongly upon the actual structure of *X*, and it is possible that several different reaction mechanisms account for the fragmentation. In a formal sense the OH loss starts with a hydrogen transfer onto the nitro group leading to a reactive intermediate of unknown structure. The direct hydroxyl elimination from this species can be effectively suppressed by competing reactions involving other parts of the isomerized molecular ion. Hence it seems impossible to predict with certainty the primary decomposition modes of the molecular ions. The examples shown in (12) indicate the variety of different reaction channels starting from structurally similar compounds, *54*.

In spite of detailed studies of labelled precursors and the application of ingenious methods, only in very rare cases are the details of the reaction mechanisms as well as the structures of the resulting ions of (12) known. Only the loss of CH_3OH, H_2O, HCN and HNO from *54* could be elucidated using CA[69]. In agreement with photochemical transformations, the electron impact induced CH_3OH elimination from *54* [X = $CH(OCH_3)_2$] leads to nitrosobenzoic acid methylester. Contrary to previous conclusions, however, the loss of H_2O, HCN and HNO upon electron impact does not produce *ortho* nitrosobenzaldehyde *58*. Rather, the fragment *m/e* 135 (13) has the structure of 2.1-benzisoxazolinone-(3), *59*. Isomerizations of the heterocycles *59*, *60* or *61* which occur under thermolytic[71] or photochemical[72] conditions in the case of the neutral molecules, or which have been discussed

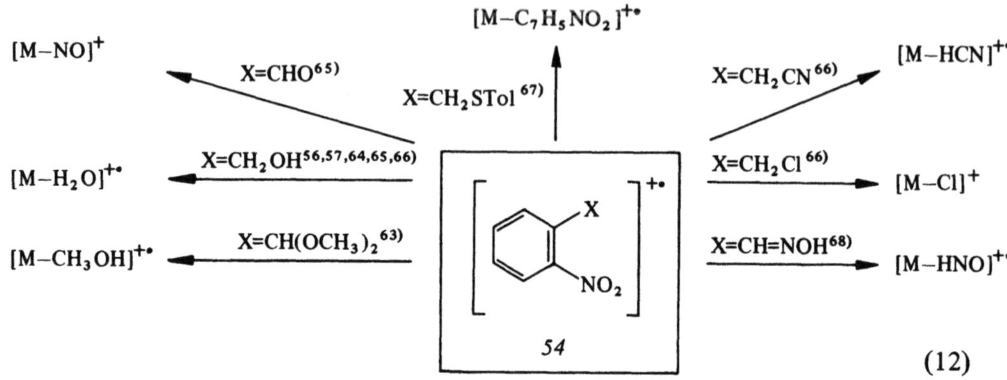

(12)

for short living reactive species *m/e* 135,[68] could not be established for ions *m/e* 135 with a lifetime of about 10^{-5} sec. In spite of the unequivocal assignment of the structure of *m/e* 135 (*59*, based on the CA studies), the detailed mechanisms concerning hydrogen transfer, elimination of neutral species and reorganization of the ionic skeleton are still unknown. It can only be stated that elimination reactions according to (13) do not belong to the 1.1. eliminations (cheletropic reactions)[73]. Moreover from the structure of the products, one has to conclude that these processes are best described by a series of intramolecular redox reactions involving hydrogen as well as oxygen migrations.

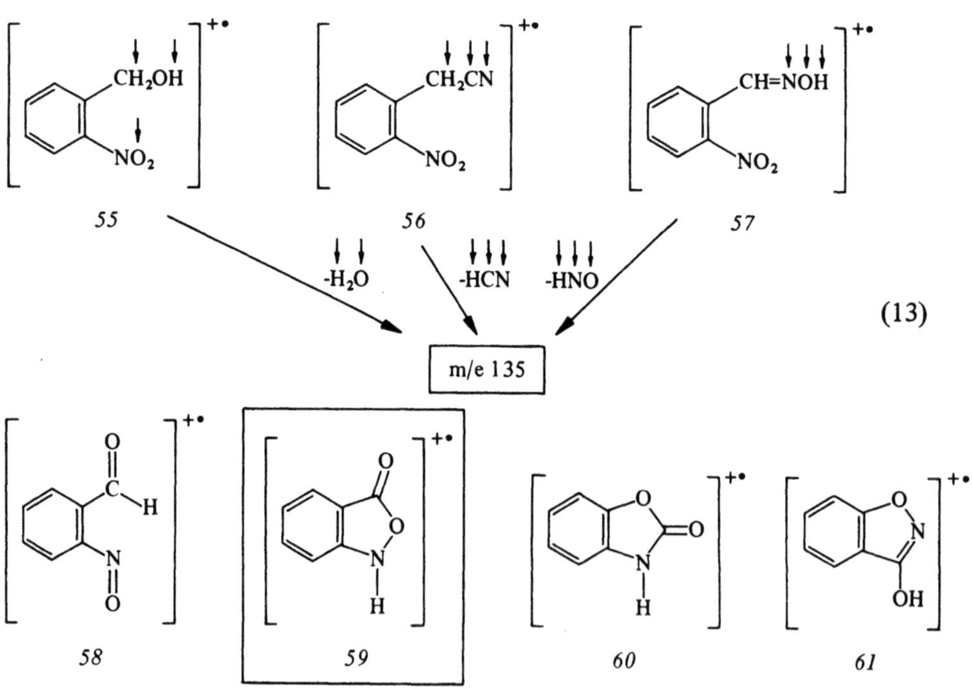

(13)

2. Oxygen Transfer

The first example for electron impact induced intramolecular redox reactions *via* oxygen transfer was published and discussed in detail by Seibl[34, 74]. He established that the fragmentation pattern of nitrophenylhydrazones of aromatic and aliphatic aldehydes and ketones as well as the decompositions of nitrostilbenes is always strongly influenced by the position of the nitro group. The decisive step of the redox reaction includes an oxygen transfer from the nitro function onto the N=C moiety followed by several rearrangements. The structural reorganization is completed by heterolytic cleavages of the intermediates *65* and *67* (14). High resolution mass spectrometry, the investigation of deuterated derivatives, and the analysis of metastable ion decompositions support the proposed mechanism.

In the past years formally related processes have been studied for derivatives of nitrostilbene[75] and nitroaniline[76]. In spite of some differences concerning the formulation of the mechanism for oxygen transfer, this redox reaction may be viewed as a classical example of a complex ortho effect. It should be mentioned that the analytical significance of this rearrangement/elimination process is relatively high.

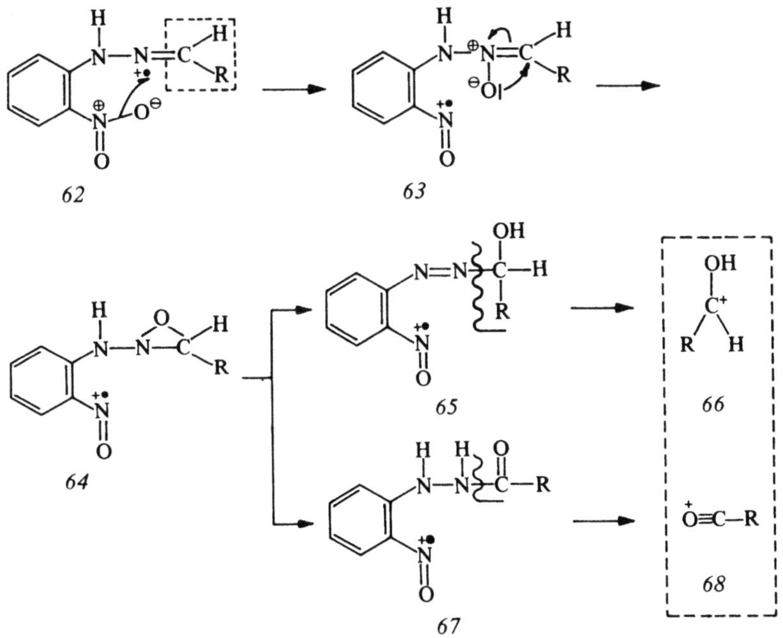

(14)

A partial insight in the complexity of such *ortho* effects (even in the case of simple compounds) permitted the thorough investigations of Middleton[77]. The data for ^{2}H and ^{13}C labelled ortho nitrostyrene show for example, that the loss of HCO from the molecular ion proceeds *via* the epoxide *70*. The carbon atom of the neutral particle, however, originates from both side chain carbons, and the elimination probably corresponds to the well-known epoxide rearrangement/fragmentation[78].

Furthermore it was concluded from the spectra that *70* is a possible intermediate for the CH_2O loss from the molecular ion of *69*. It is remarkable, however, and not yet understood why the carbon of the neutral species originates exclusively from the β position while many other positions operate as a hydrogen source.

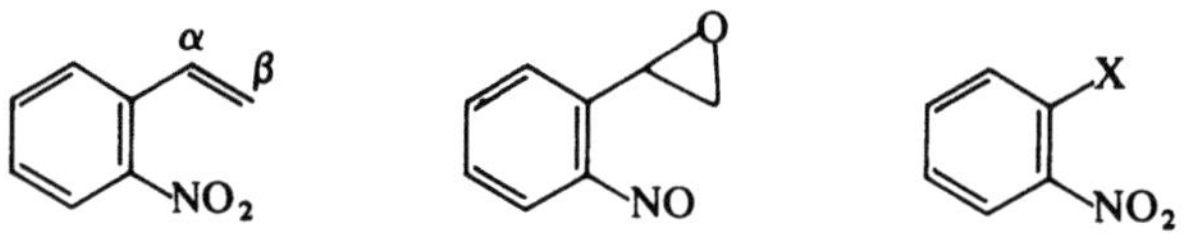

69 *70* *71*

Unusual interactions between the NO_2 function and the *ortho* substituent X, *71*, occur if this function contains several heteroatoms. Holmes[79] reported the loss of H_2O as well as H_2O_2 (!) from *71*, X = $NHNH_2$, and Hammerum[80] observed that the elimination of NO_2 from the molecular ion of *71*, X = $NHNO_2$, is followed by a consecutive loss of *two* NO radicals. All processes mentioned indicate intramolecular oxygen transfer.

Systematic investigations have been performed on compounds of the general structure *72* with the scope to elucidate the influence of the structural element X–Y on the oxygen transfer according to (15). In contrast to Seibl's oxygen migrations (14), this redox reaction represents a direct analogon to the 6-membered hydrogen rearrangement (1).

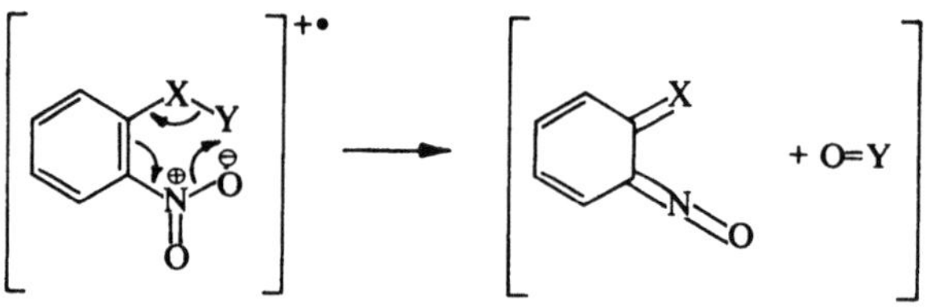

72 *73* *74* (15)

If one analyses the mass spectra of the nitrogen[80, 81], sulfur[82], selenium[83] or phosphorous derivatives[84] of *72* (X = CO), one observes a considerable variation of the intensities for the fragments *73* or *74*, which are formed *via* oxygen transfer. The probable reasons for these effects may be the different oxygen affinities as well as the abilities of *Y* to stabilize positive charges. While the *meta/para* isomers of *72* do not contain any fragment of such a redox process one observes, for the sulfur and phosphorous derivatives, that the fixed geometry in *72* is not a necessary condition for the oxygen transfer from the NO_2 group onto the acceptor. Conformationally flexible systems, for instance *75*, produce intense signals of the type $O{=}Y^+$ (16). This transformation, however, occurs preferentially if a 6-membered transition state is involved (*75*, $n = 2$). The high oxygen affinities of S and P are probably responsible for this anomalous effect[83–86]. The fixed arrangement of the oxygen donator (NO_2) and the oxygen acceptor (Y) in *72*, however, accelerates the redox reaction according to (15) with the consequence that the intensity of the molecular ions of *72* is extremely low. OH transfer which is typical for some aliphatic systems

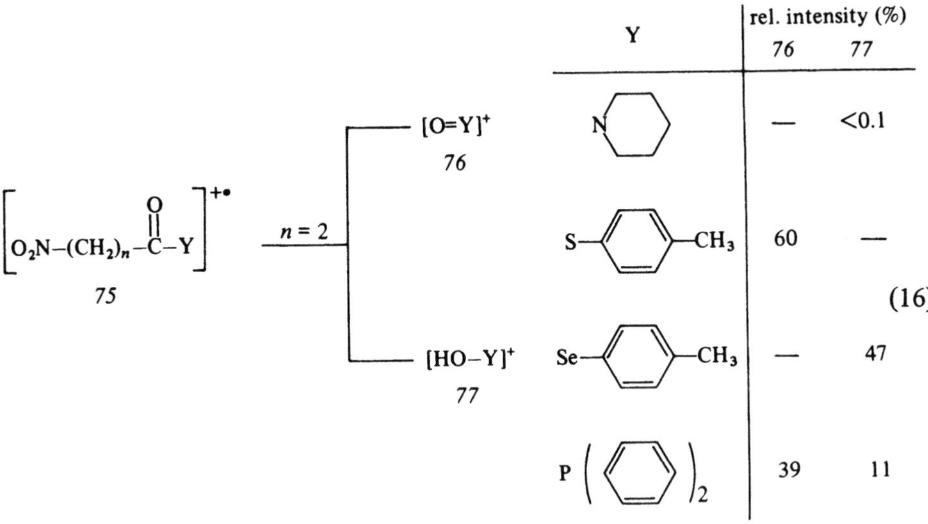

(16)

(*75* → *77*) is restricted to such compounds and has never been observed for aromatic derivatives.

A number of *multiple* oxygen transfers onto *one* acceptor are induced by electron impact during fragmentation of compounds of the structure *78*. While the mass spectra of the selenide *78* (X = Se) contain signals resulting from a *single* oxygen migration, *e.g.*, $C_7H_7O^+$ (7%) and $C_6H_4NOSe^+$ (81%)[83], the molecular ion of the sulfide (*78*, X = S) eliminates SO_2 (30%). The $(M-SO_2)^{+\cdot}$ fragment decomposes further by hydrogen elimination (100%)[67]. The very different redox potentials of the neutral particles (SO_2 is much more stable than SeO_2) are probably responsible for the different behaviour. The decomposition pathways for the oxides (*78*, X = SO, SeO) correspond very well with that for the sulfides. A single oxygen transfer is observed for the selenium oxide whereas the sulfoxide eliminates SO_2 as well as SO_3 (!). The last-mentioned reaction is terminated by a hydrogen loss leading to *79* (100%)[83].

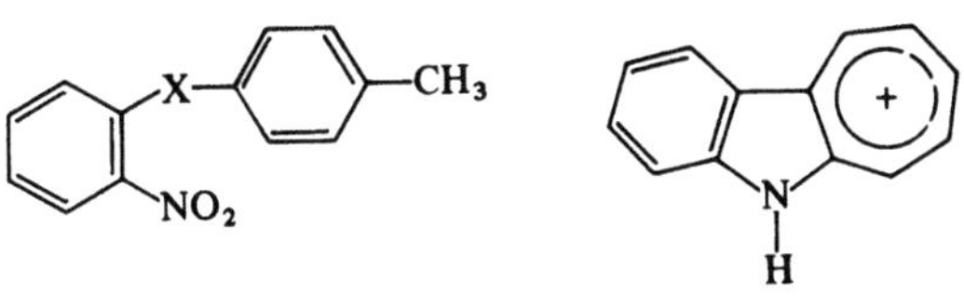

78 (X = S, Se, SO, SeO) *79*

Oxygen transfer reactions have been sporadically observed for *radical anions*[87, 88, 89]. The elimination of ^{16}OH from *80*, the formation of $(M-PhO)^-$ ions from *81* or the genesis of $PhC^{18}OO^-$ resp. $(M-PhC^{18}O-CO_2)^-$ from *82* are a few examples. The mass spectra of the radical anions of compounds *83* (*n* = 1, 2, 3; R = hydrogen, alkyl, aryl) contain some fragments as well, wherein dependence of n the oxygen transfer proceeds *via* 5-, 6- and 7-membered transition states. Typical fragments are for instance $RCOO^-$ or $RCOCHO^{-\cdot}$.

A few details of the very complex mechanisms of these rearrangements could be elucidated by the investigation of ^2H and ^{18}O labelled compounds.

It is worth mentioning that there exists no evidence for transition states with more than seven members for oxygen transfer reactions of negative ions in the gas phase[89]. Furthermore, replacement of the NO_2 function in *80–83* by another oxygen containing substituent (*e.g.*, COOR) suppresses all redox reactions. For *radical cations*, however, at least one example is known, establishing that functions other than NO_2 can operate as oxygen donators[67]. The molecular ions of *84* decomposes in $(Ph)_2P{=}O^+$ (2%) as well as $C_8H_7O_2S^+$ (15%). Both species may be caused by the high oxygen affinity of the diphenylphosphino moiety[84].

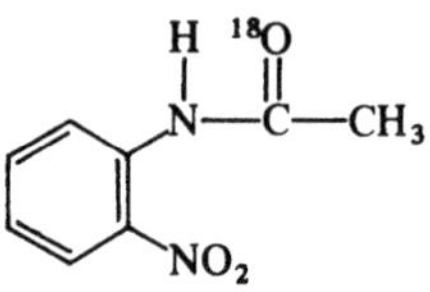

80

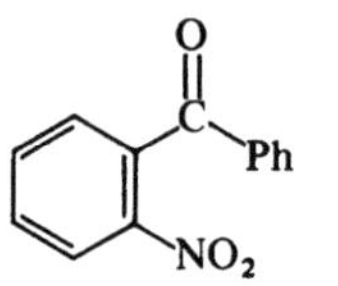

81

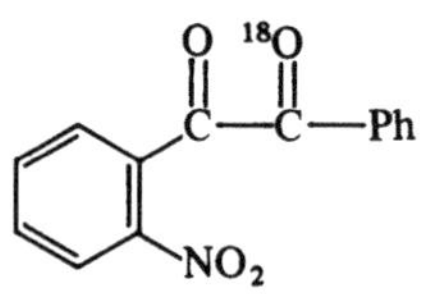

82

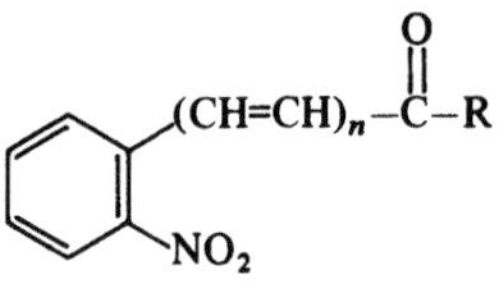

83

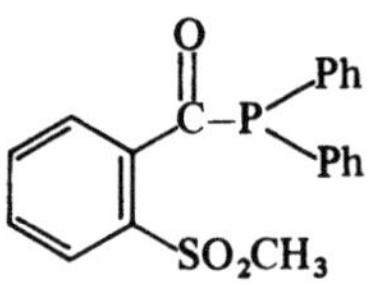

84

VI. Cyclization *via Ortho* Effect

Very different types of cyclization processes of isolated ionic species have been discussed in detail by Cooks[90] and Hesse[91]. In the following chapter only the interactions of vicinally substituted functions will be described. Such an elimination/ cyclization interaction may be recognized by two facts:

(i) the ion intensities of the cleavage products are always higher for the *ortho* isomers than for the *meta/para* derivatives,

(ii) the activation energy for the fragmentation process is considerably reduced for the *ortho* compound.

Examples for cyclization reactions *via* extrusion of partial constituents of the *ortho* function are the well-known decompositions of phthalic acid derivatives, *85*, [29, 92] or the primary decompositions of suitably substituted silyl derivatives[93].

Reactions of this type (17) will not be discussed in more detail. It should be stated, however, that in most cases it could not be established with certainty whether the elimination of X proceeds in a one step reaction, *85* → *86*, or whether a tetrahedral intermediate, *87*, is involved in the reaction sequence.

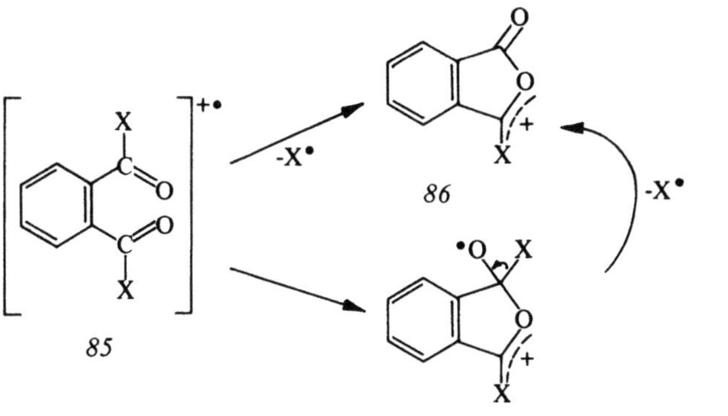

85 86 87 (17)

1. Aromatic Substitution Reactions

1.1. Five-membered Ring Systems

Extensive studies by Grützmacher and co-workers[94–96] clearly established that the elimination of *ortho* substituents X (X = hydrogen, halide) from *88* is *not* the result of a one step intramolecular substitution reaction. Instead of this, an addition/dissociation mechanism operates leading to *90* (18). The rate determining step of the reaction sequence is the formation of the σ-complex, *89*, which itself has the typical characteristics of a reactive intermediate.

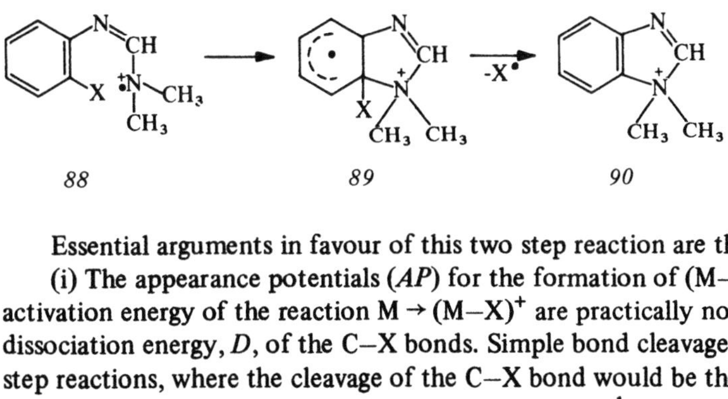

(18)

88 89 90

Essential arguments in favour of this two step reaction are the following ones:

(i) The appearance potentials (*AP*) for the formation of $(M-X)^+$ as well as the activation energy of the reaction $M \rightarrow (M-X)^+$ are practically not influenced by the dissociation energy, *D*, of the C–X bonds. Simple bond cleavages as well as multistep reactions, where the cleavage of the C–X bond would be the rate determining step, should cause differences of about 50 kcal · mol^{-1}.

(ii) The kinetic energy release, T_{kin}, is in a first approximation inversely proportional to the dissociation energy of the C–X bond.

The interdependence of *AP*, *D* and T_{kin} is shown in Fig. 3, and the published data support the proposed mechanism.

It is worth mentioning that the loss of X from *91* is strongly suppressed by electron donating substituents Y. Detailed investigations of sterically hindered biphenyl derivatives, *92*, demonstrate that a mesomeric interaction between Y and the formamidine group is responsible for the decreased cyclization abilities. The in-

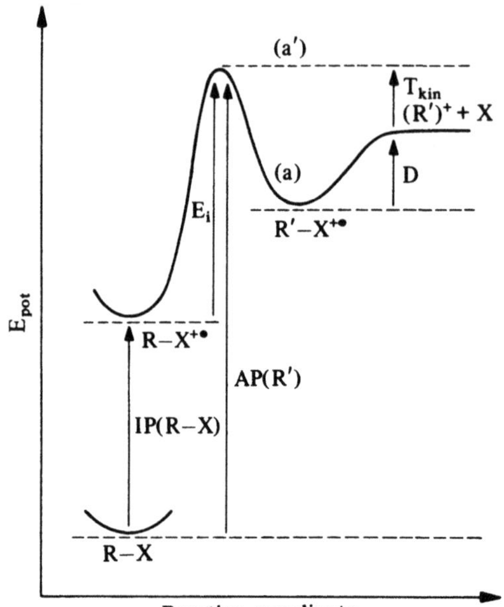

Fig. 3. *Energy profile of* (18): R–X = *88*,
(a') = transition state *88* → *89*, (a) = reactive intermediate *89*, $(R')^+$ = *90*

corporation of two CH_3 groups in the 2 and 2'position, however, is already sufficient to prevent planarity, *e.g., 93,* which is necessary for mesomeric effects. Consequently, in such a case (*91*, R = R' = CH_3), a fragmentation behaviour according to *88* is observed[96].

The electron impact induced elimination of X from the molecular ions of *94* (X = H, halide; R = $NHCOR_1$, $NHCSR_1$, $NHCSNR_2R_3$, $NHCONR_2R_3$; R_1 = alkyl, R_2, R_3 = H, alkyl) has been investigated in detail by Loudon *et al.*[97–100]. In agreement with the above mentioned study Grützmacher, the $(M-X)^+$ ions have a cyclic structure, probable *95*. The reaction mechanism, however, may only formally correspond to (18).

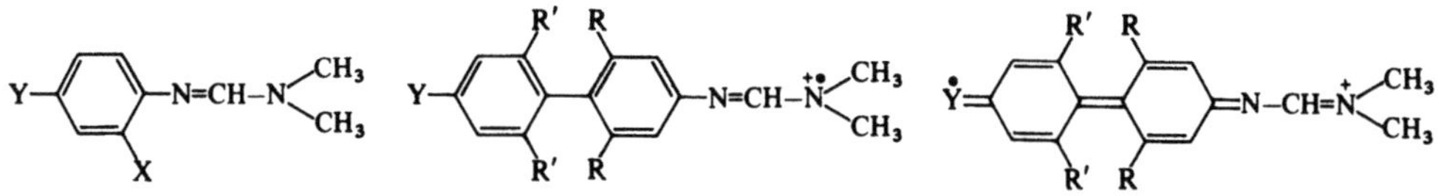

91 *92* *93*

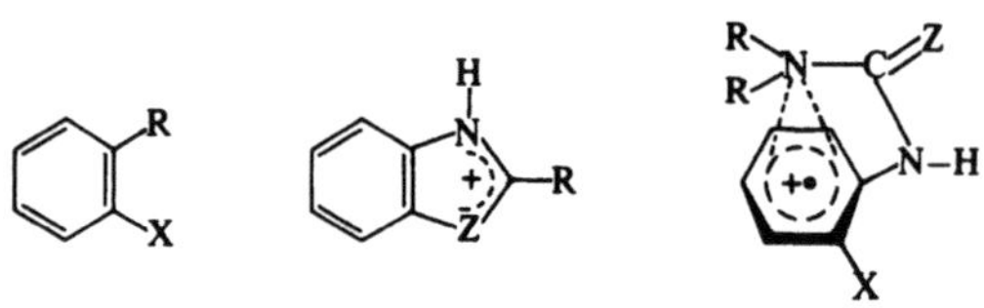

94 *95* *96*

An important difference concerning the modes of cyclization may be that the molecular ions of *94* prefer the conformation *96* from which a cyclization cannot occur. The relative rate for the generation of *96* depends very strongly on the nature of the substituents. The very complex correlations between energetic factors (*AP* and *D*) on the one hand and the dependence of the intensities of the $(M-X)^+$ ions from the type of the substituents, *94*, on the other hand possibly result from competing reactions in which the formation of *96* may play a decisive part.
Many other examples leading to 5-membered heterocyclic ions have been cited in the literature[43, 101–105]. In most cases, however, no details concerning the reaction mechanisms are known.

Another type of cyclization is described by Eguchi[106]. Isoflavones, *97*, yield intense $(M-OR')^+$ fragments, and the reaction seems to be of analytical value because it determines whether or not an alkoxy substituent is located at the 2′(6′) position (19).

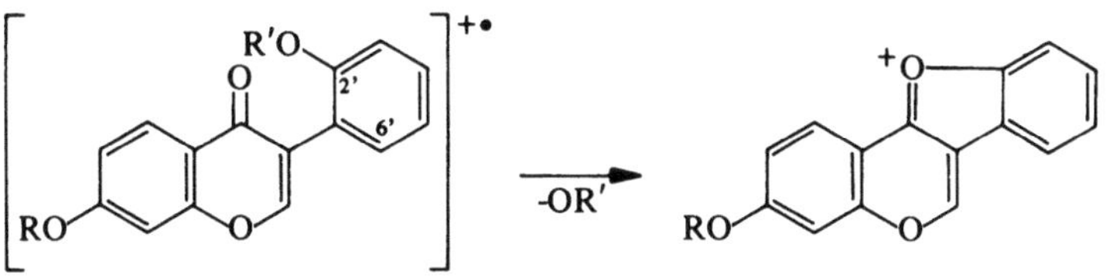

97 *98* (19)

1.2. Six-membered Ring Systems

Very different classes of organic compounds are able to yield 6-membered heterocyclic products upon electron impact induced cyclization/elimination reactions. In spite of the formal similarity of these processes with (19), it cannot be ruled out that the reaction mechanism of these decompositions is quite distinct. The structurally important parameters are the nature of the leaving group X as well as the ring forming heteroatom. The examples *99–103* may give a small impression of which structural features may be involved in the formation of $(M-X)^+$ fragments.

The elimination of $OH^{\cdot}$ and H_2 (20) from *ortho* and *peri* substituted alkoxy aldehydes, ketones or quinones is of diagnostic value for the location of functional groups[113, 114]. The reaction sequence is restricted to systems containing such a substitution pattern. Investigations of ^2H and ^{18}O labelled model compounds establish the origin of neutral species (OH and H_2) without clarifying the details of the complex skeletal reorganization. From the results it can only be concluded that the formation of stable oxonium ions, *e. g.*, *105*, is of importance.

The elimination of water from the molecular ion of *106* is of diagnostic value only for simple systems. The mass spectra of compounds containing a larger number of methoxy substituents show $(M-H_2O)^{+\cdot}$ fragments even for systems with nonadjacent substituents. For the radical cations formed by loss of water from *106*, the structure of benzofurane, *107*, was proposed based on the metastable ion characteristics[115].

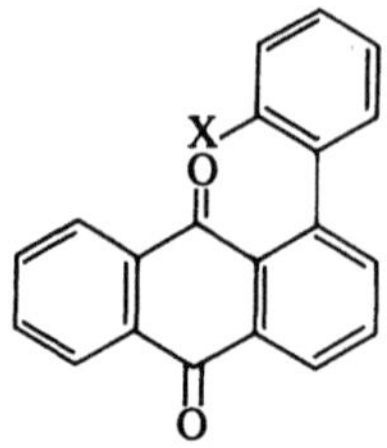

99 (X = Br, NO_2, Cl, MeO, H, F)[107]

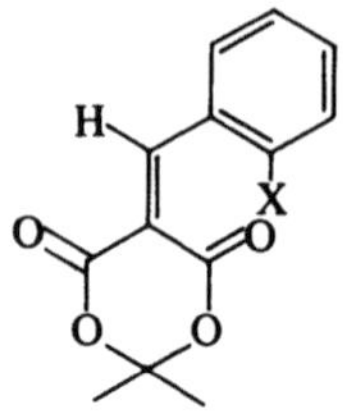

100 (X = Cl)[108]

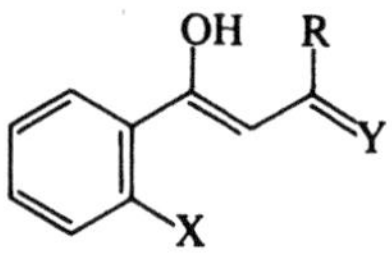

101 (X = H; R = CH_3, Ph; Y = S[109], Y = O[110])

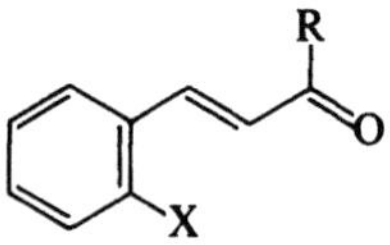

102 (X = Cl; R = Alkyl[111]; X = H, D; R = CH_3, Ph, OH, OCH_3[112])

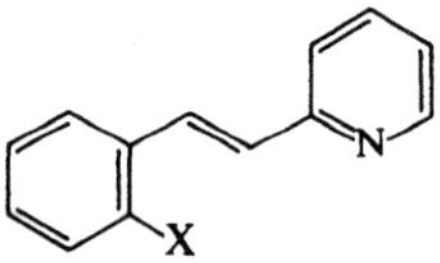

103 (X = H, NO_2)[75]

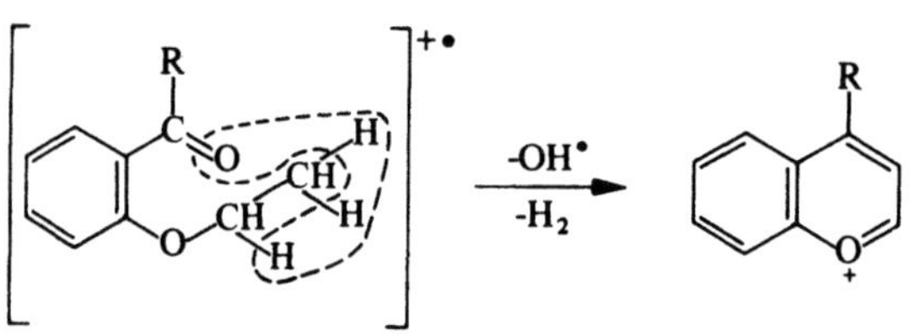

104 *105*

(20)

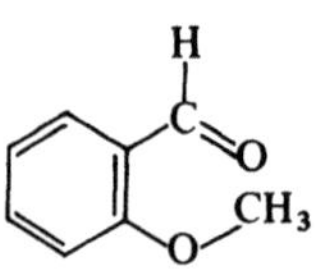

106

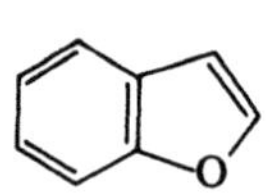

107

2. Cyclizations *via* Hydrogen Migration

As already mentioned [(3), (10), (13), *79, 82, 83*] many other types of electron impact induced reactions exist which do not belong to the classification used in this chapter. Among these are reactions which proceed *via* cyclic reactive intermediates or cyclization processes accompanied by hydrogen transfer at one step of the reaction sequence. This last-mentioned decomposition shall be discussed briefly. Compounds of structure *108* yield intense fragments at *m/e* 133, *110*[116]. ^{2}H labelling of the carboxyl function does not shift the signal *m/e* 133, and it was argued by the authors that the reaction possibly proceeds *via* the phthalide, *109* (21).

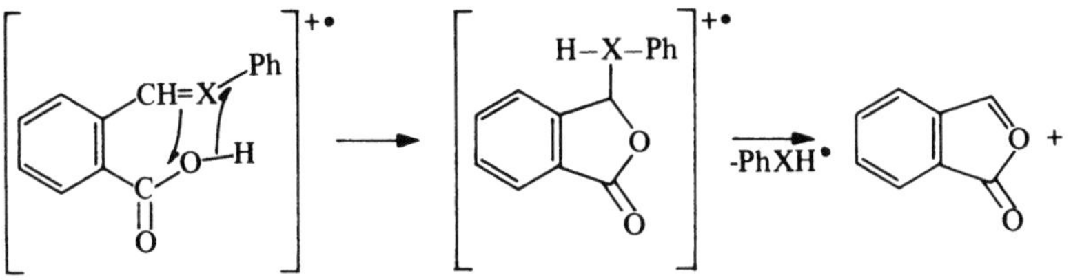

(21)

108 (X = CH, N) *109* *110*, m/e 133

The OH eliminations from *ortho* substituted aldoximes, *111* (X = CH_2, NH, O) may at least partially be the result of a hydrogen transfer/cyclization process, whereby the heterocycles *112* are formed[117]. A peak shape analysis as well as the investigation of ^{2}H labelled derivatives indicates that in addition to *112* the protonated isocyanide, *113*, is formed. This fragment, however, is generated without any measurable interaction with the *ortho* substituent.

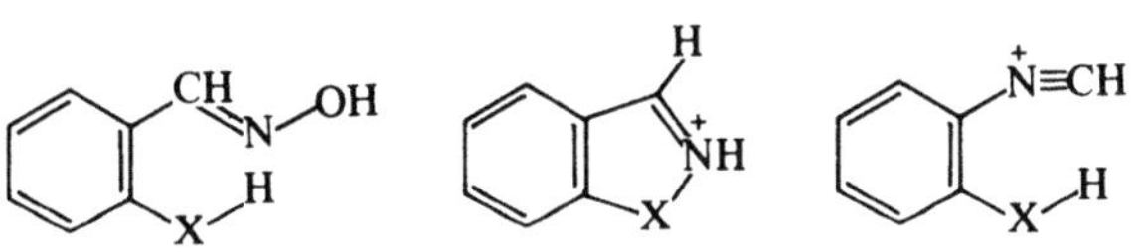

111 *112* *113*

A very unusual cyclization is observed for N(*ortho*-nitrobenzyl)amines, *114* (*n* = 4,5 or dimethylamine derivatives)[62]. While the regiospecific OH and the following NO eliminations may be regarded as typical decompositions of aromatic nitrosubstituted derivatives, this is certainly not the case for the hydrogen elimination from $(M-OH-NO)^{+\cdot}$ (22). This reaction formally corresponds to a radical induced cyclization[62, 90, 118], followed by loss of hydrogen *via* a Grob[119] fragmentation, *116* → *117*. The alternative process, *116* → *118*, is of minor importance.

A similar reaction sequence (consecutive loss of OH, NO and H) has been established for the molecular ions of dinitrophenylhydrazones of α, β-unsaturated aldehydes[34, 74].

Whether the intense Cl˙ loss from *119*[120] yields species of the structure *120* or *121* is not yet clarified. The chlorine elimination, however, has among all other competing decompositions from *119* the lowest activation energy.

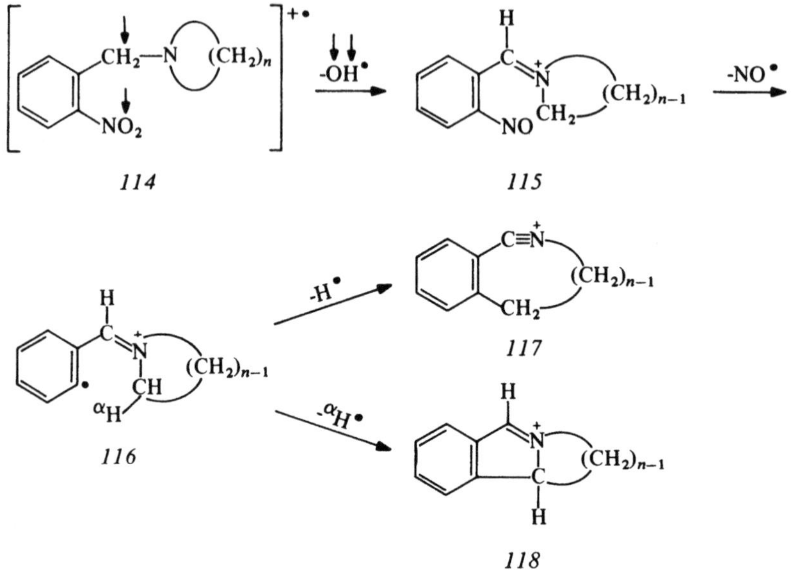

(22)

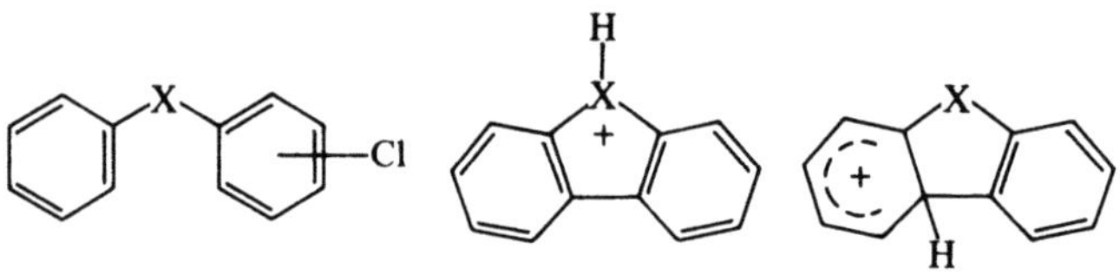

119 (X = O, S, SO) *120* *121*

VII. Exchange Processes

Detailed investigations of labelled compounds as well as the application of improved and more precise methods [*e.g.*, ion kinetic energy spectra (IKE)[37]] have demonstrated in recent years that a lot of "simple" cleavage reactions are preceded by extensive exchange reactions. A fine example is the hydroxyl elimination from ionized benzoic acid, *122*[121, 122].

An analysis of the unimolecular decompositions of metastable ions establishes that the hydroxyl loss is preceded by a hydrogen exchange of the *ortho* and carboxyl hydrogens, which occurs *via 124*. Furthermore it is noteworthy that the equilibration of both *oxygen* atoms is not the result of an intramolecular (1.3) hydrogen shift, *122* → *123*. Instead, a cascade of reversible hydrogen rearrangements takes place including hydrogens from both *ortho* positions[123–126]. Obviously this multistep reaction is energetically much more favoured than the intramolecular symmetry forbidden (1.3) hydrogen migration[127]. A similar situation is responsible for the hydroxyl loss from pyridine-4-carboxylic acid[128].

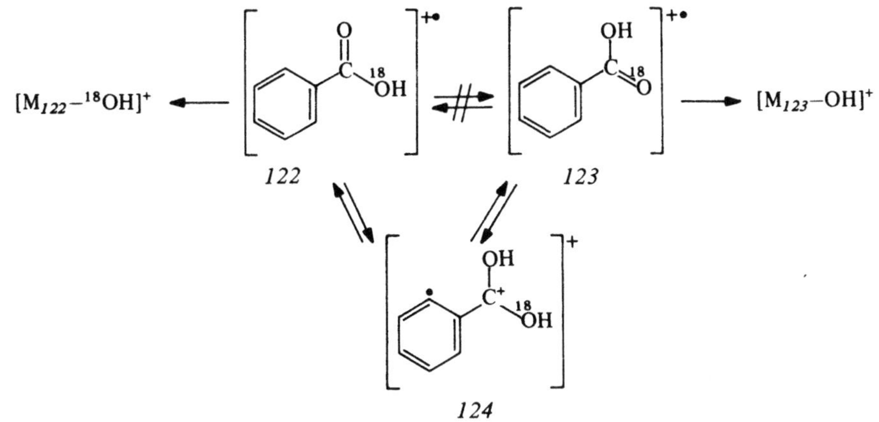

(23)

Further insight into the mechanism of exchange reactions yields the detailed investigations of Holmes and co-workers[129−131] on *cis*-configurated dicarboxylic acids, for instance maleic acid, cyclohexene-1.2-dicarboxylic acid or phthalic acid. Interestingly the molecular ions of ^{2}H labelled phthalic acid, *125*, eliminate neither OH nor HDO. This result demonstrates that the second function completely suppresses all those reactions typical for benzoic acid, *122*. Reactions of this type, however, take place exclusively for those $(M-CO_2)^{+\cdot}$ fragments, where the hydrogen has been transferred onto the carbonyl function, *125* → *127* → *129*. The acyl oxonium ions, *126* and *128*, do not participate in any exchange reaction. Furtheron it is striking, that neither the isomerized molecular ions, *126*, *127*, nor the fragments *128*, *129* interconvert via hydrogen migration (24).

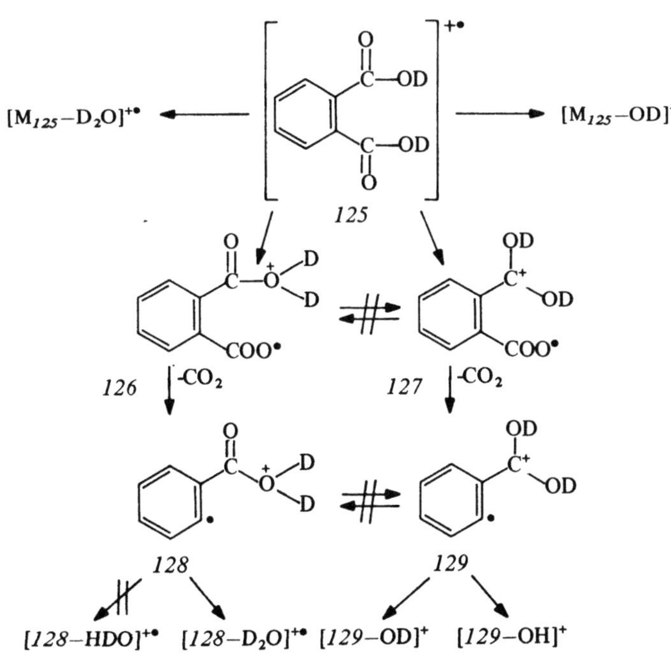

(24)

Based on the results of a combined ^{18}O and ^{2}H labelling study, the unusual water elimination from benzylbenzoate, *130*, has to be formulated as a complex multistep reaction including extensive hydrogen and skeletal rearrangements (25). A convincing mechanism for the experimentally observed equivalence of the five starred hydrogens and the two oxygens of the intermediate *132* for water elimination is not known[132]. Evidence for the proposed mechanism (25) can be provided by comparison of the IKE spectra of *130* and *134*. The identical spectra establish a common reactive intermediate. The fact that the molecular ions of *135* as well as *136* do not eliminate water at all is a further indication for *132*.

Reversible exchange reactions between the ester function and the ring protons have been reported for the water loss from isonicotinic acid methylester, *137*, as well[133].

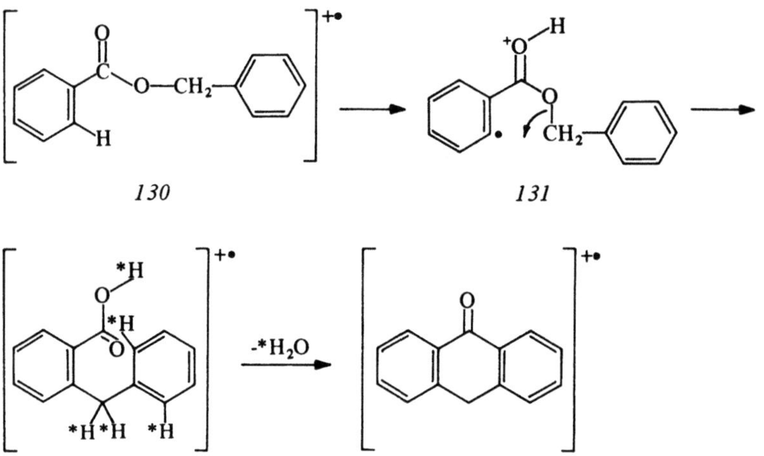

130 *131* *132* *133* (25)

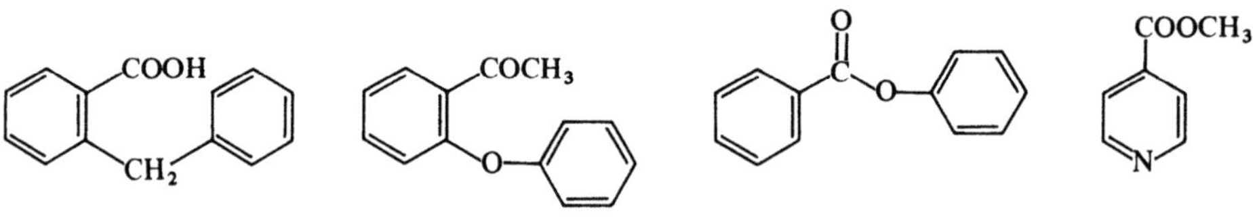

134 *135* *136* *137*

A number of exchange processes have been studied by Nibbering *et al.*[134, 135]. The authors demonstrated that in the molecular ions of *138*, reversible hydrogen migrations takes place which lead to an equilibration of the hydroxyl-, the γ- and the *ortho*-hydrogens. In the structurally related phenylpropylbromides, *139*, an exchange between $^{\alpha}H$ and *one ortho* hydrogen precedes the ethylene elimination.

The relative rates for intramolecular hydrogen migration and competing ethylene elimination allow just *one* exchange process. The decision whether the exchange processes are concerted reactions (proceeding *via* bicyclic transition states) or whether reactive intermediates come into existence (multistep reaction) is open to question.

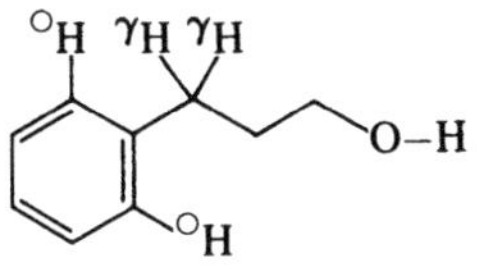

138

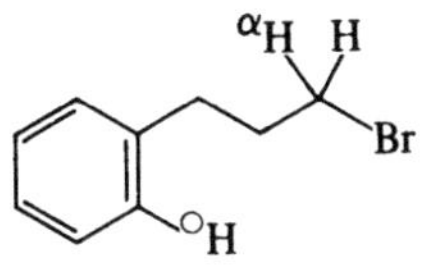

139

Complex hydrogen migrations precede also the ethylene elimination from aryl- and alkine substituted dimethylcarbenium ions[136–138]. Investigations of ^2H and ^{13}C labelled compounds, *140,* established that the carbon and hydrogen atoms of the side chain loose their positional identity before or during ethylene elimination. In the case of *140,* X = N, exchange reactions with the pyridine ring do not occur. For X = CH, however, the ethylene loss is preceded by a partial side chain/*ortho* position exchange of hydrogens. Lifetime studies demonstrate that the exchange processes increase with increasing lifetime of the $(M-CH_3)^+$ ions from *140.* An analogous behaviour has been observed for the $(M-Cl)^+$ fragments of *141*[138].

The complexity of such processes and the principal difficulties to obtain relevant informations concerning the reaction mechanism may be visualized by the C_3H_4 loss from the $(M-Cl)^+$ ions from *141* and *142.* Starting from *141*, the carbon of C_3H_4 originates nearly exclusively from the isopropylfunction (> 97%), whereas the *ortho* hydrogens are involved up to about 20% in the formation of the neutral species. In the case of *142,* however, the whole side chain contributes to the carbons of the C_3H_4 moiety, whereas the participation of hydrogens from the phenyl ring is strongly reduced[138].

These data as well as the fact that the CA spectra of the $(M-Cl)^+$ ions from *141* and *142* show distinct differences demonstrate that the hydrogen migration and the reorganization of the carbon skeleton occur at least partially independently[138].

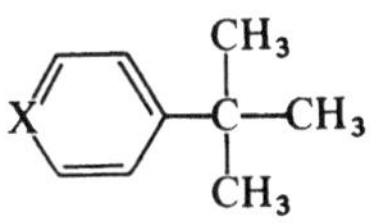

140 (X = CH, N)

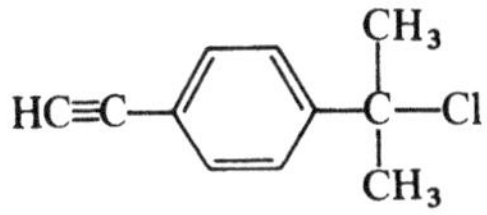

141

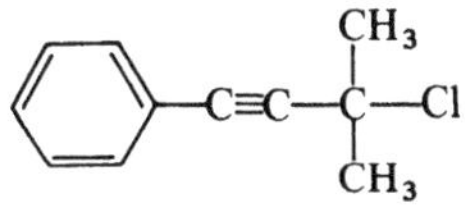

142

VIII. Interactions of *Ortho* Substituent in Biphenyl Derivatives

Sometimes the interactions typical for 1.2-disubstituted arenes have a formal correspondence in the decompositions of 2.2′substituted biphenyl derivatives. Thus the highly specific piperidine eliminations from *143*[30] as well as the formation of *146* correspond to the fragmentations of structurally analogous benzene derivative.

The amides, *147*, however, prove that, in addition to the usual behaviour, unexpected reactions may occur. Among these the sequence M → (M–H) → (M–H-piperidine)$^+$ is of special importance, and the data of ^2H labelled compounds show that both steps are highly selective. The carbonyl function obviously acts as an intramolecular catalyst for the hydrogen transfer in (26). In spite of the fact that all reactions in (26) require energetically expensive changes of the ground state conformations, they are so fast that almost all molecular ions decompose within less than 10^{-6} sec.

The biphenyl derivatives *149* and *150* give under electron impact anomalous reaction products as well. Thus an intense ion at *m/e* 370, *151*, is generated from the bisacylphosphine, *149*[84]. The diester *150* represents the first compound where

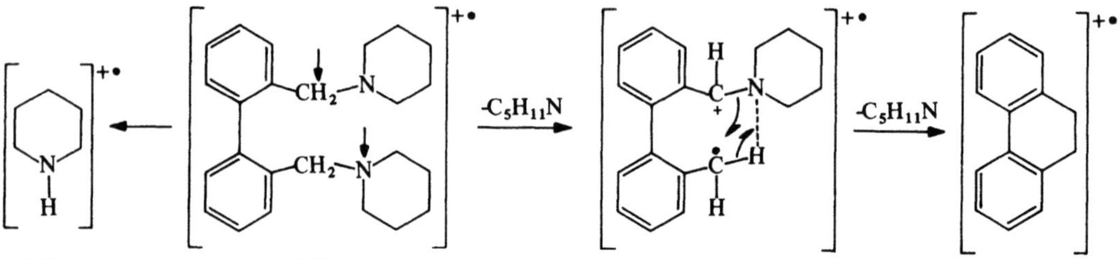

146 *143* *144* *145*

(26)

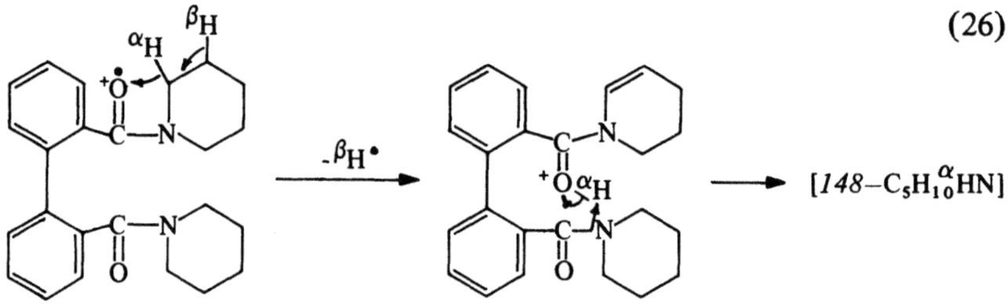

147 *148*

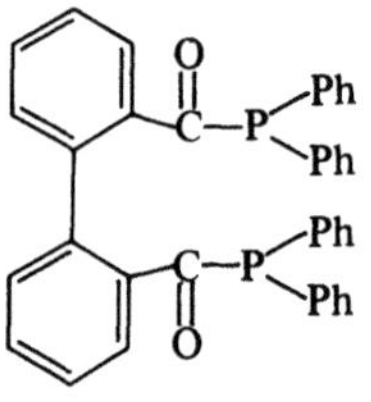

149

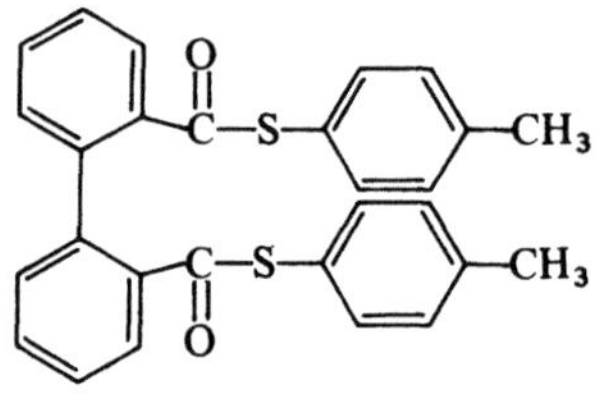

150

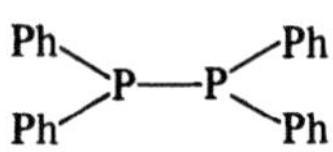

151 (mle 370)

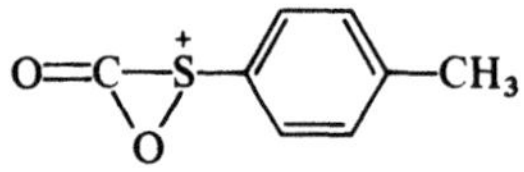

152

an oxygen transfer from the *ester* function takes place. The primary decomposition product of this redox reaction may have structure *152*, from which CO as well as CO_2 is eliminated[67, 139].

Cyclizations via aromatic substitution could be verified for suitably functionalyzed diaryl derivatives as well[140]. One of the many examples presented in this and related studies is formulated in (27).

Cyclization processes of this type also take place at appropriately substituted biphenyl derivatives, *155*. Here oxonium ions are formed, the structure of which might be described by *156*[141]. Higher substituted derivatives of *155*, however, yield mainly ions of the general structure *157*. This fragment is the result of a very interesting reciprocal methyl/hydrogen migration[141, 142].

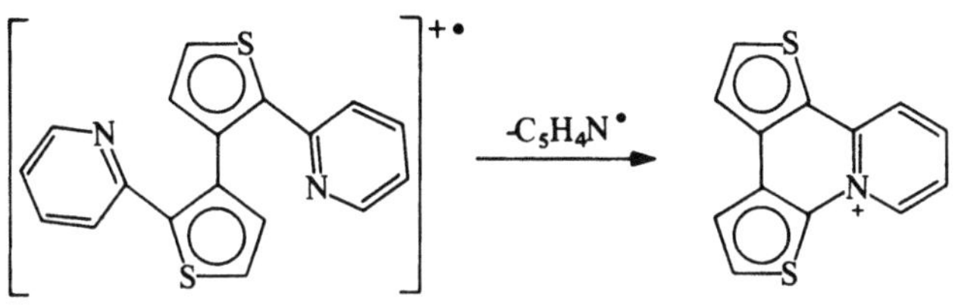

(27)

153 *154*

The investigation of many model compounds leads to the conclusions, that the following conditions have to be fulfilled in order to generate *157:*

(i) Both ring systems have to be twisted against each other. As a result the formation of a 5-membered transition state (transfer of CH_3 from the methoxy function onto C(1)) is significantly enhanced. This conformation can be preferentially populated, if all four *ortho* positions are substituted.

(ii) At least one of these substituents must have electron withdrawing properties and, moreover, should not induce competitive decompositions.

(iii) Substituents in 3(3′), 4(4′) or 5(5′) position, which may stabilize the molecular ions, should be absent.

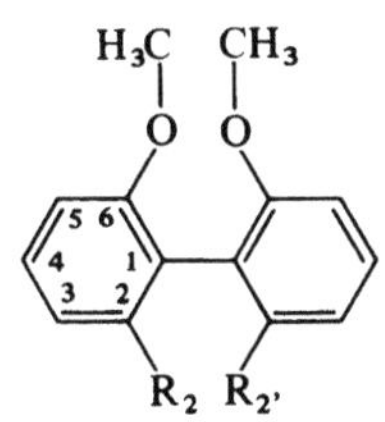

155 (R_2, $R_{2'}$: COOH, $COOCH_3$, OCH_3)

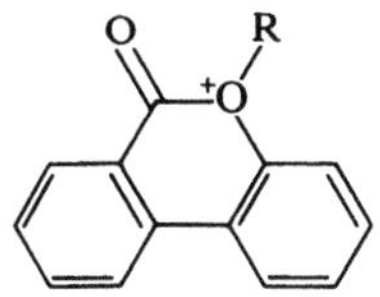

156

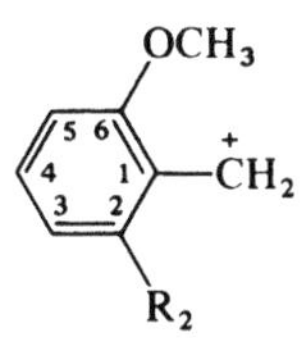

157

IX. Summary

While in the early days of organic mass spectrometry the *ortho* effect was preferentially used as a diagnostic tool for structure elucidation, the recent research activities have concentrated on mechanisms and related aspects. Among these the following were the most important topics: hydrogen transfer and especially the question of concerted or multistep reaction; the influence of the structure on the selectivity of hydrogen migration *via* higher-membered transition states; the apparent cleavage of carbon/heteroatom bonds induced by hydrogen transfer; the redox processes which formally correspond to photochemical transformations, but which are completely different according to the reaction mechanisms; the different types of cyclization reactions; and the field of exchange processes preceding or accompanying unimolecular decompositions of ions in the gas phase. If a prognosis concerning the future activities in the field of electron impact induced *ortho* effects may be risked, the following aspects will probably become more important in addition to the topics discussed in this review:

(i) Kinetic studies in the time range of nanoseconds to elucidate the time dependence of the different *ortho* effect reactions;

(ii) ion/molecule reactions with the scope to obtain more information concerning the reactivities of open and closed shell systems;

(iii) comparative studies in the fields of mass spectrometry, photochemistry and electrochemistry.

In principle all these studies may extend our knowledge concerning the inherent properties of isolated ionic species.

Acknowledgements. Grateful acknowledgement for the promotion of this work is due to the following institutions: Gesellschaft von Freunden der Technischen Universität Berlin, Fonds der Chemischen Industrie, Frankfurt, and Technische Universität Berlin (research project "Massenspektrometrie" FPS 5/1). – I would like to express my sincerest gratitude to my coworker Johannes Respondek for his assistance in preparing the English version of this paper.

X. References

1) Spiteller, G.: Massenspektrometrische Strukturanalyse organischer Verbindungen. Weinheim/Bergstr.: Verlag Chemie 1966

2) Budzikiewicz, H., Djerassi, C., Williams, D. H.: Mass spectrometry of organic compounds. San Francisco: Holden-Day 1967

3) Seibl, J.: Massenspektrometrie. Frankfurt/Main: Akademische Verlagsanstalt 1970

4) Porter, Q. N., Baldas, J.: Mass spectrometry of heterocyclic compounds. New York: Wiley-Interscience 1971

5) McLafferty, F. W.: Interpretation of mass spectra. London: W. A. Benjamin 1971

6) Bursey, J. T., Bursey, M. M., Kingston, D. G. I.: Chem. Rev. *73*, 191 (1973) and contributions cited in this journal

7) Bowie, J. H.: Aust. J. Chem. *26*, 195 (1973)

8) Smith, J. G., Wilson, G. L., Miller, J. M.: Org. Mass Spectrom. *10*, 5 (1975)

9) Schwarz, H., Köppel, C., Bohlmann, F.: Tetrahedron *30*, 689 (1974)

10) Schwarz, H., Wesdemiotis, C., Bohlmann, F.: Org. Mass Spectrom. *9*, 1226 (1974)

11) Levsen, K., Schwarz, H.: Org. Mass Spectrom. *10*, 752 (1975)

12) For a review concerning this new method see: Levsen, K., Schwarz, H.: Angew. Chem. *88*, 589 (1976); Angew. Chem. Int. Ed. Engl. *15*, 509 (1976)

13) Schwarz, H., Sezi, R., Rapp, U., Kaufmann, H., Meier, S.: Org. Mass Spectrom. *12*, 39 (1977)
14) Bowie, J. H., Ho, A. C.: Org. Mass Spectrom. *9*, 1006 (1974)
15) Rullkötter, J., Budzikiewicz, H.: Org. Mass Spectrom. *11*, 44 (1976)
16) McLafferty, F. W., Gohlke, R. S.: Anal. Chem. *31*, 2076 (1959)
17) Lacey, M. J., Macdonald, C. G., Shannon, J. S.: Org. Mass Spectrom. *5*, 1391 (1971)
18) Ballantine, J. A., Pillinger, C. T.: Org. Mass Spectrom. *1*, 425 (1968)
19) Grimshaw, J., Sell, C. S., Haslett, R. J.: Org. Mass Spectrom. *8*. 381 (1974)
20) Ion/molecule reactions are not considered. For a review concerning the method, examples and references see: Richter, W. J., Schwarz, H.: Angew. Chem. in press
21) Martens, J., Praefcke, K., Schwarz, H.: Z. Naturforsch. *30b*, 259 (1975)
22) Lumpkin, H. E., Nicholson, D. E.: Anal. Chem. *32*, 74 (1960)
23) Aczel, T., Lumpkin, H. E.: Anal. Chem. *34*, 33 (1962)
24) Spiteller, G.: Monatsh. Chem. *92*, 1147 (1961)
25) Willhalm, B., Thomas, A. F., Gautschi, F.: Tetrahedron *20*, 1185 (1964)
26) Schwarz, H., Wolfschütz, R.: Org. Mass Spectrom. *11*, 773 (1976)
27) Levi, E. M., Mao, C. L., Hauser, C. R.: Can. J. Chem. *47*, 3671 (1969)
28) Ramana, D. V., Sundaram, N.: Org. Mass Spectrom. *10*, 761 (1975)
29) Bohlmann, F., Herrmann, R., Mathar, W., Schwarz, H.: Chem. Ber. *107*, 1081 (1974)
30) Schwarz, H.: Org. Mass Spectrom, *9*, 988 (1974)
31) Schwarz, H., Wesdemiotis, C., Bohlmann, F.: Org. Mass. Spectrom. *9*, 1223 (1974)
32a) Wünsche, C., Ege, G., Beisiegel, E., Pasedach, F.: Tetrahedron *25*, 5869 (1969);
32b) Kuster, Th., Robbiani, R., Seibl, J.: Annual Meeting of the "Arbeitsgemeinschaft Massenspektrometrie", Bad Kissingen, 1977
33) Roberts, D. T.,Jr., Little, W. F., Bursey, M. M.: J. Amer. Chem. Soc. *89*, 4917 (1967)
34) Seibl, J., Völlmin, J.: Org. Mass Spectrom. *1*, 713 (1968)
35) Budzikiewicz, H., Rullkötter, J., Schiebel, H. M.: Org. Mass Spectrom. *6*, 251 (1972)
36) Herrmann, R., Schwarz, H.: Z. Naturforsch. *31b*, 1667 (1976)
37) Cooks, R. G., Beynon, J. H., Caprioli, R. M., Lester, G. R.: Metastable ions. Amsterdam: Elsevier 1973
38) Nakata, H., Tatematsu, A.: Org. Mass Spectrom. *5*, 1343 (1971)
39) Nibbering, N. M. M.: Tetrahedron *29*, 385 (1973)
40) Schwarz, H., Wesdemiotis, C., Bohlmann, F.: Org. Mass Spectrom. *9*, 1226 (1974)
41) Borchers, F., Levsen, K., Beckey, H. D.: Int. J. Mass Spectrom. Ion Phys. *21*, 125 (1976)
42) Benoit, F. M., Harrison, A. G.: Org. Mass Spectrom. *11*, 599 (1976)
43) Benezra, S. A., Bursey, M. M.: J. Chem. Soc. (B) *1971*, 1515
44) Schwarz, H., Kliment, M., Reetz, M. T., Holzmann, G.: Org. Mass Spectrom. *11*, 989 (1976)
45) Schwarz, H., Reetz, M. T.: Angew. Chem. *88*, 726 (1976); Angew. Chem. Int. Ed. Engl. *15*, 705 (1976)
46) Naito, T.: Tetrahedron *24*, 6237 (1968)
47) Budzikiewicz, H., Djerassi, C., Jackson, A. H., Kenner, G. W., Newman, D. J., Wilson, J. M.: J. Chem. Soc. *1964*, 1949
48) Grigg, R., Sargent, M. V., Williams, D. H., Knight, J. A.: Tetrahedron *21*, 3441 (1965)
49) Schwarz, H., Bornowski, H., Petruck, G.-M.: Org. Mass Spectrom. *10*, 469 (1975)
50) Kingston, D. G. I., Bursey, J. T., Bursey, M. M.: Chem. Rev. *74*, 215 (1974)
51) Bohlmann, F., Herrmann, R., Schwarz, H., Schiebel, H. M., Schröder, N.: Tetrahedron *33*, 357 (1977)
52) Tajima, E., Seibl, J.: Int. J. Mass Spectrom. Ion Phys. *3*, 245 (1969)
53) Herrmann, R., Schwarz, H.: Z. Naturforsch. *31b*, 870 (1976)
54) Reetz, M. T.: Tetrahedron *29*, 2189 (1973)
55) Borchers, F., Heimbach, H., Levsen, K., Morhenn, H., Schwarz, H.: Org. Mass Spectrom., *12*, 573 (1977)
56) Benoit, F., Holmes, J. L.: Org. Mass Spectrom. *3*, 993 (1970)
57) Meyerson, S., Puskas, I., Fields, E. K.: J. Amer. Chem. Soc. *88*, 4974 (1966)
58) Beynon, J. H., Job, B. E., Williams, A. E.: Z. Naturforsch. *21a*, 210 (1966)

59) Robinson, G. E., Thomas, C. B., Vernon, J. M.: J. Chem. Soc. (B) *1971*, 1273
60) Beynon, J. H., Lester, G. R., Williams, A. E.: J. Phys. Chem. *63*, 1861 (1959)
61) Bowie, J. H., Cooks, R. G., Fisher, J. W., Spotswood, T. M.: Aust. J. Chem. *21*, 2021 (1968)
62) Schwarz, H.: Org. Mass Spectrom. *9*, 826 (1974)
63) Bursey, M. M.: Tetrahedron Letters *1968*, 981
64) Ward, R. S., Williams, D. H.: J. Org. Chem. *34*, 3373 (1969)
65) Harley-Mason, J., Tonte, T. P., Williams, D. H.: J. Chem. Soc. (B) *1966*, 396
66) Kalir, A., Sali, E., Mandelbaum, A.: J. Chem. Soc. Perkin Trans. II, 2262 (1972)
67) Martens, J., Praefcke, K., Schwarz, H.: Liebigs Ann. Chem. *1975*, 62
68) Vijfhuizen, P. C., Heerma, W., Nibbering, N. M. M.: Org. Mass Spectrom. *11*, 787 (1976)
69) Schwarz, H., Sezi, R., Levsen, K., Heimbach, H., Borchers, F.: Org. Mass Spectrom., in press
70) Bamberger, E., Elger, F.: Liebigs Ann. Chem. *371*, 319 (1910)
71) Kinstle, T. H., Darlage, L. J.: J. Heterocycl. Chem. *6*, 123 (1969)
72) Darlage, L. J., Kinstle, T. H., McIntosh, C. L.: J. Org. Chem. *36*, 1088 (1971)
73) For further aspects see: H. Schwarz, Zon, G., Tsui, F. P.: Org. Mass Spectrom. *10*, 1160 (1975)
74) Seibl, J.: Org. Mass Spectrom. *3*, 417 (1970)
75) Ramana, D. V., Vairamani, M.: Org. Mass Spectrom. *9*, 1158 (1974)
76) Johnston, D., Smith, D. M.: Org. Mass Spectrom. *9*, 789 (1974)
77) Middleton, S., Butcher, M., Mathews, R. J.: Aust. J. Chem. *27*, 2583 (1974)
78) Audier, H. E., Dupin, J. F., Fetizon, M., Hoppilliard, Y.: Tetrahedron Letters *1966*, 2077
79) Benoit, F., Holmes, J. L.: Can. J. Chem. *47*, 3611 (1969)
80) Larsen, B. S., Becher, J., Hammerum, S., Schroll, G.: Acta Chem. Scand. *29B*, 41 (1975)
81) Schwarz, H., Bohlmann, F.: Org. Mass Spectrom. *9*, 283 (1974)
82) Martens, J., Praefcke, K., Schwarz, H.: Org. Mass Spectrom. *8*, 317 (1974)
83) Martens, J., Praefcke, K., Schulze, U., Schwarz, H., Simon, H.: Tetrahedron *32*, 2467 (1976)
84) Martens, J., Praefcke, K., Schwarz, H., Simon, H.: Phosphorus *6*, 247 (1976)
85) Martens, J., Praefcke, K., Schwarz, H.: Tetrahedron *31*, 1119 (1975)
86) Schwarz, H., Morhenn, H.: unpublished results
87) Bowie, J. H., Blumenthal, T., Walsh, I.: Org. Mass Spectrom. *5*, 777 (1971)
88) Bowie, J. H., Janposri, S.: Aust. J. Chem. *28*, 2169 (1975)
89) Bowie, J. H., Janposri, S.: J. Chem. Soc. Perkin Trans. *II*, 1656 (1976)
90) Cooks, R. G.: Org. Mass Spectrom. *2*, 481 (1969)
91) Bosshardt, H., Hesse, M.: Angew. Chem. *86*, 256 (1974); Angew. Chem. Int. Ed. Engl. *13*, 252 (1974)
92) Schwarz, H., Praefcke, K., Martens, J.: Tetrahedron *29*, 2877 (1973)
93a) Dube, G.: J. Prakt. Chem. 313, 905 (1971)
93b) Müller, D.: Annual Meeting of the "Arbeitsgemeinschaft Massenspektrometrie", Bad Kissingen 1977
94) Grützmacher, H. F., Kuschel, H., Adamietz, P.: Advanc. Mass Spectrom. *5*, 654 (1971)
95) Grützmacher, H. F., Kuschel, H.: Org. Mass Spectrom. *3*, 605 (1970)
96) Kuschel, H., Grützmacher, H. F.: Org. Mass Spectrom. *9*, 395, 403, 408 (1974)
97) Loudon, A. G., Maccoll, A., Smith, D., Webb, K. S., Ribera, A.: Chem. Commun. *1967*, 350
98) Baldwin, M. A., Cardnell, P. C., Loudon, A. G., Maccoll, A., Webb, K. S.: Advan. Mass Spectrom. *5*, 666 (1971)
99) Baldwin, M. A., Loudon, A. G.: Org. Mass Spectrom. *2*, 549 (1969)
100) Baldwin, M. A., Loudon, A. G., Maccoll, A., Webb, K. S.: Org. Mass Spectrom. *11*, 1181 (1976)
101) Cooks, R. G., Tam, S. W.: Org. Mass Spectrom. *1*, 583 (1968)
102) Shapiro, R. H., Serum, J. W., Duffield, A. M.: J. Org. Chem. *33*, 243 (1968)
103) Walter, W., Becker, R. F., Grützmacher, H. F.: Tetrahedron Letters *1968*, 3515
104) Williams, L. R.: Aust. J. Chem. *21*, 2311 (1968)
105) Cooks, R. G., Varvoglis, A. G.: Org. Mass Spectrom. *5*, 687 (1971)
106) Eguchi, S., Haze, M., Nakayama, M., Hayashi, S.: Org. Mass Spectrom. *12*, 51 (1977)

107) Buckle, D. R., Waight, S. E.: Org. Mass Spectrom. *1,* 273 (1968)
108) Egger, H.: Monatsh. Chem. *98,* 1245 (1967)
109) Chaston, S. H. H., Livingstone, S. E., Lockyer, P. N., Pickles, V. A., Shannon, J. S.: Aust. J. Chem. *18,* 673 (1965)
110) Bowie, J. H., Williams, D. H., Lawesson, S.-O., Schroll, G.: J. Org. Chem. *31,* 1384 (1966)
111) Smith, P. J., Dimmock, J. R., Taylor, W. G.: Can. J. Chem. *50,* 871 (1972)
112) Ronayne, J., Williams, D. H., Bowie, J. H.: J. Amer. Chem. Soc. *88,* 4980 (1966)
113) Bowie, J. H., Cooks, R. G., Prager, R. H., Thredgold, H. M.: Aust. J. Chem. *20,* 1179 (1967)
114) Bowie, J. H., White, P. Y., Hoffmann, P. J.: Tetrahedron *25,* 1629 (1969)
115) Bowie, J. H., White, P. Y.: J. Chem. Soc. (B) *1969,* 89
116) Ramana, D. V., Vairamani, M.: Org. Mass Spectrom. *11,* 525 (1976)
117) Vijfhuizen, P. C., Terlouw, J. K.: Org. Mass Spectrom. *11,* 888 (1976); *12,* 63, 245 (1977)
118) Richter, W. J., Liehr, J. G., Burlingame, A. L.: Org. Mass Spectrom. *7,* 499 (1973)
119) Grob, C. A.: Angew. Chem. *81,* 543 (1969)
120) Granoth, I.: J. Chem. Soc. Perkin Trans. *II,* 1503 (1972)
121) Meyerson, S., Corbin, A. L.: J. Amer. Chem. Soc. *87,* 3045 (1965)
122) Beynon, J. H., Job, B. E., Williams, A. E.: Z. Naturforsch. *20a,* 180 (1965)
123) Howe, I., McLafferty, F. W.: J. Amer. Chem. Soc. *92,* 3797 (1970)
124) Holmes, J. L., Benoit, F.: Org. Mass Spectrom. *4,* 97 (1970)
125) Shapiro, R. H., Tomer, K. B., Caprioli, R. M., Beynon, J. H.: Org. Mass Spectrom. *3,* 1333 (1970)
126) Shapiro, R. H., Tomer, K. B., Beynon, J. H., Caprioli, R. M.: Org. Mass Spectrom. *3,* 1593 (1970)
127) Woodward, R. B., Hoffmann, R.: The conversation of orbital symmetry. Weinheim: Verlag Chemie, New York: Academic Press 1970
128) Neeter, R., Nibbering, N. M. M.: Org. Mass Spectrom. *5,* 735 (1971)
129) Benoit, F., Holmes, J. L., Isaacs, N. S.: Org. Mass Spectrom. *2,* 591(1969)
130) Holmes, J. L., Benoit, F.: Org. Mass Spectrom. *4,* 97 (1970)
131) Benoit, F., Holmes, J. L.:Org. Mass Spectrom. *6,* 549 (1972)
132) Beynon, J. H., Caprioli, R. M., Shapiro, R. H., Tomer, K. B., Chang, C. W. J.: Org. Mass Spectrom. *6,* 863 (1972)
133) Neeter, R., Nibbering, N. M. M.: Tetrahedron *28,* 2575 (1972)
134) Nibbering, N. M. M., de Boer, Th. J.: Tetrahedron *24,* 1415 (1968)
135) Nibbering, N. M. M., de Boer, Th. J.: Tetrahedron *24,* 1427 (1968)
136) Rylander, P. N., Meyerson, S.: J. Amer. Chem. Soc. *78,* 5799 (1956)
137) Neeter, R., Nibbering, N. M. M.: Org. Mass Spectrom. *7,* 1091 (1973)
138) Bobrich, M., Schwarz, H., Levsen, K., Schmitz, P.: Org. Mass Spectrom., *12,* 549 (1977)
139) Martens, J., Praefcke, K., Schwarz, H.: Tetrahedron Letters *1973,* 3707
140) Kauffmann, T., Mitschken, A.: Tetrahedron Letters *1977,* 393
141) Rapp, U., Sachs, A., Wünsche, C., Mayer, W.: Chem. Ber. *105,* 217 (1972)
142) Wünsche, C., Sachs, A., Einwiller, A., Mayer, W.: Tetrahedron *24,* 3407 (1968)

Received June 21, 1977

Author Index Volumes 26-73

The volume numbers are printed in italics

GPSR Compliance

The European Union's (EU) General Product Safety Regulation (GPSR) is a set of rules that requires consumer products to be safe and our obligations to ensure this.

If you have any concerns about our products, you can contact us on ProductSafety@springernature.com

In case Publisher is established outside the EU, the EU authorized representative is:

Springer Nature Customer Service Center GmbH
Europaplatz 3
69115 Heidelberg, Germany

Batch number: 10683525

Printed by Printforce, the Netherlands